Engineering Flow and Heat Exchange

T0177961

Octave Levenspiel

Engineering Flow
and Heat Exchange

Third Edition

Springer

Octave Levenspiel
Department of Chemical Engineering
Oregon State University
Corvallis, OR, USA

ISBN 978-1-4899-7715-1   ISBN 978-1-4899-7454-9 (eBook)
DOI 10.1007/978-1-4899-7454-9
Springer New York Heidelberg Dordrecht London

© Springer Science+Business Media New York 2014
Softcover reprint of the hardcover 3rd edition 2014
This work is subject to copyright. All rights are reserved by the Publisher, whether the whole or part of the material is concerned, specifically the rights of translation, reprinting, reuse of illustrations, recitation, broadcasting, reproduction on microfilms or in any other physical way, and transmission or information storage and retrieval, electronic adaptation, computer software, or by similar or dissimilar methodology now known or hereafter developed. Exempted from this legal reservation are brief excerpts in connection with reviews or scholarly analysis or material supplied specifically for the purpose of being entered and executed on a computer system, for exclusive use by the purchaser of the work. Duplication of this publication or parts thereof is permitted only under the provisions of the Copyright Law of the Publisher's location, in its current version, and permission for use must always be obtained from Springer. Permissions for use may be obtained through RightsLink at the Copyright Clearance Center. Violations are liable to prosecution under the respective Copyright Law.
The use of general descriptive names, registered names, trademarks, service marks, etc. in this publication does not imply, even in the absence of a specific statement, that such names are exempt from the relevant protective laws and regulations and therefore free for general use.
While the advice and information in this book are believed to be true and accurate at the date of publication, neither the authors nor the editors nor the publisher can accept any legal responsibility for any errors or omissions that may be made. The publisher makes no warranty, express or implied, with respect to the material contained herein.

Printed on acid-free paper

Springer is part of Springer Science+Business Media (www.springer.com)

Preface

This volume presents an overview of fluid flow and heat exchange.

In the broad sense, fluids are materials that are able to flow under the right conditions. These include all sorts of things: pipeline gases, coal slurries, toothpaste, gases in high-vacuum systems, metallic gold, soups and paints, and, of course, air and water. These materials are very different types of fluids, and so it is important to know the different classifications of fluids, how each is to be analyzed (and these methods can be quite different), and where a particular fluid fits into this broad picture.

This book treats fluids in this broad sense, including flows in *packed beds* and *fluidized beds*. Naturally, in so small a volume, we do not go deeply into the study of any particular type of flow; however, we do show how to make a start with each. We avoid supersonic flow and the complex subject of multiphase flow, where each of the phases must be treated separately.

The approach here differs from most introductory books on fluids, which focus on the Newtonian fluid and treat it thoroughly, to the exclusion of all else. I feel that the student engineer or technologist preparing for the real world should be introduced to these other topics.

Introductory heat transfer books are devoted primarily to the study of the basic rate phenomena of conduction, convection, and radiation, showing how to evaluate "h," "U," and "k" for this and that geometry and situation. Again, this book's approach is different. We rapidly summarize the basic equations of heat transfer, including the numerous correlations for h. Then we go straight to the problem of how to get heat from here to there and from one stream to another.

The recuperator (or through-the-wall exchanger), the direct contact exchanger, the heat-storing accumulator (or regenerator), and the exchanger, which uses a third go-between stream—these are distinctly different ways of transferring heat from one stream to another, and this is what we concentrate on. It is surprising how much creativity may be needed to develop a good design for the transfer of heat from a stream of hot solid particles to a stream of cold solid particles. The flavor of this presentation of heat exchange is that of Kern's unique book; certainly simpler, but at the same time broader in approach.

Wrestling with problems is the key to learning, and each of the chapters has illustrative examples and a number of practice problems. Teaching and learning should be interesting, so I have included a wide variety of problems, some whimsical, others directly from industrial applications. Usually the information given in these practice problems has been designed so as to fall on unique points on the design charts, making it easy for the student and also for the instructor who is checking the details of a student's solution.

I think that this book will interest the practicing engineer or technologist who wants a broad picture of the subject or, on having a particular problem to solve, wants to know what approach to take.

In the university it could well form the basis for an undergraduate course in engineering or applied fluids and heat transfer, after the principles have been introduced in a basic engineering course such as transport phenomena. At present, such a course is rarely taught; however, I feel it should be an integral part of the curriculum, at least for the chemical engineer and the food technologist.

My thanks to Richard Turton, who coaxed our idiot computer into drawing charts for this book, and to Eric Swenson, who so kindly consented to put his skilled hand to the creation of drawing and sketch to enliven and complement the text. Finally, many thanks to Bekki and Keith Levien, who without their help this new revision would never have made it to print.

Corvallis, OR, USA   Octave Levenspiel

Contents

Nomenclature

a	specific surface, surface of solid/volume of vessel $[m^{-1}]$
A	area normal to flow, exterior surface area of a particle, area of exchanger $[m^2]$
A_t	cross-sectional area of flow channel $[m^2]$
Ar	Archimedes number, for fluidized beds $[-]$; see equation (7.4) and Appendix A.20
Bi	Biot number, for heat transfer $[-]$; see equation (11.4) and Appendix A.20
c	speed of sound in the fluid $[m/s]$; see equation (3.2)
C_D	drag coefficients for falling particles $[-]$; see equation (8.2) and Appendix A.20
C_g, C_l, C_s	specific heat of gas, liquid, or solid at constant pressure $[J/kg \cdot K]$
C_p	specific heat at constant pressure $[J/kg \cdot K]$; see Appendices A.16 and A.21
C_v	specific heat at constant volume $[J/kg \cdot K]$
C_{12}	conductance for flow between points 1 and 2 in a flow channel $[m^3/s]$; see equation (4.3)
d	diameter $[m]$
d_e	equivalent diameter of noncircular channel $[m]$; see equation (2.16) or (9.15)
d_p	characteristic diameter of particles to use in flow problems $[m]$; see equation (6.3)
d_{scr}	screen diameter of particles $[m]$; see discussion in text between equations (6.3) and (6.4)
d_{sph}	equivalent spherical diameter of particles $[m]$; see equation (6.2)
f_D	Darcy friction factor, for flow in pipes $[-]$; see text after equation (2.5)

f_f	friction factor, for flow in packed beds [–]; see equation (6.10) and Appendix A.20
f_F	Fanning friction factor, for flow in pipes [–]; see equation (2.1), Figs. 2.4 and 2.5, and Appendix A.20
F	force [N]
Fo	Fourier number, for unsteady state heat conduction [–]; see equation (11.2) and Appendix A.20
$F_{12}, F'_{12},$ $\overline{F}_{12}, \mathscr{F}_{12}$	various view factors for radiation between two surfaces; fraction of radiation leaving surface 1 that is intercepted by surface 2 [–]; see equations (9.74), (9.79), (9.81), and (9.83)
$\mathscr{F}$	efficiency factor for shell-and-tube heat exchangers [–]; see equation (13.17a)
F_d	drag force on a falling particle [N]; see equation (8.1)
ΣF	lost mechanical work of a flowing fluid due to friction [J/kg]; see equation (1.5)
g	acceleration of gravity, about 9.8 m/s^2 at sea level [m/s^2]
$G = u\rho = G_0/\varepsilon$	mass velocity of flowing fluid based on the mean cross-sectional area available for the flowing fluid in the packed bed [kg/m^2 open·s]
G_{nz}	mass velocity of gas through a well-rounded orifice [kg/m^2·s]; see equation (3.24)
G_{nz}^*	maximum mass velocity of gas through a well-rounded orifice [kg/m^2 · s]; see equation (3.27)
$G_0 = u_0\rho = G\varepsilon$	mass velocity of flowing fluid based on the total cross-sectional area of packed bed [kg/m^2 bed·s]
Gr	Grashof number, for natural convection [–]; see text above equation (11.31) and Appendix A.20
G_z	Graetz number [-], see equation (9.20)
h	heat transfer coefficient, for convection [W/m^2 · K]; see text above equation (9.11)
h_L	head loss of fluid resulting from frictional effects [m]; see equation (1.6) and the figure below equation (2.2)
H	enthalpy [J/kg]
He	Hedstrom number, for flow of Bingham plastics [–]; see equation (5.8) and Appendix A.20
i.d.	inside diameter [m]
k	thermal conductivity [W/m · K]; see equation (9.1) and Appendices A.15
$k = C_p/C_v$	ratio of specific heats of fluid [–]; $k \cong 1.67$ for monotonic gases; $k \cong 1.40$ for diatomic gases; $k \cong 1.32$ for triatomic gases; $k \cong 1$ for liquids

K	fluid consistency index of power law fluids and general plastics, a measure of viscosity [kg/m · s^{2-n}]; see equations (5.3) and (5.4)
$KE = u^2/2$	The kinetic energy of the flowing fluid [J/kg]
Kn	Knudsen number, for molecular flow [–]; see beginning of Chap. 4 and see Appendix A.20
l	length or distance [m]
L	length of flow channel or vessel [m]
L, L_p	characteristic length of particle [m]; see equation (11.3) and text after equation (15.13)
m	mass of particle [kg]
$\dot{m}$	mass flow rate [kg/s]
M	time for one standard deviation at spread of tracer/mean residence time in the vessel [–]; see equation (15.13)
$Ma = u/c$	Mach number, for compressible flow of gas [–]; see equations (3.1) and (3.2)
(mfp)	mean free path of molecules [m]; see Chap. 4
(mw)	molecular weight [kg/mol]; see Appendix A.10; $(mw) = 0.0289$ kg/mol, for air
n	flow behavior index for power law fluids and general plastics [–]; see equations (5.3) and (5.4)
$\dot{n}$	molar flow rate [mol/s]
N	number of stages in a multistage head exchanger [–]; see Chap. 14
$N = 4f_F L/d$	pipe resistance term [–]; see equation (3.7)
N	rotational rate of a bob of a rotary viscometer [s^{-1}]; see equation (5.15)
NNs	shorthand notation for non-Newtonian fluids
$NTU = UA/\dot{M}C$	number of transfer units [–]; see Fig. 11.4
Nu	Nusselt number, for convective heat transfer [–]; see equation (9.11) and Appendix A.20
p	pressure [Pa = N/m^2]; see Appendix A.7
$P = \Delta T_i/\Delta T_{max}$	temperature change of phase i compared to the maximum possible [–]; see Fig. 13.4
$PE = zg$	potential energy of the flowing fluid [J/kg]
Pr	Prandtl number for fluids [–]; see Appendix A.20
q	heat added to a flowing fluid [J/kg]
$\dot{q}$	heat transfer rate [W]
$\dot{q}_{12}$	flow rate of energy from surface 1 to surface 2 [W]; see equation (9.65)
Q	heat lost or gained by a fluid up to a given point in the exchanger [J/kg of a particular flowing phase]; see Fig. 13.4
$R = 8.314$ J/mol·K	gas constant for ideal gases; see Appendix A.11

R	ratio of temperature changes of the two fluids in an exchanger [–]; see Fig. 13.4
Re	Reynolds number for flowing fluids [–]; see text after equation (2.4) and Appendix A.20
$Re = du\rho/\mu$	for flow of Newtonians in pipes; see equation (2.4)
$Re = du\rho/\eta$	for flow of Bingham plastics in pipes; see equation (5.8)
$Re_{gen} = \left(\frac{d^n u^{2-n}\rho}{8^{n-1}K}\right)$ $\left(\frac{4n}{1+3n}\right)^n$	for flow of power law fluids in circular pipes; see equation (5.10)
$Re_p = d_p u_0 \rho/\mu$	for flow in packed and fluidized beds; see equation (6.9)
$Re_t = d_p u_t \rho/\mu$	at the terminal velocity of a falling particle; see equation (8.6)
S	entropy of an element of flowing fluid [J/kg K]
S	pumping speed, volumetric flow rate of gas at a given location in a pipe [m^3/s]; see equation (4.4)
t	time [s]
T	temperature [K]
$\overline{\Delta T}$	proper mean temperature difference between the two fluids in an exchanger [K]
u	velocity or mean velocity [m/s]
u_{mf}	minimum fluidizing velocity [m/s]
u_0	superficial velocity for a packed or fluidized bed, thus the velocity of fluid if the bed contained no solids [m/s]; see equation (6.9)
U	internal energy [J/kg]; see equation (1.1)
U	overall heat transfer coefficient [$W/m^2 \cdot K$]; see text after equation (10.4)
u_t	terminal velocity of a particle in a fluid [m/s]
$\dot{v}$ or v	volumetric flow rate of fluid [m^3/s]
V	volume [m^3]
W	work done [J]
W_{flow}	work done by fluid in pushing back the atmosphere; this work is not recoverable as useful work [J]
Ws	shaft work; this is the mechanical work produced by the fluid which is transmitted to the surroundings [J]
$\dot{W}s$	pumping power; that produced by fluid and transmitted to the surroundings $\left[\dot{W} = \frac{J}{s}\right]$
y	distance from the wall of a flow channel [m]; first sketch in Chap. 2
z	height above some arbitrarily selected level [m]
Z	compressibility factor, correction factor to the ideal gas law [–]; see text after equation (3.15)

Greek Symbols

α	kinetic energy correction factor [–]; see equation (2.12)
$\alpha = k/\rho C_p$	thermal diffusivity [m²/s]; see equation (11.1) and Appendices A.17 and A.21
α	the absorptivity or the fraction of incident radiation absorbed by a surface [–]; see equation (9.59)
ϵ	pipe roughness [m]; see Table 2.1
ε	voidage in packed and fluidized beds [–]; see Figs. 6.3 and 6.4
ε	emissivity of a surface [–]; see equation (9.60)
ε_f	voidage of a bubbling fluidized bed [–]
ε_m	minimum bed voidage, thus at packed bed conditions [–]
ε_{mf}	bed voidage at minimum fluidizing conditions [–]
η	pump efficiency [–]; see equation (1.14)
η	plastic viscosity of Bingham plastic non-Newtonian fluids [kg/ms]; see equation (5.2)
$\eta_i = \Delta T_i/\Delta T_{max}$	efficiency or effectiveness of heat utilization of stream i, or fractional temperature change of stream i [–]; see equation (13.15), (14.4), or (15.2)
π	atmospheric pressure
μ	viscosity of a Newtonian fluid [kg/ms]; see equation (5.1) and Appendix A.13
ρ	density [kg/m³]; see Appendices A.12 and A.21
σ	standard deviation in the spread of the temperature front in a packed bed regenerator; see equation (15.12)
$\sigma = 5.67 \times 10^{-8}$	Stefan–Boltzmann radiation constant [W/m² · K⁴]; see equation (9.63)
τ	shear stress (Pa = N/m²); see text above equation (2.1) and beginning of Chap. 5
τ	transmittance of surface [-]; see equation (9.61)
τ_w	shear stress at the wall [Pa]
τ_0	yield stress of Bingham plastics [Pa]; see equation (5.2)
ϕ	sphericity of particles [–]; see equation (6.1)
$\phi = \dot{m}_g C_g/\dot{m}_s C_s$	heat flow ratio of two contacting streams [–]; see equation (14.3)
ϕ'	heat flow ratio for each stage of a multistage contacting unit [–]; see equation (14.10)

Subscripts

f	property of the fluid at the film temperature, considered to be the average between the bulk and wall temperatures, or at $T_f = (T_{wall} + T_{bulk})/2$; see equation (9.24)
g	gas
l	liquid

lm   logarithmic mean
s   solid

The SI System of Units

Basic Quantities

Mass $M = \text{kg}$
Distance $D = \text{m}$
Time   s
Temperature $\text{Kelvin} = \text{K}$

Their combinations

Area $a = \text{m}^2$
Volume $v = \text{m}^3$
Velocity $V = \text{m/s}$
Acceleration $A = \text{m/s}^2$
Volumetric flow rate m^3/s
Mass velocity $G = \text{kg/m}^2 \cdot \text{s}$
Density $\text{kg/m}^3 = \rho$
Momentum $M \cdot v = \text{kg} \cdot \text{m/s}$
Dynamic or absolute viscosity $\text{Poiseuille, Pl} = \text{Pa} \cdot \text{s} = \text{kg/(m} \cdot \text{s)} = \mu$
Kinematic viscosity $\mu/\rho = [\text{kg/(m} \cdot \text{s)}]/(\text{kg/m}^3) = \text{m}^2/\text{s}$
When two masses collide elastically $M_1V_1 + M_2V_2 = M_1V_2 + M_2V_1$
When two masses collide non-elastically $M_1V_1 + M_2V_2 = (M_1M_2)\,V_{\text{final}}$

Note: With these SI units we have dropped the hundreds of g_c-terms in the equations used in previous versions of this book. The present version is simpler to use.

Torque for rotating bob

Q = (stress) (length of arm turning the rotating bob) (wetted perimeter of bob
        ignoring its bottom) (rotations per sec)
   = (stress) $(R)(2\pi rL)(\text{N/s})$

SI Units and their Prefixes

For larger measures:

$\text{J} \;= \text{N} \cdot \text{m}$
$\text{KJ} = \text{Kilo J} = 10^3\,\text{J}$
$\text{MJ} = \text{Mega J} = 10^6\,\text{J}$

GJ $=$ Giga J $= 10^9$ J
TJ $=$ Tera J $= 10^{12}$ J
PJ $=$ Peta J $=$ quadrillion J $= 10^{15}$ J
Q $=$ Quad $= 1015$ Btu $= 1.055 \cdot 10^{18}$ J [a crazy unit]
EJ $=$ Exa J $=$ quintillion J $= 10^{18}$ J
Zj $=$ Zeta J $= 10^{21}$ J

For smaller measures

mJ $=$ milli J $= 10^{-3}$ J
μJ $=$ micro J $= 10^{-6}$ J
nJ $=$ nano J $= 10^{-9}$ J
pJ $=$ pico J $= 10^{-12}$ J
fJ $=$ femto J $= 10^{-15}$ J
aJ $=$ atto J $= 10^{-18}$ J
zJ $=$ zepto J $= 10^{-21}$ J

From Newton's Law

Force, Newton, $N = M \cdot A = kg \cdot m/s^2 = J/m$
Pressure, Pascal, $Pa = N/a = kg \cdot m^2/s^2 = J$, joule
Work/kg of mass $= m^2/s^2$
Internal energy, $J = kg \cdot m^2/s^2$
Potential energy, $PE = J$
Kinetic energy, $KE = J$
Quantity of heat, $q = J$
Power, $W = $ work/time $= kg \cdot m^2/s^3 = J/s$
Rate of heat transfer, $q = J/s = W$
Heat transfer coefficient, $h = W/m^2 \cdot K$
Thermal conductivity $= W/m \cdot K$
Specific heat, $C_p, C_v = J/kg \cdot K$

From the ideal gas law: $Pa \cdot v = n \cdot R \cdot T$

$Pa = Na = J/m^3$
$v = m^3$
$N = $ moles of ideal gas
$T = $ degrees kelvin $= K$
$R = Pa \cdot v/n \cdot T = 8.314$ J/mol·K
$\quad = (101325)(0.0224)/(1)(273)$, J/mol·K

About the Author

Octave Levenspiel is Professor Emeritus of Chemical Engineering at Oregon State University, with primary interests in the design of chemical reactors. He was born in Shanghai, China, in 1926, where he attended a German grade school, an English high school, and a French Jesuit university. He started out wanting to study astronomy, but that was not in the stars, and he somehow found himself in chemical engineering. He studied at U.C. Berkeley and at Oregon State University, where he received his Ph.D. in 1952.

His pioneering book *Chemical Reaction Engineering* was the very first in the field, has numerous foreign editions, and has been translated into 13 foreign languages.

Some of his other books are *The Chemical Reactor Omnibook*, *Fluidization Engineering* (with co-author D. Kunii), *Engineering Flow and Heat Exchange*, *Understanding Engineering Thermo*, *Tracer Technology*, and *Rambling Through Science and Technology*.

He has received major awards from A.I.Ch.E. and A.S.E.E., has three honorary doctorates: from Nancy, France; from Belgrade, Serbia; from the Colorado School of Mines; and has been elected to the National Academy of Engineering. Of his numerous writings and research papers, two have been selected as Citation Classics by the Institute of Scientific Information. But what pleases him most is being called the "Doctor Seuss" of chemical engineering.

Part I
Flow of Fluids and Mixtures

Although the first part of this little volume deals primarily with the flow of fluids and mixtures through pipes, it also considers the flow of fluids through packed beds and through swarms of suspended solids called fluidized solids, as well as the flow of single particles through fluids. The term "fluids and mixtures" includes all sorts of materials under a wide range of conditions, such as Newtonians (e.g., air, water, whiskey), non-Newtonians (e.g., peanut butter, toothpaste), gases approaching the speed of sound, and gas flow under high vacuum where collisions between molecules are rare.

Chapter 1 presents the two basic equations which are the starting point for all analyses of fluid flow, the total energy balance and the mechanical energy balance. Chapters 2, 3, 4, 5, 6, 7, and 8 then take up the different kinds of flow.

Chapter 1
Basic Equations for Flowing Streams

1.1 Total Energy Balance

Consider the energy interactions as a stream of material passes in steady flow between points 1 and 2 of a piping system, as shown in Fig. 1.1. From the first law of thermodynamics, we have for each unit mass of flowing fluid:

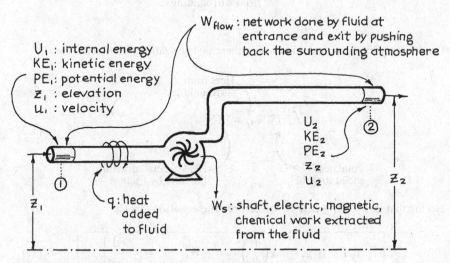

W_{flow} : net work done by fluid at entrance and exit by pushing back the surrounding atmosphere

U_1 : internal energy
KE_1: kinetic energy
PE_1: potential energy
z_1 : elevation
u_1 : velocity

U_2
KE_2
PE_2
z_2
u_2

②

z_1

z_2

q: heat added to fluid

W_s: shaft, electric, magnetic, chemical work extracted from the fluid

Fig. 1.1 Energy aspects of a single-stream piping system

© Springer Science+Business Media New York 2014
O. Levenspiel, *Engineering Flow and Heat Exchange*,
DOI 10.1007/978-1-4899-7454-9_1

$$\underset{\substack{\big\uparrow \\ \text{Potential} \\ \text{energy, PE}}}{\overset{\substack{\text{internal} \\ \text{energy} \\ \big\downarrow}}{\Delta U}} + \Delta\big(gz\big) + \underset{\substack{\big\uparrow \\ \text{Flow work} \\ \text{gained by fluid}}}{\overset{\substack{\text{Kinetic} \\ \text{energy, KE} \\ \big\downarrow}}{\Delta\left(\frac{u^2}{2}\right)}} + \Delta\left(\frac{p}{\rho}\right) = \underset{\substack{\big\downarrow \\ \text{Work received by} \\ \text{surroundings from fluid}}}{\overset{\substack{\text{Heat added to fluid} \\ \text{from surroundings} \\ \big\downarrow}}{q - W_s}} \qquad \left[\frac{J}{kg}\right] \qquad (1.1)$$

Consider the internal energy term in the above expression. From the second law of thermodynamics,

$$\Delta U = \int T\,dS - \int p\,d\left(\frac{1}{\rho}\right) + \begin{array}{l} \text{magnetic,} \\ \text{electrical,} \\ \text{chemical work,} \\ \text{etc.} \end{array} \qquad (1.2)$$

The $\int T\,dS$ term accounts for both heat and frictional effects. Thus, in the ideal situation where there is no degradation of mechanical energy (no frictional loss, turbulence, etc.):

$$\int T\,dS = q = \begin{array}{l} \text{heat added to flowing fluid} \\ \text{from surroundings} \end{array}$$

On the other hand, in situations where there is degradation (frictional losses),

$$\underset{\substack{\big\uparrow \\ \text{Total heat} \\ \text{added to fluid}}}{\overset{\substack{\text{Heat from} \\ \text{surroundings} \\ \big\downarrow}}{\int T\,dS = q}} + \underset{\substack{\big\uparrow \\ \text{Heat generated within} \\ \text{the fluid by friction}}}{\Sigma F} \qquad (1.3)$$

Noting that $\Delta H = \Delta U + \Delta(p/\rho)$, we can rewrite equation (1.1) as

$$\boxed{\Delta H + \Delta(gz) + \Delta\left(\frac{u^2}{2}\right) = q - W_s \qquad \left[\frac{J}{kg}\right]} \qquad (1.4)$$

This is the first law of thermodynamics in its usual and useful form for steady-flow single-stream systems.

[AUTHOR'S NOTE: g_c is a conversion factor, to be used with American engineering units. In SI units, g_c is unity and drops from all equations. Since this book uses SI units throughout, g_c is dropped in the text and problems.]

1.2 Mechanical Energy Balance

For each kilogram of real flowing fluid, with its unavoidable frictional effects, with no unusual work effects (magnetic, electrical, surface, or chemical), and with constant value of g, equations (1.1) and (1.3) combined give the so-called mechanical energy balance.

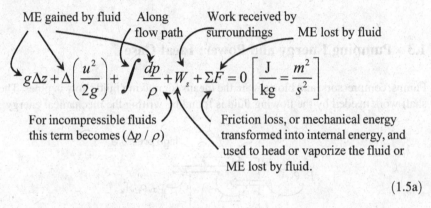

$$g\Delta z + \Delta\left(\frac{u^2}{2g}\right) + \int \frac{dp}{\rho} + W_s + \Sigma F = 0 \quad \left[\frac{J}{kg} = \frac{m^2}{s^2}\right]$$

ME gained by fluid — Along flow path — Work received by surroundings — ME lost by fluid

For incompressible fluids this term becomes ($\Delta p / \rho$)

Friction loss, or mechanical energy transformed into internal energy, and used to head or vaporize the fluid or ME lost by fluid.

$$(1.5a)$$

Multiplying by $1/g$ gives, in alternative form,

$$\Delta z + \Delta\left(\frac{u^2}{2g}\right) + \frac{1}{g}\int \frac{dp}{\rho} + \frac{1}{g}W_s + \frac{1}{g}\Sigma F = 0 \quad \left[\begin{array}{c} m \text{ of} \\ \text{fluid} \end{array}\right] \qquad (1.5b)$$

called shaft work

Lost head $= h_L$

In differential form these equations are

$$g\,dz + u\,du + \frac{dp}{\rho} + dW_s + d(\Sigma F) = 0 \qquad \left[\frac{J}{kg}\right] \qquad (1.6a)$$

and

$$dz + \frac{u\,du}{g} + \frac{1}{g}\frac{dp}{\rho} + \frac{1}{g}dW_s + d(h_L) = 0 \qquad [m] \qquad (1.6b)$$

These equations, in fact, represent not a balance, but a loss of mechanical energy (the transformation into internal energy because of friction) as the fluid flows down the piping system. In the special case where the fluid does no work on the surroundings (W = 0) and where the frictional effects are so minor that they can be completely ignored ($\Sigma F = 0$), the mechanical energy balance reduces to

$$g\Delta z + \Delta \frac{u^2}{2} + \int \frac{dp}{\rho} = 0 \tag{1.7}$$

which is called the Bernoulli equation.

The mechanical energy balance, equations (1.5a), (1.5b), (1.6a), (1.6b), and (1.7), is the starting point for finding work effects in flowing fluids—pressure drop, pumping power, limiting velocities, and so on. We apply these expressions to all types of fluids and mixtures.

1.3   Pumping Energy and Power: Ideal Case

Pumps, compressors, and blowers are the means for making fluids flow in pipes. The shaft work needed by the flowing fluid is found by writing the mechanical energy

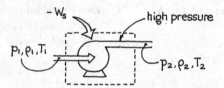

balance about the device. In the ideal case where the kinetic and potential energy changes and frictional losses are negligible, equation (1.5a and b) reduces to

$$-W_{s,\text{ideal}} = \int_{p_1}^{p_2} \frac{dp}{\rho} \quad \left[\frac{J}{kg}\right] \tag{1.8}$$

received by fluid

For *liquids* and *slurries*, ρ and $T \cong$ const, so the shaft work from the surroundings to the fluid is

$$-W_{s,\text{ideal}} = \frac{p_2 - p_1}{\rho} = \frac{\Delta p}{\rho} \quad \left[\frac{J}{kg}\right] \tag{1.9}$$

Thus the work delivered by the fluid

$$+ \dot{W}_{s,\text{ideal}} = \dot{m} W_{s,\text{ideal}} = (\rho u A) - W_{s,\text{ideal}} \qquad \left[\frac{J}{s} = W \right] \qquad (1.10)$$

Assuming ideal gas behavior and adiabatic reversible compression or expansion, the work received by each unit of flowing gas is

$$-W_{s,\text{ideal}} = \frac{k}{k-1} \frac{RT_1}{(mw)} \left[\left(\frac{p_2}{p_1} \right)^{(k-1)/k} - 1 \right] = C_p (T_2 - T_1)$$

$$= \frac{k}{k-1} \frac{RT_2}{(mw)} \left[1 - \left(\frac{p_1}{p_2} \right)^{(k-1)/k} \right] \qquad \left[\frac{J}{kg} \right]$$

$$(1.11)$$

The power produced by the flowing gas is then

$$\dot{m} W_{s,\text{ideal}} = (mw) \dot{n} W_{s,\text{ideal}} \quad [W] \qquad (1.12)$$

where

$$k = \frac{C_p}{C_v}, \quad \frac{RT_i}{(mw)} = \frac{p_i}{\rho_i}, \quad \dot{n} RT_i = p_i v_i, \quad \text{and} \quad \frac{T_2}{T_1} = \left(\frac{p_2}{p_1} \right)^{(k-1)/k} \qquad (1.13)$$

1.4   Pumping Energy and Power: Real Case Compression

For *gases* compare *ideal adiabatic compression* with the real situation with its frictional effects and heat interchange with the surroundings, both cases designed to take fluid from a low pressure p_1 to a higher pressure p_2. The p-T diagram of Fig. 1.2 shows the path taken by the fluid in these cases.

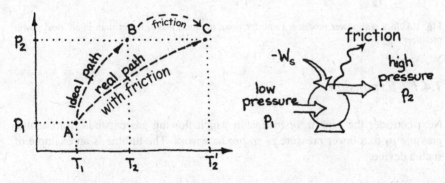

Fig. 1.2 In a real adiabatic compression (with friction), the fluid leaves hotter than in an ideal compression

For real compression some of the incoming shaft work is needed to overcome friction and ends up heating the gas. Thus, the actual incoming shaft work is greater than the ideal needed and related to it by

$$W_{s,\text{actual}} = \frac{W_{s,\text{ideal}}}{\eta}, \quad W_s < 0 \tag{1.14}$$

where η is the compressor efficiency, typically

$\eta = 0.55$–0.75 for a turbo blower
$\eta = 0.60$–0.80 for a Roots blower
$\eta = 0.80$–0.90 for an axial blower or a two-stage reciprocating compressor

The temperature rise in ideal and real compression (see Fig. 1.2) is related to the compressor efficiency by

$$\eta = \frac{T_2 - T_1}{T_2' - T_1 - (q/C_p)} \tag{1.15}$$

where q(J/kg) is the heat going from surroundings to the compressor and then to the fluid, per kg of fluid being compressed.

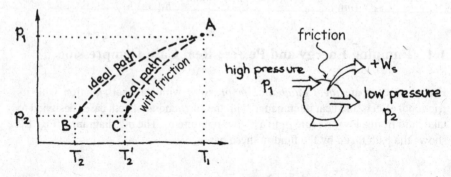

Fig. 1.3 In a real power producer (with friction), the fluid leaves hotter than in an ideal power producer

1.4.1 Expansion

Next consider the *reverse operation* in which flowing gas expands from a high pressure p_1 to a lower pressure p_2 *to produce work*. The turbine is an example of such a device.

In the real situation with its frictional losses, less work is produced than would be generated ideally. Figure 1.3 shows the p-T paths taken by the gas in the real and the ideal cases. The efficiency of the turbine relates these work terms, as follows:

$$W_{s,\text{actual}} = \eta W_{s,\text{ideal}} \tag{1.16}$$

and the outlet temperature of the fluid in the two cases is (see Fig. 1.3)

$$\eta = \frac{T_2' - T_1 - (q/C_p)}{T_2 - T_1} \tag{1.17}$$

where q again represents the heat added to each kilogram of flowing gas.

For *liquids* the relationship between actual and ideal work is given by equations (1.14) and (1.16), the same as for gases. However, since the work needed to compress a liquid is very much smaller than for the same mass of gas (by 2 or 3 orders of magnitude), the temperature change is usually quite small and often can be safely neglected when compared with the other energy terms involved. Thus, more usefully, the efficiency of operations between p_1 and p_2 is best gotten from equation (1.9), or

$$\eta_{\text{compression}} = \frac{-W_{s,\text{ideal}}}{-W_{s,\text{actual}}} = \frac{p_2 - p_1}{\rho(-W_{s,\text{actual}})} \tag{1.18}$$

and

$$\eta_{\text{expansion}} = \frac{W_{s,\text{actual}}}{W_{s,\text{ideal}}} = \frac{\rho(W_{s,\text{actual}})}{p_1 - p_2} \tag{1.19}$$

As may be seen in Figs. 1.2 and 1.3, for real compression or expansion, a portion of the mechanical energy is lost by friction to heat the flowing fluid. This is the unavoidable tax imposed by the second law of thermodynamics on all real processes.

Example 1.1. Hydrostatics and Manometers
Find the pressure p_4 in the tank from the manometer reading shown below, knowing all heights z_1, z_2, z_3, z_4, all densities ρ_A, ρ_B, ρ_C, and the surrounding pressure p_1.

(continued)

(continued)

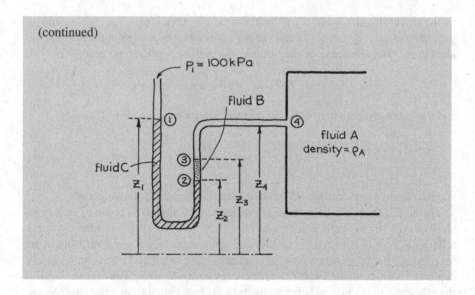

Solution

To find the pressure at point 4, apply the mechanical energy balance from a point of known pressure, point 1, around the system to point 4. Thus, from point 1 to point 2, we have

$$g\Delta z + \frac{\Delta u^2}{2} + \int \frac{dp}{\rho} + W_s + \sum F = 0$$

and for fluids of constant density (liquids), this reduces to

$$p_2 - p_1 = \rho_C g(z_1 - z_2) \quad \left[\frac{kg}{m^3} \cdot \frac{m}{s^2} \cdot m = Pa \right]$$

Repeating this procedure gives

$$p_3 - p_2 = \rho_B g(z_2 - z_3)$$
$$p_4 - p_3 = \rho_A g(z_3 - z_4)$$

Adding the above expression and noting that $p_1 = 100$ kPa gives the desired expression

(continued)

(continued)
$$p_4 = 100 + G[\rho_C(z_1 - z_2) + \rho_B(z_2 - z_3) + \rho_A(z_3 - z_4)], \qquad [\text{kPa}]$$

The same strategy of working around the system holds for other geometries and for piping loops.

Example 1.2. Counting Canaries Italian Style

Italians love birds, many homes have these happy songsters in little cages, and to supply them is a big business. Tunisian Songbirds, Inc., is a major supplier of canaries for Southern Italy, and every Wednesday a large truck carrying these chirpy feathered creatures is loaded aboard the midweek Tunis to Naples ferry. The truck's bird container is 2.4 m wide, 3.0 m high, solid on the sides and bottom, open at the top except for a restraining screen, and has a total open volume available for birds of 36 m^3. On arrival at Naples a tax of 20 lira/bird is to be charged by the customs agent, but how to determine the amount to be assessed? Since counting these thousands of birds one by one would be impractical, the Italians use the following ingenious method.

The customs agent sets up his pressure gauges and then loudly bangs the side of the van with a hammer. This scares the birds off their perches up into the air. Then he carefully records the pressure both at the bottom and at the top of the inside of the van.

If the pressure at the bottom is 103,316 Pa, the pressure at the top is 102,875 Pa, and the temperature is 25 °C, how much tax should the customs agent levy?

Additional data: Juvenile canaries have a mass of 15 g and a density estimated to be 500 kg/m^3.

Solution

With birds flying in all directions, consider the interior of the van to be a "fluid" or "slurry" or a "suspension" of mean density $\bar{\rho}$; then the mechanical energy balance of equation (1.5) reduces to

$$p_1 - p_2 = \bar{\rho}g(z_2 - z_1) \tag{i}$$

where

$$\bar{\rho} = \frac{\text{total mass}}{\text{total volume}} = \frac{m_{\text{birds}} + m_{\text{air}}}{V_{\text{total}}} = \frac{\rho_b V_b + \rho_a V_a}{V_t}$$

$$= \rho_b \left(\frac{V_b}{V_t} \right) + \rho_a \overbrace{\left(\frac{V_a}{V_t} \right)}^{\approx 0}$$

In their calculations the customs agents ignore the second term in the above expression because the density of air is so small compared to the density of the birds. Thus, on replacing all known values into equation (i) gives

$$103{,}316 - 102{,}875 = [(500)(V_b/V_t)](9.8)(3 - 0)$$

from which the volume fraction of birds in the van is found to be

$$\frac{V_b}{V_t} = 0.03$$

The number of birds being transported is then

$$N = \left(0.03 \frac{\text{m}^3 \text{ birds}}{\text{m}^3 \text{ van}} \right) (36\,\text{m}^3 \text{ van}) \left(500 \frac{\text{kg}}{\text{m}^3 \text{ bird}} \right) \left(\frac{1 \text{ bird}}{0.015 \text{ kg}} \right)$$

$$= 36{,}000 \text{ birds}$$

therefore

$$\text{Tax} = \left(20 \frac{\text{L}}{\text{bird}} \right) (36{,}000 \text{ birds}) = 720{,}000\,\text{L}$$

Example 1.3. Compressor Efficiency
An adiabatic compressor takes in 1 atm, 300 K air and produces a product stream of 3 atm, 450 K air. What is its efficiency? Take $k = C_p/C_v = 1.4$.

Solution
The efficiency of the real adiabatic compressor is given by equation (1.15), or

$$\eta = (T_2 - T_1)/\left(T_2' - T_1\right) = (T_2 - 300)/(450 - 300) \qquad \text{(i)}$$

For ideal compression, the exit temperature is given by equation (1.13), or

$$T_2/T_1 = (p_2/p_1)^{(k-1)k} = (3/1)^{0.4/1.4} = 1.36$$

and with $T_1 = 300$ K we find

$$T_2 = 1.36(300) = 408 \text{ K} \qquad \text{(ii)}$$

Combining equations (i) and (ii) gives the compressor efficiency to be

$$\eta = (408 - 300)/(450 - 300) = 0.72, \text{ or } 72\%$$

Problems on Energy Balances

1.1. Thermodynamics states that at high enough pressure diamond is the stable form for carbon. General Electric Co. and others have used this information to make diamonds commercially by implosion and various other high-pressure techniques, all complex and requiring sophisticated technology.

   Let us try something different. Take a canvas sack of lead pencils, charcoal briquettes, or coal out in a rowboat to one of the deepest parts of the ocean, the Puerto Rico trench; put some iron pipe in the sack, lower it to the ocean bottom 10 km below, wait a day, and then haul it up. Lo and behold!—a sack full of diamonds, we hope. It may work if the pressure on the ocean bottom is above the critical, or transition, pressure. Find the pressure on the ocean bottom.

   Data: Down to 10 km, seawater has an average density of 1,036 kg/m³.

1.2. *Streamline trains.* Forest Grove, Oregon, has a museum—a railroad museum—and with your admission ticket you get a free ride around their private track on either an old 1810 puffer or a streamliner of the 1950 era which was designed to speed at 160 km/h.

To measure the speed of the streamliner, we attach a pressure measuring device at the very front of the engine, at the stagnation point for the flowing air. We then measure the pressure when the train is moving and when it is standing still. We find:

(i) On a nice sunny 25 °C day, we get a pressure reading of $p = 102{,}750$ Pa when the streamliner is not moving.

(ii) When the streamliner is traveling at its museum top speed along the straightaway, we find $p = 102{,}760$ Pa.

How fast is the train barreling along?

1.3. *Artificial hearts.* The human heart is a wondrous pump, but only a pump. It has no feelings, no emotions, and its big drawback is that it only lasts one lifetime. Since it is so important to life, how about replacing it with a compact, super reliable mechanical heart which will last two lifetimes. Wouldn't that be great? The sketch below gives some pertinent details of the average relaxed human heart. From this information calculate the power requirement of an ideal replacement heart to do the job of the real thing.

Comment. Of course, the final unit should be somewhat more powerful, may be by a factor of 5, to account for pumping inefficiencies and to take care of stressful situations, such as running away from hungry lions. Also assume that blood has the properties of water.

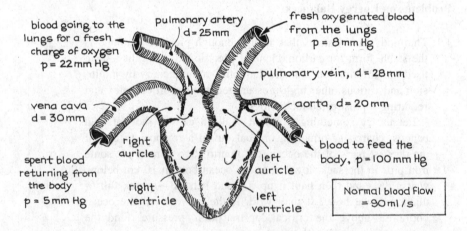

blood going to the lungs for a fresh charge of oxygen
p = 22 mm Hg

pulmonary artery
d = 25 mm

fresh oxygenated blood from the lungs
p = 8 mm Hg

pulmonary vein, d = 28 mm

vena cava
d = 30 mm

aorta, d = 20 mm

right auricle

left auricle

blood to feed the body, p = 100 mm Hg

spent blood returning from the body
p = 5 mm Hg

right ventricle

left ventricle

Normal blood Flow
= 90 ml / s

1.4. *The Great Salt Lake flood.* The water level of the Great Salt Lake in Utah has been rising steadily in recent years. In fact, it has already passed its historic high of 1283.70 m above mean sea level set in June 1873.[1] This rise has flooded many of man's works, and if it continues it will damage many more including the road bed of the main railroad line from San Francisco to the East Coast as well as long stretches of interstate highway I-80.

One idea for countering this alarming rise is to pump water out of the lake up over a dam into channels that lead to two very large evaporation ponds. Determine the annual energy cost of pumping an average of 85 m^3/s of water year round if electricity costs \$0.0385/kW · h. The six giant 85 % efficient turbines are designed to raise water from an elevation of 1,280.75–1,286.25 m above mean sea level, the lake water has a density of 1,050 kg/m^3, and the exhaust pipe is 3 m i.d.

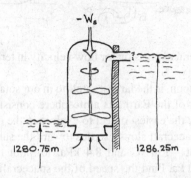

1.5. For the manometer shown on the next page, develop the expression for the pressure difference $p_1 - p_4$ as a function of the pertinent variables. In the sketch on the next page, is the fluid flowing up or down the pipe?

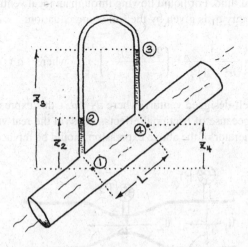

---

[1] See the *Ogden Standard Examiner*, 15 May 1986.

1.6. *Pitot tubes* are simple reliable devices for measuring the velocity of flowing fluids. They are used in the laboratory, and, if you look carefully, you will see them on all airplanes. Sketch (a) shows how the pitot tube works. Fluid flows past probe B but is brought to a stop at probe A, and according to Bernoulli's equation the difference in velocity is translated into a difference in pressure. Thus probe A, which accounts for the kinetic energy of the fluid, reads a higher pressure than does probe B. Real pitot tubes compactly combine these two probes into two concentric tubes as shown in sketch (b).

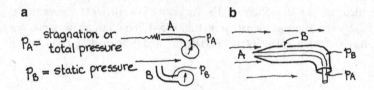

(a) Develop the general expression for flow velocity in terms of the pressures p_A and p_B.

(b) Titan, Saturn's moon, is the largest satellite in our solar system. It is roughly half the diameter of the Earth, its atmosphere consists mainly of methane, and it is probably the easiest object to explore in the outer solar system. As the Voyager 2 spacecraft slowly settles toward the surface of Titan through an atmosphere at $-130\ °C$ and 8.4 kPa, its pitot tube reads a pressure difference of 140 Pa. Find the speed of the spacecraft.

1.7. A *venturi meter* is a device for measuring the flow rate of fluid in a pipe. It consists of a smooth contraction and expansion of the flow channel, as shown below ($p_1 \gg p_2$). Pressure measurements at the throat and upstream then give the flow rate of fluid. For liquid flowing through an ideal venturi, show that the approach velocity u_1 is given by the following equation:

$$u_1 = \frac{1}{(\alpha^4 - 1)^{1/2}} \cdot \left(\frac{2(p_1 - p_2)}{\rho}\right)^{1/2} \quad \text{where} \quad \alpha = \frac{d_1}{d_2}$$

Note: For a well-designed venturi, where $d_2 < d_1/4$, this expression is off by 1–2 % at most, because of frictional effects. Thus, in the real venturi meter, the "1" in the numerator of the above expression should be replaced by 0.98–0.99.

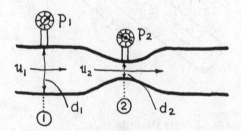

1.8. A *hydraulic ram* uses the kinetic energy of flowing fluid to raise part of that fluid to a higher derivation. The principle of operation is as follows:

When water flowing down a supply tube is abruptly stopped, the pressure at the bottom of the pipe surges, and this allows some of the water to be raised above the feed level. With proper valving this pulsing action repeats 15–200 times/min, with pumping efficiencies as high as 90 %. This type of pump requires no motor and is convenient to use in faraway places.

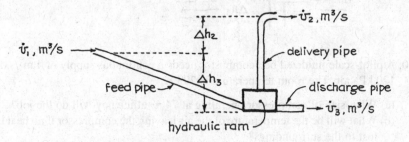

(a) Develop an expression relating the fraction of feed which can be delivered to the higher elevation to the pertinent variables shown above.

(b) A stream flows past my home, and I want to use it for my domestic water supply. I do not want to buy electricity to pump the water up to my house, so I've decided to use a hydraulic ram. If the fall of the stream is 3 m, the usable flow rate in the feed pipe is 2 lit/s, and the height from the ram to my storage tank is 8 m, what delivery rate can I expect at 50 % efficiency if the efficiency is defined as actual delivery rate divided by the theoretical?

1.9. The *air lift pump*[2] works by pumping small air bubbles into the bottom of a vertical pipe immersed in the fluid to be raised, as shown below. Ideally, the bubbles are so small that the relative velocity between air and water is negligible. Actual air lift pumps can achieve pumping efficiencies of 75 % or more. For a 60 % water and 40 % air mixture in the lift pipe, how high can an ideal air lift pump raise the water?

[2] For practical applications see F. A. Zenz, Chem. Eng. Prog. **89** 51 (August 1993).

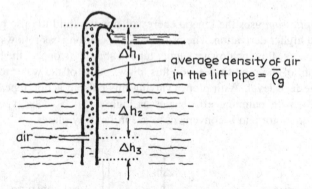

1.10. A pilot-scale fluidized bed combustor needs a continuous supply of 4 m³/s of
120 kPa air. The room temperature is 293 K.

(a) What size of compressor operating at 74 % efficiency will do the job?
(b) What will be the temperature of the air leaving the compressor if no heat is
lost to the surroundings?

1.11. I want to get into the *Guinness Book of Records* by constructing the world's
highest fountain. For this I will pump water out of a 2-cm-i.d. pipe pointing
straight up. Find the compressor power needed for the stream of water to rise
100 m. Now people tell me that I have to consider frictional effects, but I've
got an ace up my sleeve. The most honest-looking fellow at the race track sold
me a very, very special powder. One pinch into the water supply tank and
friction in the pipe and in the air magically disappears. So please don't worry
about friction, just make your calculations assuming that frictional effects can
be ignored.

1.12. Consider a very high column of gas, isothermal, and with an average compressibility factor of $\overline{Z}$.

(a) Show that the pressure change between the top (point 2) and the bottom
(point 1) of this column is given by

$$p_2 = p_1 e^{-g(mw)(z_2 - z_1)/\overline{Z}RT}$$

(b) A capped natural gas well has a pressure of 15 bar at ground level.
Assuming this natural gas to be pure methane, calculate the pressure of
gas 5,000 m underground if the temperature is 300 K and $\overline{Z} = 0.95$
throughout.

1.13. What is the air pressure on top of Mt. Hood (~3,430 m) assuming that the
temperature all about the mountain is 7 °C? Assume 1 atm at sea level.

1.14. Should we look for an adiabatic gas compressor or an isothermal? Which
needs less power? To find out, please compare the power needed to compress
a stream of room temperature air from 1 bar to 10 bar in these two ways.

1.15. An adiabatic 82 % efficient power turbine takes in hot 3 atm air and produces 1,000 kW of work and an exhaust air stream at 300 K and 1 atm. What is the temperature of the hot incoming air?

1.16. This chapter notes that much less work is needed to compress liquid than gas; consequently, the associated temperature rise for the liquid is much smaller than for the gas and often can safely be neglected. Let us check this statement by comparing the ideal work and temperature rise associated with the compression from 1 to 2 atm of 1 kg of flowing streams of air and of water entering at 20 °C. Note that per unit mass $\Delta H = C_p \Delta T$, where C_p is given for both air and water in the appendix.

1.17. Lots of artificial heart pumps are actually left ventricular assist devices. To help design such devices, calculate the fraction of the total work that is done by the left side of the heart. See Problem 1.3 for additional details or contact Professor Carlos A. Ramirez at University of Puerto Rico.

1.18. An adiabatic 82 % efficient power turbine takes in hot 3 atm air and produces 1,000 kw of work and an exhaust air stream at 300 k and 1 atm. What is the temperature of the hot incoming air?

1.19. Air at 3.3 atm and 7.62 mol/min enters and flows through some experimental equipment for which the frictional loss is $\Sigma F = 10^5$ J/kg of flowing air. The whole setup is immersed in boiling water and thus can be taken to be at 100 °C.
What is the pressure of the air leaving the equipment? Ignore any possible kinetic energy contribution.

1.20. A combustion gas (0.1 m³/s, 208 °C, 1 bar, mw = 0.030 kg/mol, Cp = 36 J/mol • K) from a reactor enters a compressor and leaves at 408 °C and 5 bar. The whole unit is hot, and its heat loss to the surroundings is 9 kw. What is the power consumption of the compressor, and what is its efficiency?

1.21. Here are two ways of getting a stream of air (1 mol/s) from 20 °C 1 atm to 200 °C and 10 atm:

   (1) Compress adiabatically then heat at constant pressure.
   (2) Heat first, then compress adiabatically. For each of these two ways, determine the heating requirement and the compressor requirement for the adiabatic 100 % efficient compressor. Take $C_{p, \text{air}} = 29$ J/mol K.

   Which arrangement would you choose?

1.22. For a high-temperature catalyst test cell, 1 mol/s of air at 300 K and 1 atm is first compressed in a well-insulated 50 % efficient compressor, then heated to 900 k in a heat exchanger for which the pressure drop is negligible. This air then flows through the isothermal 900 K test cell for which the frictional loss is estimated to be $F = 179{,}500$ J/kg of flowing air and finally exits at 1 atm. Find the duty of the heat exchanger, and assume $\overline{C}_{p,air} = 30$ J/mol K.

1.23. A reactor in our chemical plant is presently being fed a stream of hot high-pressure air (0.07 m3/s, 5 bar, 200 °C). We produce this stream by feeding room temperature air (20 °C, 1 bar) to a 60-kw adiabatic compressor and then passing the high pressure stream (5 bar) through a heat exchanger having negligible pressure drop to get the desired temperature.

What is the efficiency of the compressor?
What is the duty of the heat exchanger?

1.24. This chapter notes that much less work is needed to compress liquid in gas; consequently the associated temperature rise for the liquid is much smaller than for the gas and often can safely be neglected. Let us change this statement by comparing the ideal work and temperature rise associated with the compression from 1 to 2 atm of 1 kg of flowing stream of air and water entering at 20 °C.

1.25. From CEN I read about Joseph Gay-Lussac's ballooning ventures long ago, and it reminded me of the Great American Lead Balloon Contest of 1977 held by the staff of the Arthur D. Little Co. of Boston. The winning balloon was close to spherical, about 2 m in circumference, fabricated with a 1-mil lead foil skin, and filled with helium.

In Sept. 2021 Alma Ata in Russia (the world's largest center of lead processing as we all well know) is planning to hold a contest to see who could build the world's smallest spherical lead balloon that could float on air in their giant stadium. The winning prize is 3 million rubles, roughly equivalent to 200,000 euros.

NASA of the USA, the Russian Space agency, and other organizations are planning to compete, so I'm thinking, why not me too, with my grad students.

I wonder what would be the diameter of the smallest perfectly spherical hydrogen-filled balloon with a 0.4-mm-thick lead skin which would just float in air on a quiet 20 °C day at sea level and how does it compare with what was reported in CEN, pg 47, July 19, 2004.

Chapter 2
Flow of Incompressible Newtonian Fluids in Pipes

Newtonians are fluids in which the relative slip of fluid elements past each other is proportional to the shear on the fluid, as shown in Fig. 2.1. All gases, liquid water, and liquids of simple molecules (ammonia, alcohol, benzene, oil, chloroform, butane, etc.) are Newtonians. Pastes, emulsions, biological fluids, polymers, suspensions of solids, and other mixtures are likely to be non-Newtonian.[1] This chapter deals with Newtonians.

When a fluid flows in a pipe, some of its mechanical energy is dissipated by friction. The ratio of this frictional loss to the kinetic energy of the flowing fluid is defined as the Fanning friction factor, f_F. Thus[2]

$$f_F = \left(\frac{\frac{\text{frictional}}{\text{drag force}} \Big/ \frac{\text{area of pipe}}{\text{surface}}}{\text{kinetic energy/m}^3 \text{ of fluid}} \right) = \frac{\tau_w}{\rho \frac{u^2}{2}} \quad [-] \qquad (2.1)$$

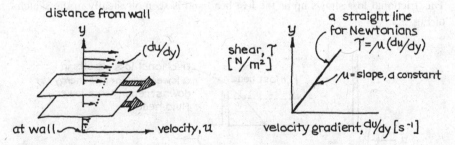

Fig. 2.1 Representation of a Newtonian fluid

---

[1] Strictly speaking, we should use the terms *Newtonian fluids* and *non-Newtonian fluids*. However, there should not be much confusion or distress if we drop the word "fluid" and simply call these materials *Newtonians* and *non-Newtonians*.

[2] SI readers may ignore g_c in all equations, if they wish.

© Springer Science+Business Media New York 2014
O. Levenspiel, *Engineering Flow and Heat Exchange*,
DOI 10.1007/978-1-4899-7454-9_2

Making a force balance about a section of pipe, as sketched in Fig. 2.2, relates the wall shear τ_w to the frictional loss ΣF (or the frictional pressure drop Δp_{fr}). In words then, for a length of pipe L,

$$\begin{pmatrix} \text{Force transmitted} \\ \text{to the walls} \end{pmatrix} = \begin{pmatrix} \text{frictional energy} \\ \text{loss by the fluid} \end{pmatrix}$$

Fig. 2.2 A force balance on a section of pipe

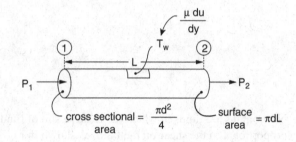

or in symbols

$$(L\pi d)\tau_w = \frac{\pi d^2}{4}\Delta p_{\text{fr}} = \frac{\pi d^2}{4}\rho\Sigma F \quad [\text{N}]$$

Substituting in equation (2.1) and rearranging gives

$$\Sigma F = \frac{2f_F L u^2}{d} = gh_L \qquad \left[\frac{\text{J}}{\text{kg}} = \frac{\text{m}^2}{\text{s}^2}\right] \tag{2.2}$$

This frictional loss shows up as the lost head and is seen physically in the sketch of Fig. 2.3.

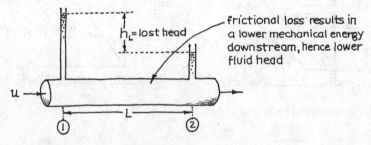

Fig. 2.3 Physical representation of the lost head

We would expect the friction factor to depend on the velocity of flow u, the fluid properties of density ρ and viscosity μ, the pipe size d, and its roughness, and so it does. Thus, we have

$$f_F = f\left[\left(\begin{array}{c}\text{Reynolds number :}\\ \text{a combination of } d,\ u,\ \mu,\ \rho\end{array}\right),\ \left(\begin{array}{c}\text{pipe}\\ \text{roughness, } \in\end{array}\right)\right]$$

where the Reynolds number is defined as

$$\text{Re} = \frac{du\rho}{\mu}$$

Figures 2.4 and 2.5, prepared from both theory and experiment, represent this relationship for Newtonians in two alternative ways. Each figure is useful for

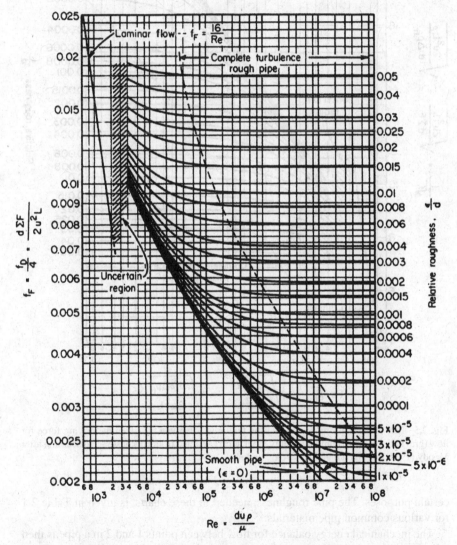

Fig. 2.4 This figure is useful for finding the pumping requirement or frictional loss when you are given the flow rate of fluid in a pipe (Adapted from Moody (1944)) or given d and u, find W_s or Δp

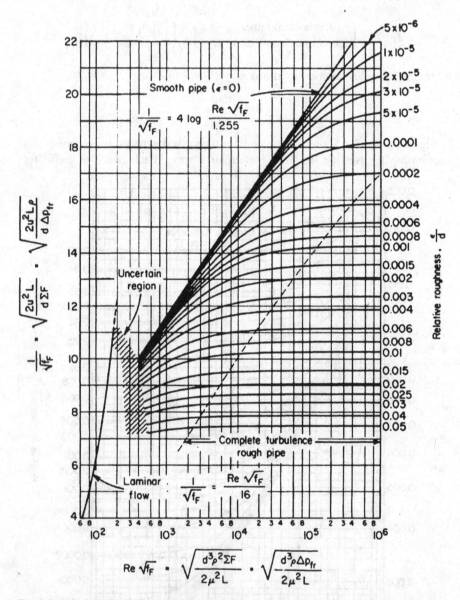

Fig. 2.5 This figure is useful for finding the flow rate when you are given the driving force for flow (gravitational head, pumping energy input, etc.) (Adapted from H. Rouse; see discussion after Moody (1944)) or given d and Δp or W_s, find u

certain purposes. The pipe roughness, needed in these charts, is given in Table 2.1 for various common pipe materials.

The mechanical energy balance for flow between points 1 and 2 in a pipe is then represented by equation (1.5); thus, referring to Fig. 2.6 we have

Table 2.1 Roughness of clean pipe[a]

Pipe material	ϵ, mm
Riveted steel	1–10
Concrete	0.3–3
Wood stave	0.2–1
Cast iron	0.26 $(0.25)^b$
Galvanized iron	0.15 $(0.15)^b$
Asphalted cast iron	0.12 $(0.13)^b$
Commercial steel or wrought iron	0.046 $(0.043)^b$
Drawn tubing	0.0015
Glass	0
Plastic (PVC, ABS, polyethylene)	0

[a]Adapted from Fischer and Porter Co., Hatboro, PA, catalogs section 98-A (1947)
[b]Values in parentheses from Colebrook (1939)

Fig. 2.6 Development of the mechanical energy balance for flow in pipes

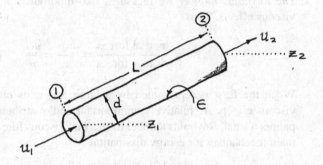

$$= 0 \text{ for no pump or turbine}$$
$$\text{in the line}$$

where

$$g\Delta z + \Delta\left(\frac{u^2}{2}\right) + \int_1^2 \frac{dp}{\rho} + W_s + \sum F = 0 \left. \begin{array}{c} \\ \\ \\ \\ \end{array} \right\} \quad \left[\frac{J}{kg}\right] \qquad (2.3)$$

$$\sum F = \frac{2 f_F L u^2}{d}$$

or in terms of head of fluid

where

$$\Delta z + \Delta\left(\frac{u^2}{2g}\right) + \frac{1}{g}\int\frac{dp}{\rho} + \frac{1}{g}W_s + h_L = 0$$

$$\quad\quad\quad\quad\quad\quad\quad \uparrow \begin{array}{l}=0 \text{ for no pump or turbine}\\ \text{ in the line}\end{array}$$

$$h_L = \frac{1}{g}\sum F = \frac{2f_F Lu^2}{gd}$$

$$\left.\right\}\quad [\text{m}]$$

$$(2.4)$$

These equations are used with Figs. 2.4 and 2.5 to solve pipe flow problems.

2.1 Comments

1. *The Reynolds number Re* measures the importance of energy dissipation by viscous effects. Thus

$$\text{Re} = \frac{\text{inertial forces}}{\text{viscous forces}} = \frac{du\rho}{\mu} = \frac{dG}{\mu}\quad [\text{-}]$$

When the flow is represented by a large Reynolds number, this means that viscous effects are relatively unimportant and contribute little to energy dissipation; a small Reynolds number means that viscous forces dominate and are the main mechanism for energy dissipation.

2. *Flow regimes* (see Fig. 2.7). Newtonians flowing in pipes exhibit two distinct types of flow, *laminar* (or streamline) when Re < 2,100 and *turbulent* when Re > 4,000. Between Re = 2,100 and Re = 4,000, we observe a transition regime with uncertain and sometimes fluctuating flow.

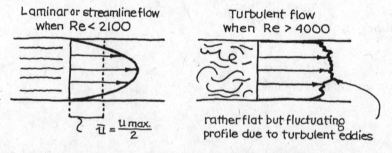

Fig. 2.7 The difference between laminar and turbulent flow in pipes

3. *For laminar flow* (Re < 2,100), the friction factor and the frictional loss can be found either from Fig. 2.4 or 2.5 or from the following simple theoretical expressions derived by Poiseuille:

$$f_F = \frac{16}{Re} \quad \text{or} \quad \Sigma F = \frac{32u\mu L}{d^2\rho} \quad \text{or} \quad u = \frac{d^2\rho\Sigma F}{32\mu L} \tag{2.5}$$

4. *Two different friction factors* are in common use today:

(i) f_F, the Fanning friction factor, defined in equation (2.1)
(ii) f_D, the Darcy friction factor

Chemical engineers favor f_F; most others prefer f_D. Don't confuse these two definitions; also, note that

$$f_D = 4f_F \quad \text{or} \quad f_F = \frac{f_D}{4}$$

The simplest way to tell which one is being used (when no subscript is shown) is to look at the laminar flow regime in the f vs. Re chart. There

$$f_F = \frac{16}{Re} \quad \text{while} \quad f_D = \frac{64}{Re}$$

5. *In the turbulent regime* (Re > 4,000), the friction factor and frictional loss are found from Fig. 2.4 or 2.5 or from the experimentally tested expressions of Nikuradse, which closely approximate the theoretical expressions of Nikuradse, Prandtl, and von Karman, as discussed in Schlichting (1979).

For the range of flows from Re = 4,000 to 10^8, these expressions were cleverly combined by Colebrook (1939) to give

$$\frac{1}{\sqrt{f_F}} = -4\log\left(\frac{1}{3.7}\frac{\in}{d} + \frac{1.255}{Re\sqrt{f_F}}\right) \tag{2.6}$$

In a form useful for calculating Re given the value of f_F

$$\frac{1}{Re} = \frac{\sqrt{f_F}}{1.255}\left[10^{-0.25/\sqrt{f_F}} - \frac{1}{3.7}\frac{\in}{d}\right] \tag{2.7}$$

and in a form useful for calculating f_F from Re Pavlov et al. (1981) give the very good approximation

$$\frac{1}{\sqrt{f_F}} \cong -4\log\left[\frac{1}{3.7}\frac{\in}{d} + \left(\frac{6.81}{Re}\right)^{0.9}\right] \tag{2.8}$$

The above expressions reduce to a number of special cases. Thus for *fully developed turbulence in rough pipes*, where f_F is independent of Re, equation (2.6) or (2.8) becomes

$$\frac{1}{\sqrt{f}_F} = 4 \log \left(3.7 \frac{d}{\epsilon} \right) \tag{2.9}$$

For *smooth pipes* ($\epsilon/d = 0$) equation (2.8) simplifies to

$$\frac{1}{\sqrt{f}_F} = 3.6 \log \frac{\text{Re}}{6.81} \tag{2.10}$$

6. *Transition regime* (Re = 2,100 ~ 4,000). Here we have an uncertain situation where the flow may be turbulent, or laminar, or even fluctuating.

7. *Piping systems* have contractions, expansions, valves, elbows, and all sorts of fittings. Each has its own particular frictional loss. A convenient way to account for this is to put this loss in terms of an equivalent length of straight pipe. Thus, the equivalent length of a piping system as a whole is given by

$$L_{\substack{\text{equiv} \\ \text{total}}} = L_{\substack{\text{straight} \\ \text{pipe}}} + \sum \left(L_{\text{equiv}} \right) \atop {\substack{\text{all fittings} \\ \text{contractions,} \\ \text{expansions, etc.}}} \tag{2.11}$$

In turbulent flow the equivalent lengths of pipe fittings are independent of the Reynolds number, and Table 2.2 shows these values for various fittings. Unfortunately, in laminar flow the equivalent length varies strongly with the Reynolds number is distinctive for each fitting. Thus, simple generalizations such as Table 2.2 cannot be prepared for the laminar flow regime.

8. The kinetic and potential energy terms of flowing fluids. In solving flow problems and replacing values in the mechanical energy balance, we often find

  • For liquids—the kinetic energy terms are negligible and can be ignored.
  • For gases—the potential energy terms are negligible and can be ignored.

When in doubt, evaluate all the terms and then drop those which are small compared to the others.

In cases where the kinetic energy must be considered, the sketches of Fig. 2.8 show how to account for this effect. In general, then

Table 2.2 Equivalent pipe length for various pipe fittings (turbulent flow only)[a]

Pipe fitting	L_{equiv}/d
Globe valve, wide open	~300
Angle valve, wide open	~170
Gate valve, wide open	~7
3/4 open	~40
1/2 open	~200
1/4 open	~900
90° elbow, standard	30
long radius	20
45° elbow, standard	15
Tee, used as elbow, entering the stem	90
Tee, used as elbow, entering one of two side arms	60
Tee, straight through	20
180° close return bend	75
Ordinary entrance (pipe flush with wall of vessel)	16
Borda entrance (pipe protruding into vessel)	30
Rounded entrance, union, coupling	Negligible
Sudden enlargement from d to D	
Laminar flow in d:	$\frac{Re}{32}\left[1-\left(\frac{d^2}{D^2}\right)\right]^2$
Turbulent flow in d:	$\frac{1}{4}f_{F,\,in\,d}\left[1-\left(\frac{d^2}{D^2}\right)\right]^2$
Sudden contraction from D to d; all conditions except high-speed gas flow where $p_1/p_2 \geq 2$. For this see Chap. 3.	
Laminar flow in d:	$\frac{Re}{160}\left[1.25-\left(\frac{d^2}{D^2}\right)\right]$
Turbulent flow in d:	$\frac{1}{10}f_{F,\,in\,d}\left[1.25-\left(\frac{d^2}{D^2}\right)\right]$

[a]Adapted in part from Crane (1982) and from Perry (1950)

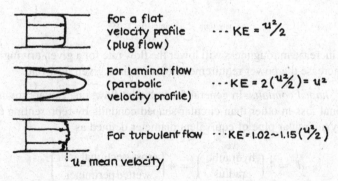

For a flat velocity profile (plug flow) $\cdots KE = u^2/2$

For laminar flow (parabolic velocity profile) $\cdots KE = 2(u^2/2) = u^2$

For turbulent flow $\cdots KE = 1.02\sim1.15(u^2/2)$

u = mean velocity

Fig. 2.8 An accounting of the kinetic energy term in the mechanical energy balance

$$KE = \frac{u^2}{\alpha 2} \quad \text{where} \quad \alpha \begin{cases} = \frac{1}{2} \text{ for laminar flow} \\ = 1 \text{ for plug flow} \\ \rightarrow 1 \text{ for turbulent flow} \end{cases} \tag{2.12}$$

Since the kinetic energy of flowing fluids only contributes significantly at high velocities where flow is turbulent, α usually is close to 1. Only for gases can the flow be both laminar and at high velocity. This situation occurs only rarely.

9. *Evaluation of the $\int (dp/\rho)$ term in the mechanical energy balance:*

   • For liquids, $\rho \simeq$ constant so

$$\int_1^2 \frac{dp}{\rho} = \frac{1}{\rho}(p_2 - p_1) = \frac{\Delta p}{\rho} \tag{2.13}$$

   • For ideal gases with small density changes, we can use an average density and then treat the gas as an incompressible fluid; thus

$$\bar{\rho} = \frac{1}{\bar{v}} = \frac{\bar{p}(mw)}{R\bar{T}} \quad \text{where} \quad \begin{cases} \bar{p} = \frac{p_1 + p_2}{2} \\ \bar{T} = \frac{T_1 + T_2}{2} \end{cases} \tag{2.14}$$

   • For large fractional changes in pressure or density, and by this we mean when $p_1/p_2 > 2$ or $\rho_1/\rho_2 > 2$, we must use the treatment of Chap. 3.

10. *Aging of pipes.* The value of pipe roughness given in Table 2.1 is for clean pipe. With time, however, roughness may increase because of corrosion and scale deposition. Colebrook (1939) found that a simple linear expression can reason-ably represent such a change

$$\epsilon_{\text{any time}} = \epsilon_{\text{time}=0} + \alpha t \tag{2.15}$$

   An increase in roughness will lower the flow rate for a given driving force or will increase the power requirement to maintain a given flow rate.

11. *Other shaped conduits.* In general for turbulent flow, one can approximate the frictional loss in other than circular-shaped conduits by representing the conduit by a circular pipe of equivalent diameter defined as

$$d_e = 4 \left(\frac{\text{hydraulic}}{\text{radius}} \right) = 4 \left(\frac{\text{cross-sectional area}}{\text{wetted perimeter}} \right) \tag{2.16}$$

For certain shapes,

- Plain and eccentric annuli
- Pipes containing various shapes of internals, including finned tubes
- Parallel plates
- Rectangular, triangular, and trapezoidal conduits

—experiments have been made and frictional losses reported [see Knudsen and Katz (1958), Chaps. 4 and 7].

Example 2.1. Tomato Growing in Absentia

Every summer I carefully grow a giant tomato plant because I love the taste of its fresh-picked fruit. Since these plants need 2 lit of water each day of the growing season to produce these delectable and irresistible fruits, how do I grow my plant next summer when I will be away for 4 weeks with no way to water it?

One solution would be to connect a long plastic tube 0.4 mm i.d. to the faucet at my home where the water pressure is 100 kPa above atmospheric and lead it to the plant. Determine how long the tube would have to be to deliver 2 lit/day of water. Of course, everything is on the level.

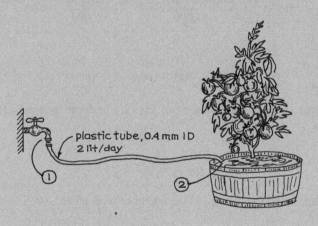

plastic tube, 0.4 mm ID
2 lit/day

Solution

Knowing the volumetric flow rate of water and the tube diameter will allow calculation of the Reynolds number for the flow in the plastic tube. Thus

$$u_2 = \frac{v}{A} = \frac{\left(2\frac{\text{lit}}{\text{day}}\right)\left(\frac{1\,\text{m}^3}{1,000\,\text{lit}}\right)\left(\frac{1\,\text{day}}{24\times3,600\,\text{s}}\right)}{\frac{\pi}{4}(0.0004\,\text{m})^2} = 0.184\frac{\text{m}}{\text{s}}$$

(continued)

(continued)

and

$$Re_2 = \frac{d_2 u_2 \rho}{\mu} = \frac{(0.0004)(0.184)(1,000)}{10^{-3}} = 73.7 \quad \text{thus, laminar flow}$$

After these preliminaries let us write the mechanical energy balance between points 1 and 2

$$g\overbrace{(z_2 - z_1)}^{=0} + u_2^2 - u_1^2 + \frac{p_2 - p_1}{\rho} + \overbrace{W_s}^{=0} + \frac{2 f_F L u^2}{d} = 0 \quad \left[\frac{J}{kg}\right]$$

Note that "2" is absent from the denominator of the kinetic energy term. This is because the fluid is in laminar flow. Also, since the diameter of the faucet opening is large compared to that of the tube, we can reasonably assume that the velocity therein is negligible, or $u_1 = 0$. Thus, on replacing values we find

$$(0.184)^2 + \frac{101,325 - 201,325}{1,000} + \frac{2 f_F L (0.184)^2}{0.0004} = 0$$

or

$$0.034 - 100 + 169.6 f_F L = 0 \tag{i}$$

Then, either from an extrapolation of Fig. 2.4 or from equation (2.5), we find that

$$f_F = \frac{16}{Re} = \frac{16}{73.7} = 0.2171 \tag{ii}$$

Combining (i) and (ii) then gives the length of tube needed, or

$$L = 2.715 \text{ m}$$

NOTE ON THE KE CONTRIBUTION: The numbers in equation (i) show that the kinetic energy contributes less than 0.04 % to the total energy loss. Thus, the kinetic energy term could very well have been ignored in this problem. This frequently is the case, especially when flow velocities are not great and when frictional losses are severe.

(continued)

(continued)

NOTE ON THE ENTRANCE LOSSES: In the above solution we ignored the entrance losses. Let us see if this is reasonable. From Table 2.2 we find that the extra length of tubing representing this loss is given by

$$\frac{L_{eq}}{d} = \frac{1.25(\text{Re})}{160} = \frac{1.25(74)}{160} = 1.57, \quad \text{or} \quad L_{eq} = 0.2 \text{ mm}$$

This is quite negligible.

Example 2.2. Overflow Pipe for a Dam

Next summer I plan to dam Dope Creek to form a little lake. Building the dam is straightforward; however, I haven't yet figured out what size of galvanized pipe to use for the water overflow. The dry season is no problem; it is the wet season with its thunderstorms and flash floods that worries me. My personal meteorologist estimates that in the very worst conceivable situation, flow in Dope Creek can reach $\pi/2$ m^3/s.

If the pipe diameter is too small, the water level will rise too high (more than 1 m above the water intake) and will overflow and damage the dam. How large a pipe is needed to guarantee that the water level does not ever rise above this danger point? The equivalent length of this overflow pipe is 19.6 m, and its discharge is located 3 m below its intake.

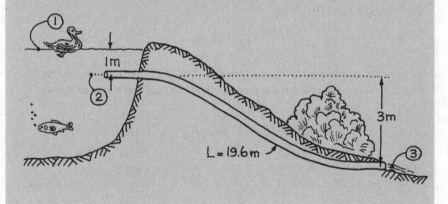

Solution

We can choose to write the mechanical energy balance between points 1 and 3 or between points 2 and 3. The former pair seems simpler because then $\Delta p = 0$. Thus, between points 1 and 3, we have

$$g\Delta z + \underbrace{\frac{\Delta u^2}{2}}_{=0} + \underbrace{\frac{\Delta p}{\rho}}_{=0} + \underbrace{W_s}_{=0} + \Sigma F = 0$$

In Example 2.1 we found that the kinetic energy contribution and pipe entrance losses were negligible, so let us start by assuming that they are negligible here as well (we will check this later), and with no pump or turbine in the line, the shaft work term disappears. On replacing values we find that

$$9.8(-4) + 0 + 0 + 0 + \frac{2f_F u_2^2 (19.6)}{d} = 0 \tag{i}$$

Next, relate the flow velocity and Reynolds number with the pipe diameter as follows:

$$u_2 = \frac{\dot{v}}{A} = \frac{\pi/2}{(\pi/4)d^2} = \frac{2}{d^2} \tag{ii}$$

$$Re = \frac{d u_2 \rho}{\mu} = \frac{d(2/d^2)(1{,}000)}{10^{-3}} = \frac{2 \times 10^6}{d} \tag{iii}$$

Combining (i) and (ii) gives

$$f_F = d^5/4 \tag{iv}$$

Now as d changes, so does Re and f_F, and these in turn are related by Fig. 2.4. So let us solve for the pipe diameter by trial and error, as shown.

Guess d	Re [from (iii)]	d/d [from Table 2.1]	f_F [from Fig. 2.4]	f_F [from (iv)]
0.1	2×10^7	0.0015	5.4×10^{-3}	0.0025×10^{-3}
0.4	5×10^6	0.00038	4.0×10^{-3}	2.6×10^{-3}
0.44	4.6×10^6	0.00034	3.9×10^{-3}	4.1×10^{-3} ... close enough

Therefore, the pipe diameter needed is $d = 0.44$ m.

Finally, let us check to see if we are justified in assuming that the kinetic energy contribution and pipe entrance losses can be ignored. So let us evaluate these contributions and include them in equation (i)

(continued)

(continued)

KE contribution. From equations (2.12) and (ii)

$$\frac{\Delta u^2}{2} = \frac{u_2^2}{2} = \frac{2}{d^4}$$

Entrance loss. Referring to Table 2.2 and the sketch of the dam for Dope Creek above shows that the Borda entrance most closely represents the pipe; thus,

$$\frac{L_{eq}}{d} = 30 \quad \text{or} \quad L_{eq} = 30d$$

Including both these terms in (i) gives

$$9.8(-4) + \frac{2}{d^4} + 0 + 0 + \left[\frac{2f_F(4/d^4)(19.6 + 30d)}{d}\right] = 0 \qquad \text{(v)}$$

To solve this for the pipe diameter, we repeat the trial-and-error procedure, but with equation (v) in place of equation (iv). This gives the pipe diameter of

$$d = 0.56 \text{ m}$$

In which case equation (v) becomes

$$-39.2 \quad + \; 20.3 \quad + \; 0 \; + 0 + \quad [10.3 \; + \; 8.8] \; = \; 0$$

PE : 100%   KE : 52%   Pipe : 26% Entrance : 21%

These numbers show that only 26 % of the potential energy is dissipated by pipe friction, not 100 %, and that the major transformation, over 50 %, is to kinetic energy of the flowing water.

NOTES: These two examples illustrate the general finding that for flow in rather long small-diameter pipes, the frictional resistance at the pipe walls dominates, while kinetic energy and entrance effects can be ignored. On the contrary, in short large-diameter pipes, kinetic energy and pipe entrance losses should not be ignored and can actually dominate.

Figures 2.4 and 2.5 allow us to solve flow problems without trial and error whenever frictional losses, flow rate, or pipe length are the unknowns. Unfortunately, when the pipe diameter is unknown, one needs to resort to trial-and-error procedures, as we have seen in this problem. Laminar flow is the exception; for then no trial and error is needed.

Problems on Incompressible Flow in Pipes

2.1. A 100 % efficient 1-kW pump–motor lifts water at 1.6 lit/s from a lake through 80 m of flexible hose into a tank 32 m up on a hill. A second pump with the same length of hose will be used to pump water from the lake at the same rate into a reservoir at lake level. What size of 100 %-efficient pump–motor is needed?

2.2. *California's savior.* "Big projects define a civilization. So why war, why not big projects," said Governor Walter Hickel of Alaska, referring to the following scheme.

California will run out of excess water by year 2000, Alaska has lots of excess, so why not build a suboceanic pipeline to pump freshwater from Alaska to California. It would be enormous, the cost would be very large (about $150 billion), and it would take about 15 years to build, but it would supply about 10 % of the whole of California's needs.

As proposed it would pump a trillion gallons of water per year through a 30-foot-diameter plastic pipe 1,700 miles long, buried in the ocean floor. What would be the pumping cost per year and per m^3 of water delivered if energy costs 3¢/kW · h?

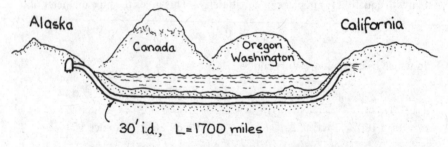

2.3. *Acid pump.* 3 kg/s of 75 % H_2SO_4 ($\rho = 1,650$ kg/m^3, $\mu = 8.6 \times 10^{-3}$ kg/m s) is to be pumped from one open tank to another through 600 m (total equivalent length which includes bends, fittings, etc.) of 50-mm-i.d. pipe ($\epsilon = 0.046$ mm). The outlet of the first tank is 7 m below its surface, and the inlet of the second tank is 2 m below its surface and 13 m above the surface of the first tank. Find the power required for this operation if the pump–motor is 50 % efficient.

2.4. Water at 20 °C flows horizontally at 3 m/s from the base of a tank through 100 m of 100-mm-i.d. PVC plastic pipe. How high is the level of water in the tank?

(a) Solve ignoring kinetic energy and pipe entrance effects.
(b) Solve accounting for these effects.

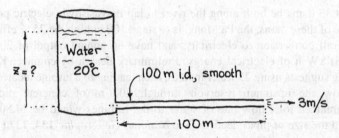

2.5. Water at 20 °C is to flow out of a settling pond to a drain trench through 100-m equivalent length of 100-mm-i.d. galvanized pipe. The level of the pond is 10 m above the discharge end of the pipe. Find the flow rate of water in m^3/min.

2.6. Water at 20 °C flows from the base of a large storage tank through a smooth horizontal pipe (100 mm i.d., 1 km in length) at a velocity of 1 m/s. That is not fast enough. How much pipe must I saw off to get 2.5 times the velocity through the pipe? Ignore kinetic energy and entrance effects.

2.7. *The Sweat-hose.* We are testing a new type of soaker garden hose. It has porous walls through which water seeps. Calculate the seepage rate of this hose in liters/hour.

    Data. The hose is 15 m long, 3 cm o.d., and 2 cm i.d. It is connected to a water faucet at one end, and it is sealed at the other end. It has 100 pores/cm^2 based on the outside of the hose surface. Each pore is tubular, 0.5 cm long and 10 µm in diameter. The water pressure at the faucet feeding the hose is 100 kPa above atmospheric pressure.

2.8. *The incredible osmotic pump.* Unbelievable though it may seem, if you could sink a pipe about 10 km down into the ocean capped with an ideal semiper-meable membrane at its bottom, *fresh*water would pass through the membrane, rise up the pipe, and burble out above the surface of the ocean ... all by itself and with no expenditure of energy! However, there are not many places in the world where one could go that far down into the ocean.

    A somewhat more practical alternative for getting freshwater from the ocean requires sinking the pipe only about 250 m down. Freshwater would pass through the membrane into the pipe and stop at the 231-m depth. Then all you need to do is pump this freshwater to the surface.

    What would be the pumping cost/m^3 of water to pump 1 lit/s of 20 °C freshwater up 240 m to an inland storage tank, through a 50.8-mm-i.d. pipe 300 m in equivalent length?

2.9. *Taming the Mekong.* The giant Mekong River runs from the Himalayas through Southeast Asia, and the Mekong Development Project proposed that 35 dams be built along the river to tap its vast hydroelectric potential. One of these dams, the Pa Mong, is to stand 100 m high, be 25 % efficient in overall conversion to electricity, and have an annual output of 20 billion (US) kW h of electrical energy. Preliminary design by engineer Kumnith Ping suggests using 25 water intakes, each leading to a turbine located 100 m below the upstream reservoir through 200 m of concrete pipe. The guaranteed total flow rate of water to the turbines would be 14,800 m^3/s. Find the size of pipes needed. [See *National Geographic* **134**, 737 (1968).]

2.10. The *Alaska Pipeline*. Oil is pumped right across Alaska from Prudhoe Bay to Valdez in a 1.22-m-i.d. pipe, 1,270 km long at pressures as high as 8 MPa. The crude is at 50 °C, and the flow rate of the line is 2.2 m^3/s. Calculate:

(a) The theoretical number of pumping stations needed
(b) The kW rating of the pump–motor set

NOTE: This data is from *National Geographic* **150**, 684 (1976). In addition, estimate for crude oil at this temperature: $\rho = 910\,kg/m^3$, $\mu = 6 \times 10^{-3}\,kg/m\,s$.

Geothermal energy. Northwest Natural Gas Co. is doing exploratory work for a possible $50,000,000 geothermal development to supply Portland, Oregon, with hot water from Mt. Hood. Wells would be drilled on the slopes of Mt. Hood, 760 m above Portland to obtain 74 °C hot water at 1 atm. This would then be piped to Portland at 1.6 m^3/s in a 1.1-m-i.d. pipe 70 km in length.

2.11. Calculate the required size of motor and the pumping cost assuming 50 % efficiency for the pump–motor and 2¢/kW h for electricity, or else, if no pump is needed, find the water pressure in the pipeline at Portland.

2.12. We would like the water pressure at Portland to be 700 kPa, and this can be obtained without pumps by choosing the proper pipe size. What size would do the job? [See *Corvallis Gazette-Times* (September 28, 1977).]

2.13. Water at 10 °C is to flow out of a large tank through a 30-gauge 8-in. commercial steel (0.205-m-i.d.) piping system with valve open, as shown below. What length of pipe could be used while still maintaining a flow rate of 0.2 m^3/s?

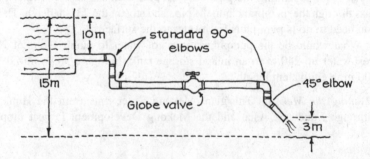

2.14. In the above problem replace the globe valve with a gate valve. What maximum length can now be used?

2.15. *Swiss ingenuity.* The village of Schaffzell high in the Alps operates its own modest hydroelectric plant which produces electricity continuously whether it is used or not. When not needed the 240 kW of electricity runs a motor-turbine at 75 % efficiency which pumps 5 °C water at 0.1 m^3/s through an equivalent length of 780 m of pipe to a little lake located 153 m uphill.

When extra electricity is needed, the flow is reversed, and water runs downhill at the same flow rate, 0.1 m^3/s, from the little lake through the turbine to generate the needed electricity, again at 75 % efficiency. How much power can be generated with this downflow from the little lake?

2.16. What size of pipe was used by the village of Schaffzell in their system that pumps water up to the little lake (see previous problem)?

2.17. *Drip irrigation* is a means of getting water directly to growing plants with very little waste. One method uses a large-diameter polyethylene "mother" tube (10–15 mm i.d.) from which lead many small-diameter polyethylene feeder tubes, called drippers, which go directly to the individual plants. What length of 0.5-mm tubing will give a flow rate of 20 °C water of 6 lit/day if the pressure is 2 bar in the mother tube and 1 bar in the surroundings?

2.18. *More on drip irrigation.* Referring to the above problem, what diameter and length of dripper should we use if we want a flow rate of 4 lit/h to each plant and if the desired dripper length is to be somewhere between 0.5 and 1.5 m? *Data*: The temperature is 20 °C; mother tube pressure = 200 kPa; ambient pressure = 100 kPa; drippers are manufactured in three diameters: 0.5 mm, 1.0 mm, and 1.5 mm.

2.19. *Jet printers.* The heart of jet printers for computers is a cluster of very thin ink-filled tubes, each about the thickness of a human hair or 100 μm. These spit out tiny droplets of ink to form desired characters on paper. About 4,000 drops/s are produced by each tube, and each drop has a mass of 15 μg and is ejected in about 3 μs.

These drops are ejected from the tube by passing pulses of electricity, about 15 μJ, through a heater located very close to the end of the tube, or 50 μm. This heat vaporizes a bit of ink which raises the pressure high enough to force a drop out of the tube. The vapor then condenses, and the process is repeated rapidly enough to generate these 4,000 drops/s.

Estimate the pressure in these vapor bubbles, and ignore possible surface tension effects. Accounting for surface tension effects would only change the answer by about 13 %.

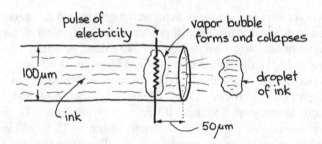

2.20. *Air-conditioned homes for little animals.* Prairie dogs, cute squirrel-like creatures minus bushy tails, live in large colonies in underground burrows connected by tunnels. A curious feature of these tunnels is that some of their exits are at ground level while others are in carefully built up mounds of earth. Is there a purpose to this?

Some speculate that these mounds which dot the colony act as lookout posts while their tunnels act as emergency getaways. You see, hawks love prairie dogs, but not vice versa. But do they need that many lookout stations, seeing that just about every second tunnel exit is in a raised mound?

Vogel and Bretz (*Science* **175**, 210 (1972)) propose that these mounds are part of the scheme for ventilating the burrows. Since air flows more slowly past a ground level exit than past one higher up, they feel that this should cause air to flow through the tunnel. Let us see if this explanation makes sense by considering the idealized situation sketched below and with values as shown. For this:

(a) Find the direction and velocity of air flow through the tunnel.
(b) Find the mean replacement time for the air in the tunnel.

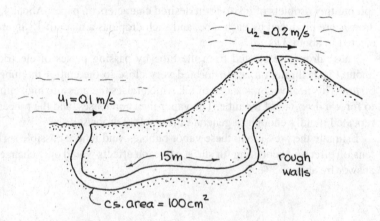

2.21. *Oil pipeline.* A steel pipe 0.5 m i.d. is to carry 18,000 m^3/day of oil from an oil field to a refinery located 1,000 km away. The difference in elevation of the two ends of the line is negligible.

(a) Calculate the power required to overcome friction in the pipeline.
(b) Since the maximum allowable pressure in any section of the line is 4 MPa (about 40 atm), it will be necessary to have pumping stations at suitable intervals along the pipeline. What is the smallest number of pumping stations required?

Data: At the temperature involved the oil has a viscosity of 0.05 kg/m s and a density of 870 kg/m^3.

2.22. Williams Brothers–CMPS Engineers are presently designing a pipeline to transport 40 m^3/h of concentrated sulfuric acid from Mt. Isa to Phosphate Hill, both in Queensland, Australia. At the head of the pipeline will be one pumping station, the line will be constructed of soft iron because this material is only very slowly attacked by the concentrated acid (at most 6 mm in the estimated 20-year life of the line), and the line is to discharge into a storage tank.

(a) Determine the highest pressure expected in the pipeline. Where should this be?
(b) Determine the pressure at the pipeline discharge; note that the pressure should not go below atmospheric anywhere in the line.
(c) Calculate the pumping power requirement for this pipeline for a 33 % pump–motor efficiency. This low value is to account for corrosion of the pump, low winter temperatures, etc.

Data

- The pipe to be used is API standard seamless 6″ pipe; 219-mm o.d. and 9.52-mm wall thickness.
- The length of the pipe is 142 km.
- Elevations above sea level are as follows:

  360 m at Mt. Isa
  265 m at Phosphate Hill
  450 m at Thunder Ridge, the highest point in the line and 50 km from Mt. Isa

- Properties of 98 weight % H$_2$SO$_4$ at 15 °C:

$\rho = 1,800$ kg/m^3, $\mu = 29 \times 10^{-3}$ kg/m-s.

2.23. Air at rest, at 20 °C and 100 kPa, is forced by means of a fan through a horizontal galvanized flue 1 m in diameter and 10 m long at a velocity of 10 m/s. What size motor should be used if the motor is 90 % efficient and the fan with its entrance resistance is 20 % efficient?

2.24. *Teaser.* Consider Problem 2.25 on the horizontal flue. If the flue were vertical and flow were upward, would you get a different answer? Calculate it please. Did you find that the size of motor needed changes from 3 to 8 kW? If so, then what would happen for downflow in the flue? Would you be generating 2 kW of useful work for free? Try to resolve the dilemma of a vertical flue.

2.25. *The Aquatrain.* Low-sulfur coal, highly prized by electric utilities because it needs no pretreatment for sulfur removal, is mined in Colorado. Los Angeles needs this coal, and the slurry pipeline is one way to transport this coal from mine to city. Some claim that this method of transportation is cheaper than by railroad. However, there is one overwhelming drawback to this method—the use of freshwater:

- Freshwater must be used in the slurry; otherwise salts will remain with the powdered coal to cause corrosion problems when the coal is burned.
- On arrival at Los Angeles the freshwater will be contaminated and not usable further.
- Freshwater is scarce in the Southwest United States and too valuable to be used this way.

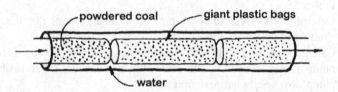

W.R. Grace and Co. has an alternative proposal called the Aquatrain: The powdered coal is placed in giant cylindrical plastic bags 5 m long and 0.75 m in diameter, and the bags are then pumped along with water in a 0.91-m-i.d. pipeline all the way from their mine near Axial, Colorado (elevation = 2,000 m), to Los Angeles (at sea level), 2,000 km away. It is reported that 15 million tons of coal can be transported per year this way.

This method has another big advantage. At Glenwood Springs some very salty water enters the Colorado River, significantly raising the saline content of the river. By using this saline water in the pipeline instead of fresh river water, one can keep away 250,000 t/yr of salt from the river, thereby reducing the load on the desalination plants presently being built downstream on the Colorado River.

Suppose the pipeline operates 360 days in the year and that the plastic bags are carried down the river in neutral buoyancy and are in end-to-end contact along the whole length of the pipeline. For a preliminary rough estimate:

(a) Calculate the mean velocity of coal in the pipeline.
(b) Estimate the volumetric flow rate of water needed to transport this coal assuming first laminar flow, then turbulent flow.
(c) Determine whether the flow of water is laminar or turbulent.

(d) Find the theoretical pumping cost per ton to transport the coal assuming a straight rundown to Los Angeles in a commercial steel pipe and 2.5¢/kW h for electrical energy.

Information for this problem is from *Christian Science Monitor* (June 4, 1982).

2.26. Water at 20 °C flows horizontally at 3 m/s from the base of a tank through a Borda entrance and 100 m of 100-m-i.d. PVC pipe. How high is the level of the water in the tank?

(a) Solve ignoring kinetic energy effects and pope entrance effects.
(b) Solve accounting for these effects.

References and Recommended Readings

C.F. Colebrook, Friction factors for pipe flow. Inst. Civil Eng. **11**, 133 (1939)

Crane Co. Technical paper 410, New York (1982)

J.G. Knudsen, D.L. Katz, *Fluid Dynamics and Heat Transfer* (McGraw-Hill, New York, 1958)

R. Lemlich, A kinetic analogy. J. Chem. Educ. **31**, 431 (1954)

L.F. Moody, Turbulent flow in pipes, with particular reference to the transition region between the smooth and rough pipe law. Trans. ASME **66**, 641 (1944)

K.F. Pavlov, P.G. Romankov, A.A. Noskov, *Problems and Examples for a Course in Basic Operations and Equipment in Chemical Technology*, Mir, Moscow (1981), translated

J.H. Perry, *Chemical Engineer's Handbook*, 3rd edn. (McGraw-Hill, New York, 1950)

H. Schlichting, *Boundary Layer Theory*, 7th edn., transl. by J. Kestin (McGraw-Hill, New York, 1979)

Chapter 3
Compressible Flow of Gases

When the density change of fluid is small ($\rho_1/\rho_2 < 2$) and the velocity not too high (Mach number, Ma < 0.3), then the mechanical energy balance reduces to the forms developed in Chap. 2. These equations represent the flow of all liquids as well as relatively slow-moving gases. This is called *incompressible flow*.

On the other hand, for gases only, when the pressure ratio is large and/or the flow very fast, then kinetic energy and compressibility effects (large density changes) may well become the dominant terms in the mechanical energy balance. In this situation the mechanical energy balance takes on quite different forms from those developed in Chap. 2. This is called *compressible gas flow*, and the equations for this flow are developed in this chapter.

Air flowing through an air conditioning duct can be considered to be incompressible, while flow from a high-pressure tank through a short discharge tube or through a pipeline whose inlet pressure is 50 atm and outlet pressure is 10 atm should be treated as compressible.

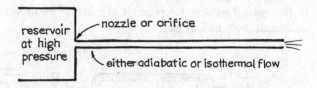

Fig. 3.1 High pressure nozzle and adiabatic or isothermal flow

The importance of these two factors, compressibility and velocity, is indicated by the Mach number, defined as

$$\text{Ma} = \left(\frac{\text{velocity of gas}}{\text{velocity of sound in the gas}} \right)_{\text{at same } T \text{ and } p} = \frac{u}{c} \qquad (3.1)$$

where the speed of sound is given by thermodynamics as

© Springer Science+Business Media New York 2014
O. Levenspiel, *Engineering Flow and Heat Exchange*,
DOI 10.1007/978-1-4899-7454-9_3

$$\begin{matrix} \curvearrowright \end{matrix} c\,\frac{\mathrm{ideal}}{\mathrm{gas}}\left[\frac{kp}{\rho}\right]^{1/2}=\left[\frac{kRT}{(mw)}\right]^{1/2}\underset{20^\circ\mathrm{C}}{\overset{\mathrm{air}}{=}}343.5\,\frac{\mathrm{m}}{\mathrm{s}}$$

$$\begin{matrix}\text{Depends on}\\ T,p\,\text{and}\,\rho\end{matrix}\quad\quad k=\frac{C_p}{C_v}\begin{matrix}\cong1.67 & \text{for monatomic gases}\\ \cong1.40 & \text{for diatomic gases}\\ \cong1.32 & \text{for triatomic gases}\end{matrix}$$

$$\tag{3.2}$$

and the mass velocity of an ideal gas is

$$G=\mu\rho\quad\left[\frac{\mathrm{kg}}{\mathrm{m^2\cdot s}}\right]$$

In this chapter we treat in turn

- Adiabatic flow in a pipe
- Isothermal flow in a pipe
- Flow through a nozzle

We then combine the equations for nozzle and pipe to treat the discharge of gases from a tank of reservoir through a pipe, as sketched in Fig. 3.1.

3.1 Adiabatic Flow in a Pipe with Friction

Consider a gas flowing through a well-insulated but real pipe, one with frictional resistance. Figure 3.2 shows how conditions change as the gas moves down the pipe.

Suppose p_1 is fixed and p_3 is adjustable. When p_3 is a bit smaller than p_1, flow through the pipe is slow and $p_2=p_3$. However, as p_3 is lowered more and more the velocity of the gas at the pipe outlet u_2 increases until it reaches the speed of sound in that gas. This speed represents the actual mean velocity of motion of the individual molecules of gas.

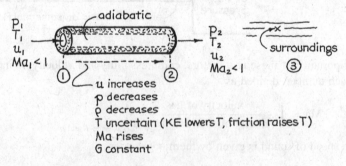

Fig. 3.2 Slow flow in a well-insulated pipe; $p_2=p_3$

If p_3 is lowered further, the gas leaving the pipe cannot go faster and the velocity remains sonic at p^*, T^*, u^*, and $\mathrm{Ma}_2 = 1$, and the flow rate remains unchanged. Thus, the maximum possible flow rate of gas in that pipe has been reached for the particular inlet pressure p_1. So for $p_3 < p^*$ we have what is called *choked flow*, as shown in Fig. 3.3.

To quantitatively characterize this adiabatic flow with friction, use both the mechanical energy and the total energy balances, equations (1.7) and (1.4), for a differential section of pipe.

$$\underbrace{g\,dz}_{\approx 0} + u\,du + \underbrace{\frac{dp}{\rho}}_{} + \underbrace{d\left(\Sigma F\right)}_{\dfrac{2f_F u^2 dL}{d} = 0} + \underbrace{dW_s}_{= 0} = 0 \quad \left[\frac{\mathrm{J}}{\mathrm{kg}}\right] \tag{3.3}$$

ρ changes greatly, cannot be assumed to be a constant, and accounts for adiabatic expansion

$$\underbrace{dh}_{C_p dT} + u\,du + \underbrace{g\,dz}_{\approx 0} = \underbrace{dq}_{= 0 \text{ adiabatic}} - \underbrace{dW_s}_{= 0} = 0 \quad \left[\frac{\mathrm{J}}{\mathrm{kg}}\right] \tag{3.4}$$

Combining these expressions, writing everything in terms of the Mach number

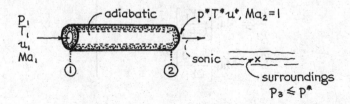

Fig. 3.3 The fastest possible flow in a well-insulated pipe gives sonic velocity at the exit

with the help of equations (3.1) and (3.2), assuming a constant value for f_F throughout the length of the pipe, we obtain, after considerable manipulation and integration,

$$\left.\begin{array}{l} \dfrac{T_2}{T_1} = \dfrac{Y_1}{Y_2}, \quad \text{where } Y_i = 1 + \dfrac{k-1}{2}\text{Ma}_i^2 \\[3mm] \dfrac{p_2}{p_1} = \dfrac{\text{Ma}_1}{\text{Ma}_2}\left[\dfrac{Y_1}{Y_2}\right]^{1/2} \\[3mm] \dfrac{\rho_2}{\rho_1} = \dfrac{\text{Ma}_1}{\text{Ma}_2}\left[\dfrac{Y_2}{Y_1}\right]^{1/2} \end{array}\right\}$$

(3.5)

$$G = \text{Ma}_1 p_1 \left[\dfrac{(mw)k}{RT_1}\right]^{1/2} = \text{Ma}_2 p_2 \left[\dfrac{(mw)k}{RT_2}\right]^{1/2} \quad \left[\dfrac{\text{kg}}{\text{m}^2\text{s}}\right]$$

(3.6)

and

$$\underbrace{\dfrac{k+1}{2}\ln\left[\dfrac{\text{Ma}_2^2 Y_1}{\text{Ma}_1^2 Y_2}\right]}_{\text{KE term}} - \underbrace{\left(\dfrac{1}{\text{Ma}_1^2} - \dfrac{1}{\text{Ma}_2^2}\right)}_{\substack{\text{Compressibility} \\ \text{term}}} + \underbrace{k\left(\dfrac{4f_F L}{d}\right)}_{\substack{\text{Pipe} \\ \text{resistance} \\ \text{term}}} = 0$$

(3.7)

which relates the flow to the frictional resistance.

The *largest possible flow rate* in an adiabatic pipe with friction (choked flow) comes with sonic velocity at the exit. Hence, putting $\text{Ma}_2 = 1$ we find

$$\left.\begin{array}{l} \dfrac{T*}{T_1} = \dfrac{2Y_1}{k+1} \underset{k=1.4}{\overset{\text{for Ma}_1 \approx 0}{=\!=\!=\!=\!=}} 0.833 \\[3mm] \dfrac{p*}{p_1} = \text{Ma}_1\left[\dfrac{2Y_1}{k+1}\right]^{1/2} \\[3mm] \dfrac{\rho*}{\rho_1} = \text{Ma}_1\left[\dfrac{k+1}{2Y_1}\right]^{1/2} \end{array}\right\}$$

(3.8)

$$G* = \text{Ma}_1 p_1 \left[\dfrac{(mw)k}{RT_1}\right]^{1/2} = p*\left[\dfrac{(mw)k}{RT*}\right]^{1/2} \quad \left[\dfrac{\text{kg}}{\text{m}^2\text{s}}\right]$$

(3.9)

and

$$\dfrac{k+1}{2}\ln\left[\dfrac{2Y_1}{(k+1)\text{Ma}_1^2}\right] - \left(\dfrac{1}{\text{Ma}_1^2} - 1\right) + k\left(\dfrac{4f_F L}{d}\right) = 0$$

(3.10)

Note that the outlets T, ρ, and p are dependent only on the Mach numbers and that it is equation (3.7) or (3.10) which relates the Mach numbers with the frictional loss in the length of pipe.

3.2   Isothermal Flow in a Pipe with Friction

This represents a long pipeline (see Fig. 3.4). Consider first the flow through a pipe with fixed inlet conditions. As p_3 is lowered the flow rate increases until choking conditions are reached. However, for isothermal flow we will show that the limiting exit Mach number is $1/\sqrt{k}$ instead of 1 which was found for adiabatic flow. So

$$\left.\begin{array}{l} \text{Ma}_{\text{choking, isothermal}} = 1/\sqrt{k} \\ \text{Ma}_{\text{choking, adiabatic}} = 1 \end{array}\right\} \tag{3.11}$$

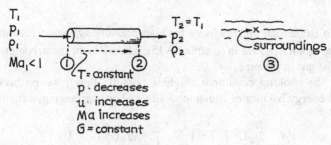

Fig. 3.4 Slow isothermal flow in a pipe; $p_2 = p_3$

Applying the mechanical energy balance to a differential slice of pipe gives

$$\underbrace{g\,dz}_{\cong 0} + u\,du + \frac{dp}{\rho} + d\left(\Sigma F\right) + \underbrace{d\,W_s}_{=0} = 0 \qquad \left[\frac{\text{J}}{\text{kg}}\right] \tag{3.12}$$

Ignoring the total energy balance since the gas is isothermal, dividing by u^2, writing everything in terms of the Mach number, assuming constant f_F, and integrating gives

$$\underbrace{2\ln\frac{\text{Ma}_2}{\text{Ma}_1}}_{\text{KE term}} - \underbrace{\frac{1}{k}\left(\frac{1}{\text{Ma}_1^2} - \frac{1}{\text{Ma}_2^2}\right)}_{\substack{\text{Compressibility} \\ \text{term}}} + \underbrace{\frac{4f_F L}{d}}_{\substack{\text{Pipe} \\ \text{resistance}}} = 0 \tag{3.13}$$

with

$$\left.\begin{array}{c} T_2 = T_1 \\[4pt] \dfrac{p_2}{p_1} = \dfrac{\mathrm{Ma}_1}{\mathrm{Ma}_2} \\[8pt] \dfrac{\rho_2}{\rho_1} = \dfrac{\mathrm{Ma}_1}{\mathrm{Ma}_2} \end{array}\right\} \qquad (3.14)$$

and

$$G = \mathrm{Ma}_1 p_1 \left[\frac{(mw)k}{RT}\right]^{1/2} \qquad \left[\frac{\mathrm{kg}}{\mathrm{m^2 s}}\right] \qquad (3.15)$$

For close to isothermal conditions or for high-pressure systems where the gas is nonideal, replace RT by $\overline{Z}R\overline{T}$ in equation (3.15), where $\overline{Z}$ is the mean compressibility factor for the gas in the pipe.

To find the choking conditions (highest possible flow), we go back to the mechanical energy balance of equation (3.12), and on rearranging, write it as

$$-\frac{dp}{dL} = \frac{2f_F G^2}{\rho d}\left[\frac{1}{1-(u^2\rho/p)}\right] = \frac{2f_F G^2}{\rho d}\left[\frac{1}{1-k\mathrm{Ma}^2}\right] \qquad (3.16)$$

Now $-(dp/dL)$ can only be positive, never negative, and the highest flow rate results when $-(dp/dL) \to \infty$. Thus, for choked flow, as sketched in

Figure 3.5, equation (3.16) shows that $\mathrm{Ma}_2 = 1/\sqrt{k}$, so

$$\left.\begin{array}{c} T_{ch} = T_1 \\[4pt] p_{ch} = p_1 \mathrm{Ma}_1 \sqrt{k} \\[4pt] \rho_{ch} = \rho_1 \mathrm{Ma}_1 \sqrt{k} \\[8pt] u_{ch} = \dfrac{u_1}{\mathrm{Ma}_1 \sqrt{k}} \end{array}\right\} \qquad (3.17)$$

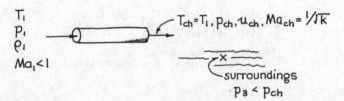

Fig. 3.5 The fastest possible does not give sonic velocity at the exit

$$G_{ch} = u_{ch}\rho_{ch} = u_1\rho_1 = G_1 = Ma_1p_1\left[\frac{(mw)k}{RT}\right]^{1/2} = p_{ch}\left[\frac{(mw)}{RT}\right]^{1/2} \qquad (3.18)$$

and

$$\ln\left[\frac{1}{k Ma_1^2}\right] - \left(\frac{1}{k Ma_1^2} - 1\right) + \frac{4f_FL}{d} = 0 \qquad (3.19)$$

Equation (3.19) relates the frictional resistance to choked flow conditions.

3.3   Working Equations for Flow in Pipes (No Reservoir or Tank Upstream)

For *adiabatic flow* equation (3.7) in alternative and probably more useful form becomes

$$\frac{k+1}{k}\ln\frac{p_1T_2}{p_2T_1} - \frac{k-1}{2k}\left(\frac{p_1^2T_2^2 - p_2^2T_1^2}{T_2 - T_1}\right)\left(\frac{1}{p_1^2T_2} - \frac{1}{p_2^2T_1}\right) + \frac{4f_FL}{d} = 0 \qquad (3.20a)$$

with

$$G = \left[\frac{2(mw)k}{(k-1)R} \cdot \frac{T_2 - T_1}{(T_1/p_1)^2 - (T_2/p_2)^2}\right]^{1/2} \qquad (3.20b)$$

With upstream conditions known and p_2 given, find T_2 from equation (3.20a), then G from equation (3.20b); see Turton (1985).

Similarly, for *isothermal flow*, equation (3.13) becomes

$$2\,\ln\frac{p_1}{p_2} - \frac{(mw)}{G^2RT}(p_1^2 - p_2^2) + \frac{4f_FL}{d} = 0 \qquad (3.21)$$

For situations where nonideal gas behavior and/or somewhat nonisothermal flow occurs, we can use the coagulated expression

$$\frac{2}{\alpha}\ln\frac{p_1}{p_2} - \frac{(mw)\left(p_1^2 - p_2^2\right)}{G^2\left(\bar{Z}R\bar{T}\right)} + \frac{4f_FL}{d} = 0$$

$$\uparrow$$

$$\text{Deviation from plug flow}$$
$$\alpha \cong 1 \text{ for turbulent flow}$$

$$(3.22a)$$

where

$$G = \frac{v\rho}{A} = \frac{vp(mw)}{AR\overline{T}} \quad \left[\frac{kg}{m^2 \cdot s}\right] \tag{3.22b}$$

3.4   Flow Through an Orifice or Nozzle

Suppose that gas flows in frictionless adiabatic flow from p_0 to p_1 through a smooth well-rounded orifice as shown in Fig. 3.6. We call this a nozzle. The mechanical and total energy balances then give

In so small a region we can reasonably assume that the frictional loss is negligible compared to other effects

$$g\cancel{dz} + u\, du + \underset{\displaystyle \curvearrowleft}{\cancel{\frac{dp}{\rho}}} + d\cancel{W_s} + d\left(\Sigma F\right) = 0$$

Changes adiabatically and reversibly

$$\int \frac{C_p dT}{dh} + u\, du + g\cancel{dz} = d\cancel{q} - d\cancel{W_s}$$

Accounting for the changing cross-sectional area of the flow channel, and writing everything in terms of the Mach number gives, on integration,

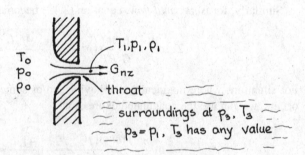

T_0
p_0
ρ_0

T_1, p_1, ρ_1
G_{nz}
throat
surroundings at p_3, T_3
$p_3 = p_1$, T_3 has any value

Fig. 3.6 Flow through a smooth orifice or nozzle

$$\left.\begin{array}{l} \mathrm{Ma}_1^2 = \dfrac{2}{k-1}(Y_1 - 1) \ \text{ or } \ Y_1 = 1 + \dfrac{k-1}{2}\mathrm{Ma}_1^2 \\[2mm] \dfrac{T_1}{T_0} = \dfrac{1}{Y_1} \\[2mm] \dfrac{p_1}{p_0} = \dfrac{1}{Y_1^{k/k-1}} \ \text{ or } \ Y_1 = \left(\dfrac{p_0}{p_1}\right)^{(k-1)/k} \\[2mm] \dfrac{\rho_1}{\rho_0} = \dfrac{1}{Y_1^{1/(k-1)}} \end{array}\right\} \tag{3.23}$$

$$G_{nz} = p_0 \mathrm{Ma}_1 \left[\frac{(mw)k}{RT_0} \frac{1}{Y_1^{(k+1)/(k-1)}}\right]^{1/2} \left[\frac{\mathrm{kg}}{\mathrm{m}^2\mathrm{s}}\right] \tag{3.24}$$

$$u_1 = \mathrm{Ma}_1 c_1 = \mathrm{Ma}_1 \left[\frac{kRT_0}{(mw)Y_1}\right]^{1/2} \tag{3.25}$$

If p_1 is lowered there comes a point where the exit gases are at sonic velocity, or $\mathrm{Ma}_1 = 1$. Further lowering of p_1 will not give an increase in flow. For these conditions, sketched in Fig. 3.7,

$$\left.\begin{array}{l} \dfrac{T*}{T_0} = \dfrac{2}{1+k} \xrightarrow{\ k=1.4\ } 0.833 \\[3mm] \dfrac{p*}{p_0} = \left(\dfrac{2}{1+k}\right)^{k/(k-1)} \xrightarrow{\ k=1.4\ } 0.528 \\[3mm] \dfrac{\rho*}{\rho_0} = \left(\dfrac{2}{1+k}\right)^{1/(k-1)} \xrightarrow{\ k=1.4\ } 0.634 \end{array}\right\} \tag{3.26}$$

At the throat the maximum flow rate is

$$\left.\begin{array}{l} G_{nz}^* = p_0 \left[\dfrac{(mw)k}{RT_0}\left(\dfrac{2}{1+k}\right)^{(k+1)/(k-1)}\right]^{1/2} \xrightarrow{\ k=1.4\ } 0.685\, p_0\left[\dfrac{(mw)}{RT_0}\right]^{1/2} \\[5mm] \hspace{3cm} \xrightarrow[\substack{20°C,\, 1\,\text{atm}}]{\text{air}} 239\dfrac{\mathrm{kg}}{\mathrm{m}^2\mathrm{s}} \end{array}\right\} \tag{3.27}$$

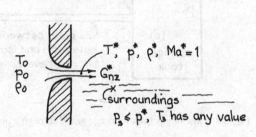

Fig. 3.7 Maximum flow
through a smooth orifice

or

$$u_0^* = c_0^* \text{Ma}_0^* = \left[\frac{2kRT_0}{(mw)(k+1)}\right]^{1/2} \underset{\substack{\text{air} \\ \overline{20\,°C}}}{\overset{k=1.4}{=\!=\!=}} 1.080 \left[\frac{RT_0}{(mw)}\right]^{1/2} \Bigg\} \qquad (3.28)$$

$$313.6\,\frac{m}{s}$$

3.4.1 Comments

1. First check to see whether sonic velocity is reached at the throat, and then use the correct equation.
2. By using $p_1/p_0 < 0.5$ we have a simple accurate flow gauge which depends on p_0 and A_{throat} alone; see equation (3.26).
3. For a nonrounded nozzle, say a square orifice, all sorts of uncertainties intrude in flow which we don't know how to treat. So probably it is best to treat the sharp-edged orifice as an ordinary pipe entrance with an equivalent length of 16 d (from Chap. 2). However, whenever possible stay away from them. Just round and smooth the opening.

3.5 Pipe Leading from a Storage Vessel

In a tank–nozzle–pipe system, such as shown in Fig. 3.8, the time needed to go through the nozzle is so short that adiabatic flow is always a good approximation, while flow in the pipe will be somewhere between adiabatic and isothermal—but which? Consider the following:

The pipe flow equations developed in this chapter all assumed one-dimensional (or plug) flow. But this is not strictly correct because a layer of slow-moving fluid always coats the inner wall of the pipe. Consequently, although an adiabatic gas progressively expands and cools as it races down the pipe, it returns to its stagnation temperature whenever it is brought to rest at the pipe wall. Thus, the pipe walls

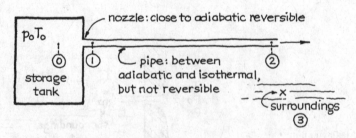

Fig. 3.8 Flow of gas from a storage vessel through a pipe

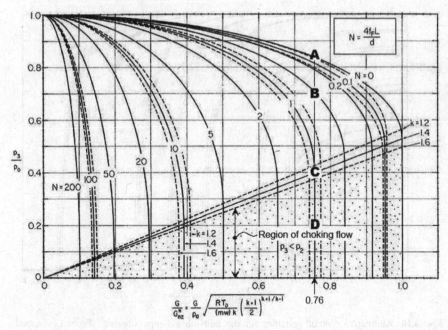

Fig. 3.9 Graph representing the flow of gas through a pipe from a high-pressure storage tank. Simultaneous solution of equations (3.5), (3.6), (3.7), (3.8), (3.9), and (3.10) with (3.21), (3.22a), (3.22b), (3.23), (3.24), (3.25), and (3.26) (From Levenspiel (1977))

experience the stagnation temperature of the gas despite the fact that the fast-moving gas in the central core may be much cooler.

For this reason, the adiabatic nozzle–adiabatic pipe equations probably better represents the real pipe with constant wall temperature. We consider this situation from now on. In any case, this is not a very serious point because the difference in predictions of the isothermal and adiabatic pipe equations is not significant and can hardly be noted on the performance charts.

Figure 3.9 is a dimensionless plot relating flow rate with overall pressure drop and frictional resistance for a smooth well-rounded orifice followed by a pipe. Note that the longer the pipe, the smaller the maximum throughput even though the velocity of gas at the pipe exit can be sonic. Figure 3.10 is a cross-plot of Fig. 3.9, useful for longer pipes. For an abrupt sharp-edged orifice, add the equivalent length of 16 pipe diameters (from Chap. 2) to the length of pipe.

For long pipes the contribution of the orifice pressure drop becomes negligible, and the graphs just represent the equations for pipe flow alone.

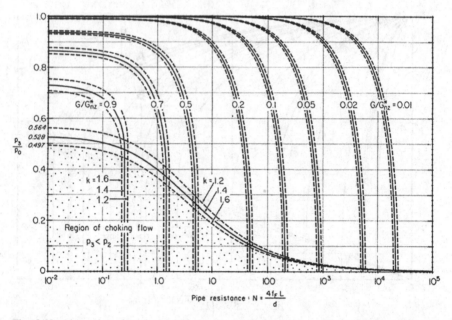

Fig. 3.10 Alternative plot of variables for the tank–nozzle–pipe situation (From Levenspiel (1977))

Example 3.1. Nitrogen to an Ammonia Plant
Nitrogen ($k = 1.4$) is to be fed through a 15-mm-i.d. steel pipe 11.5 m long to a synthetic ammonia plant. Calculate the downstream pressure in the line for a flow rate of 1.5 mol/s, an upstream pressure of 600 kPa, and a temperature of 27 °C throughout.

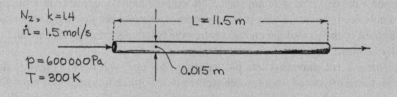

Solution
Method A. Let us use equation (3.13). For this, evaluate in turn

(continued)

(continued)

$$\dot{m} = \left(1.5\,\frac{mol}{s}\right)\left(0.028\,\frac{kg}{mol}\right) = 0.042\,\frac{kg}{s}\;\text{(mass flow rate)}$$

$$G = \frac{\dot{m}}{\text{area}} = \frac{0.042\;kg/s}{(\pi/4)(0.015)^2 m^2} = 237.7\,\frac{kg}{m^2 s}\;\text{(mass velocity)}$$

$$\mathrm{Re} = \frac{dG}{\mu} = \frac{(0.015)(237.7)}{2\times 10^{-5}} = 180{,}000$$

$\epsilon = 0.046\;mm$ (from Table 2.1, pipe roughness)

$$\frac{\epsilon}{d} = \frac{0.046\;mm}{15\;mm} = 0.003\;\text{(roughness ratio)}$$

$$f_F = 0.006\,75$$

$$N = \frac{4f_F L}{d} = \frac{4(0.00675)11.5\;m}{0.015\;m} = 20.7$$

$$\mathrm{Ma}_1 = \frac{G}{p_1}\left[\frac{RT}{(mw)k}\right]^{1/2} = \frac{237.7}{600{,}000}\left[\frac{8.314\;(300)}{0.028\;(1.4)}\right]^{1/2} = 0.1$$

Replacing in equation (3.13) gives

$$-\ln\left(\mathrm{Ma}_1^2/\mathrm{Ma}_2^2\right) - 71.54\left[1 - \left(\mathrm{Ma}_1^2/\mathrm{Ma}_2^2\right)\right] + 20.70 = 0$$

Solve this equation by trial and error.

Guess $\mathrm{Ma}_1/\mathrm{Ma}_2$	l.h.s. (left-hand side)
0.5	−31.57
0.4	−37.56
0.8	−4.61
0.87	+3.59
0.84	−0.014 (close enough)
0.841	+0.104

Therefore, from equation (3.14)

$$\frac{p_2}{p_1} = \frac{\mathrm{Ma}_1}{\mathrm{Ma}_2} = 0.84$$

Thus, the downstream pressure

(continued)

(continued)

$$p_2 = 0.84\,p_1 = 0.84(60{,}000) = 504{,}000 \ \text{Pa}$$

Method B. Let us use the recommended working expression for isothermal flow in pipes or equation (3.21). This can be written as

$$-\ln\left(\frac{p_2^2}{p_1^2}\right) - \frac{(mw)p_1^2}{G^2RT}\left(1 - \frac{p_2^2}{p_1^2}\right) + \frac{4f_FL}{d} = 0$$

and on replacing values

$$-\ln\left(\frac{p_2^2}{p_1^2}\right) - 71.54\left(1 - \frac{p_2^2}{p_1^2}\right) + 20.70 = 0$$

Noting that $p_2/p_1 = \text{Ma}_1/\text{Ma}_2$, we recognize the above expression as equation (i) of the previous method. The rest follows as with method A.

Example 3.2. Design of a Critical Orifice Flow Meter

We want the flow of helium ($k = 1.66$) in a 100-mm-i.d. pipe to be 4 m/s at 105 °C and 200 kPa. The storage tank from which we draw helium contains the gas at 1 MPa and -20 °C. How do we get the desired flow with a critical orifice?

Solution

The required flow rate is

$$\dot m = \left[\frac{\pi}{4}(0.1)^24\right]\frac{\text{m}^3}{\text{s}}\left(\frac{200{,}000}{101{,}325}\right)\left(\frac{273}{378}\right)\left(\frac{1\ \text{mol}}{0.0224\ \text{m}^3}\right)\left(\frac{0.00403\ \text{kg}}{\text{mol}}\right)$$

$$= 8\times10^{-3}\frac{\text{kg}}{\text{s}}$$

Since the pressure ratio is well over 2:1, we have critical flow through the orifice. So equation (3.27) becomes

(continued)

(continued)

$$G_{nz}^* = 10^6 \left[\frac{(0.00403)(1.66)}{(8.314)(253)} \left(\frac{2}{2.66} \right)^{2.66/0.66} \right]^{1/2} = 1,000 \frac{kg}{m^2 s}$$

To find the diameter of hole needed, note that

$$\dot{m} = AG = (\pi/4)d^2 G = \frac{kg}{s}$$

or

$$d = \left(\frac{4\dot{m}}{\pi G} \right)^{1/2} = \left(\frac{4}{\pi} \frac{8 \times 10^{-3}}{1,000} \right)^{1/2} = 3.19 \text{ mm}$$

Comment. Since we end up with a small-diameter orifice followed by a large-diameter pipe, it is safe to ignore the resistance of the pipe.

Example 3.3. Use of Design Charts for Flow of Gases
Air at 1 MPa and 20 °C in a large high-pressure tank (point 0) discharges to the atmosphere (point D) through 1.25 m of 15-mm-i.d. drawn tube with a smooth rounded inlet. What is the pressure:

  (a) Halfway down the tube or at point B
  (b) Just at the tube inlet or at point A
  (c) Just before the tube exit or at point C

Solution
Since this problem involves a high-pressure tank–nozzle–pipe discharging air to the atmosphere (point 0), we can solve it with the design charts of this chapter. So referring to the drawing below, let us evaluate the terms needed to use the charts.

(continued)

(continued)

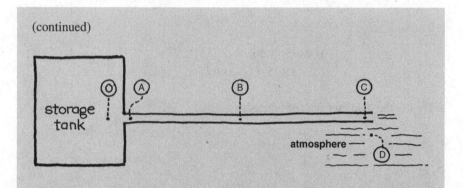

$$\epsilon = 0.0015 \text{ mm (from Table 2.1)}$$

$$\frac{\epsilon}{d} = \frac{0.0015 \text{ mm}}{15 \text{ mm}} = 0.0001$$

$$f_F = 0.003 \text{ (from Fig 2.4 assuming complete}$$

$$\text{turbulence---check below)}$$

$$N = \frac{4f_F L}{d} = \frac{4(0.003)(1.25 \text{ m})}{0.015 \text{ m}} \cong 1$$

$$k = 1.4(\text{for air})$$

$$\text{so,} \quad \frac{p_D}{p_0} = \frac{101,325}{10^6} \cong 0.1$$

With this information, locate point A in Fig. 3.9, as shown in the sketch below.

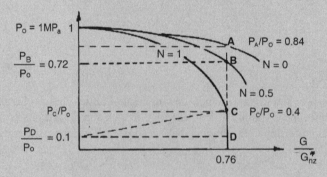

This gives $G/G^*_{nz} = 0.76$.

(continued)

(continued)

(a) Halfway down the pipe G/G_{nz}^* remains unchanged at 0.76; however, $N = 0.5$. This locates point B which in turn gives

$$\frac{p_B}{p_0} = 0.72$$

Thus, the pressure halfway down the pipe is

$$p_B = \boxed{720 \text{ kPa}}$$

Ignoring the total energy

(b) At the tube inlet $p_A/p_0 = 0.84$ or $p_A = 840$ kPa, see point A
(c) At the tube exit $p_C/p_0 = 0.40$ or $p_C = 400$ kPa, see point C.

Now check the assumption of complete turbulence. From G/G_{nz}^*, (3.27) with appendix A.13 and the data given in this problem, we find

$$\text{Re} = \frac{dG}{\mu} = \frac{(0.015)(0.76)(239)}{1.83 \times 10^{-5}} \cong 1.48 \times 10^5$$

which is well in the complete turbulence regime.

Problems on the Compressible Flow of Gases

The topics dealt with in the problems are:

Problems		
	1–2	Mach number alone
	3–9	Flow in pipes alone
	10	Nozzle
	11–30	Pipes leading from tanks or reservoirs
	31–34	Heat pipes
	35	?

3.1. *Storms on Neptune. Science*, **246** 1369 (1989) reports on spacecraft Voyager's visit to the outer planets of Jupiter, Saturn, Uranus, and Neptune. It took 12 years to get to Neptune, and there it found a surface of solid nitrogen and an atmosphere of nitrogen at 1.4 Pa and 38 K. Also, it was reported that the jet stream on Neptune blew at 1,500 mph. What Mach number does this represent? Do you think that your answer makes sense?

3.2. *Ultrasonic gas flowmeter. Chemical and Engineering News*, p. 16 (August 27, 1984) reports that physicists at the National Bureau of Standards have patented a novel instrument for measuring the flow rate of gases in pipes. Basically, it consists of two microphones plus a loudspeaker, all mounted on the flow pipe.

With such a device find the flow rate in kg/s of helium (15 °C, 200 kPa) in a 0.1-m-i.d. duct if the two microphones are located 6 pipe diameters apart and if the difference in arrival time at these microphones of a sharp "beep" from a downstream sound source is 1 ms.

3.3. It bugs me that they are so tightlipped about the production rate of their new coal gasification plant. But, perhaps we could work it out for ourselves. I noted that the gas produced ($mw = 0.013$, $\mu = 10^{-5}$ kg/m s, $k = 1.36$) is sent to neighboring industrial users through a bare 15-cm-i.d. pipe 100 m long. The pressure gauge at one end of the pipe read 1 MPa absolute. At the other end it read 500 kPa. I felt pain when I touched the pipe, but when I spat on it, it didn't sizzle, so I guess that the temperature is 87 °C. Would you estimate for me the flow rate of coal gas through the pipe, both in tons/day and in m^3/s measured at 1 atm and 0 °C?

3.4. For our project on the biochemical breakdown of grass straw, we need to oxygenate the deep fermenter vat by introducing 5 lit/s of air at 2 atm. We obtain this air from a compressor located 100 m away through a steel pipe 0.1 m i.d. What should be the pressure at the pipe inlet to guarantee this flow rate? Assume that everything, vat and feed pipe, is at 20 °C.

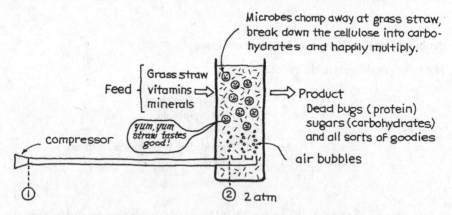

3.5. Repeat Example 3.1 with one change; the flow rate is to be doubled to 3 mol/s.

3.6. *Hydrogen pipelines.* If the United States converts to a hydrogen economy, the electricity produced by power plants and dams will be used directly to decompose water into its elements. Then, instead of pumping electricity everywhere on unsightly power lines, hydrogen will be pumped in a network of unseen underground pipelines. This hydrogen will be used to fuel autos, to heat homes, and to produce small amounts of needed electricity. If this day comes, one trunk pipeline (0.5 m i.d.) will come direct to Corvallis from Bonneville Dam 300 km away. The pressure of H_2 entering the pipeline will be 2 MPa; at Corvallis, it will be 1 MPa. The temperature throughout the line can be estimated to be 20 °C. Find the flow rate of hydrogen under these conditions in kg/s and in std. m³/s (at 1 atm, 0 °C).

Consider various aspects of the proposed hydrogen pipeline of the above problem.

3.7. If gas consumption at Corvallis rises high enough so that the pipeline pressure at Corvallis drops from 1 to 0.5 MPa, what would be the gas flow rate? The pressure at Bonneville would remain unchanged.

3.8. If a 1-m pipeline were laid instead of the 0.5-m line, how much gas in kg/s and in std. m³/s could be transported?

3.9. If the hydrogen pressure at Bonneville were doubled to 4 MPa, while the pressure at Corvallis stays at 1 MPa, what flow rate can be obtained?

3.10. Nitrogen ($k = 1.39$) at 200 kPa and 300 K flows from a large tank through a smooth nozzle (throat diameter $= 0.05$ m) into surroundings at 140 kPa. Find the mass flow rate of nitrogen and compare it with choked flow. [This problem was taken from F. A. Holland, *Fluid Flow for Chemical Engineers*, p. 125, Arnold, London (1973).]

3.11. A safety vent from a 3.5-atm reactor to the atmosphere consists of 5 m of 15-mm commercial steel pipe leading off from the reactor. Is flow choked or not?

3.12. Methane ($k = 1.2$) discharges from one tank (112 kPa) to another (101 kPa) through 2.4 m of 7.66-mm-i.d. pipe. I am afraid that the connecting pipe will snap off at the upstream end since it is just held in place there with chewing gum. If it does, what will happen to the discharge rate of methane from the upstream tank?

3.13. 25 mol/s of ethylene ($k = 1.2$, $\mu = 2 \times 10^{-5}$ kg/m s) are to be fed to a reactor operating at 250 kPa from a storage tank at 60 °C and 750 kPa. This flow is to be controlled by a discharge control tube 24 mm i.d. made of commercial iron pipe as shown below. What length of control tube is needed?

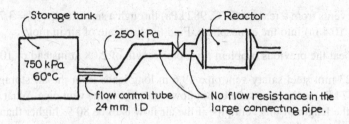

Storage tank
250 kPa
Reactor
750 kPa
60°C
flow control tube
24 mm ID
No flow resistance in the large connecting pipe.

3.14. Hydrogen ($k = 1.4$) flows from a tank at 1 Mpa to a second tank at 400 kPa through 36 m of 115-mm commercial steel pipe with a smooth entrance. What is the pressure p_1 in this connecting pipe just outside the high-pressure tank?

3.15. Here's my flowmeter—2 m of 3-cm-i.d. pipe sticking out from the gas storage tank ($-20\,°C$), ending with a cap having a 0.5-mm hole drilled in it so as to make a critical orifice. However, the size of hole is not right, because the flow is exactly 8 % too high. Even if I drill another hole, I'll most likely still be off, so no more holes for me. What should I do? How to get the right flow rate with the present orifice. Make your calculations and present your answer with a sketch.

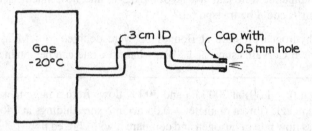

3.16. Drat! The smooth hole drilled into the wall of the tank (5 atm inside, 1 atm outside) is too big because the flow is 5 times that desired. With a 200-mm-long pipe (with same i.d. as the hole) fitted to the hole, the flow is 2.5 times that desired. What length of pipe should we fit to the hole to get just the right flow rate?

3.17. With a hole at the wall of the storage tank (200 kPa inside), the discharge to the surroundings (144 kPa) is 2.5 times that desired. But when we attach a tube 1 m long to the orifice (same diameters), the flow rate is three-fourths that desired. What length of tube should we use to get the right flow rate?

3.18. When the pressure in the tank rises to 180 kPa, the relief valve to the atmosphere (100 kPa) opens. When the pressure drops to 110 kPa, the valve closes and the pressure rises again. Estimate the ratio of flow rates of air at the end and beginning of this operation if the temperature remains at 300 K throughout the cycle.

3.19. Air vents from a reactor (0 °C, 982 kPa) through a smooth pipe ($d = 3.75$ mm, $L = 10.4$ m) into the atmosphere. Find the flow rate of air in mol/s.

3.20. Repeat the previous problem for a pipe length of 2.08 m instead of 10.4 m.

3.21. A 23-mm steel safety vent pipe 19.6 m long leads from the air storage tank (167 kPa) to the atmosphere. But flow is too low in the vent pipe. What should be the length of this vent pipe if the air flow is to be 80 % higher than in the present pipe?

Oxygen is fed to a reactor operating at 240 kPa from a storage tank at 300 kPa through 1,187.5 mm of 7.6-mm-i.d. steel pipe with a smooth inlet. Flow is highly turbulent; however, the flow rate is not high enough. By what percentage or fraction does the mass flow rate change:

3.22. If the pressure of oxygen in the storage tank is raised to 800 kPa?

3.23. If the original connecting pipe is replaced by one which is twice the diameter?

Gas flows from tank A through a pipe to tank B, and the pressures in the tanks are such that choked flow exists in the pipe. What happens to the flow rate:

3.24. If the pressure in tank A is doubled and the pressure in tank B is halved?

3.25. If the pressures in tanks A and B are both doubled?

3.26. Nitrogen is fed to our reactor operating at 280 kPa from a storage tank which is at 350 kPa through 1,187.5 mm of 7.6-mm-i.d. steel tubing with a smooth inlet. If the pressure in the reactor is lowered to 1 atm, by what percentage will the flow rate increase?

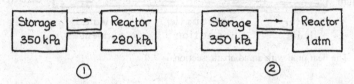

3.27. Air discharges from a large tank through 1.25 m of 15-mm-i.d. drawn tubing with smooth entrance. The pressure in the tank is 1 MPa, 1 bar outside. What is the flow rate of air from the tank in kg/s?

3.28. I. Györi, in *Chem. Eng.,* p. 55 (October 28, 1985) presents a calculator program for solving gas flow problems and as an example shows how to solve the following.

Gas ($\overline{mw} = 0.029$, $k = 1.4$, $C_p = 1.22$) flows from one tank ($p = 10^6$ Pa) to a second ($p = 10^5$ Pa) through $L = 11.88$ m of $d = 0.051$ m iron pipe which contains three 90° elbows.

Find the mass velocity and mass flow rate (kg/s) of this gas.

You may wish to compare your method of solution and answer with that given in this article.

3.29. Calculate the discharge rate of air (kg/s) to the atmosphere from a reservoir at 1.10 MPa at 20 °C through 10 m of straight 2″ sch 40 steel pipe (5.52 cm i.d.) and 3 standard 90° elbows. The pipe inlet is abrupt or sharp edged.

NOTE: Compare your answer to that given in *Perry's Chemical Engineers' Handbook*, sixth edition, pp. 5–30, McGraw-Hill, New York, 1984.

3.30. Calculate the discharge rate of air to the atmosphere from a reservoir 150 psig (150 psi + 1 atm) and 70 ° F through 39 ft of straight steel pipe with 2.067 in. i.d. and 3 standard elbows. The pipe inlet is abrupt.

NOTE: This problem statement comes from Perry's *Chemical Engineers' Handbook*, third edition, p. 381, McGraw-Hill, 1950. Your answer and the *Handbook's* answer will differ. Can you figure out why?

Information on Heat Pipes, Useful for Problems 3.31–3.34

The heat pipe is a crafty and efficient way of transferring heat from a hot place to a cold place even with a very small temperature difference. It consists of a sealed pipe with a wick going from end to end and containing just the right fluid. At the hot end (the evaporator) the vapor pressure is high, so liquid boils. At the cold end the vapor pressure is low, so vapor flows to that end and condenses. Then, by capillarity, the condensed liquid is drawn along the wick from the cold end back to the hot end. As an illustration consider the heat pipe containing water and steam shown in Fig. 3.11.

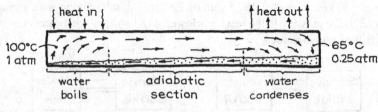

Fig. 3.11 The heat pipe with an adiabatic section

The capacity of a heat pipe is enormous and is limited by one of the following factors:

- Heat transfer rates into and out of the pipe at the hot and cold ends.
- The capillary stream may break, meaning that the hot end uses up liquid faster than the wick can suck it from the cold end.
- The vapor flow from hot to cold end under the prevailing pressure difference may be the slow step.

Here we only consider the last factor. To estimate this limiting flow, note that vapor speeds up in the evaporator section, moves fastest in the adiabatic section, and then slows down in the condenser. As an approximation we can look at this as an orifice followed by a pipe (the adiabatic section).

Sometimes there is no adiabatic section. In this situation vapor speeds up to a maximum at the boundary between evaporator and condenser, then slows down. This can be looked upon as a nozzle, or an orifice, as shown in Fig. 3.12.

Fig. 3.12 The heat pipe
with no adiabatic section

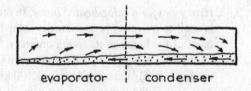

evaporator ┆ condenser

3.31. A *water heat pipe* 1.5 m long is to be used to help equalize the temperature in
two side-by-side regions. The hot portion of the pipe is 0.6 m long and
contains boiling water at 124 °C (vapor pressure = 225 kPa); the cool portion
is 0.9 m long and contains condensing steam at 90 °C (vapor
pressure = 70 kPa). Find the heat transfer rate if the diameter of the vapor
transport section is 1 cm and if vapor transport is the limiting process. Each
kilogram of water going from hot to cold end transfers 2,335 kJ of heat.

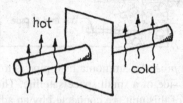

3.32. *Cryogenic medical probe.* The compact handheld cryogenic probe sketched
below is designed to freeze tumors and tissues. It consists of a small high-
pressure nitrogen heat pipe having an evaporator section (fluid at 100 K and
799 kPa) to contact the tumor, a 0.3-m adiabatic section, and a condenser (fluid
at 85 K and 229 kPa) bathed in liquid nitrogen and open to the atmosphere.
Estimate the heat removal rate of the tip if the inside diameter of the heat pipe
is 2 mm, its roughness is 0.02 mm, and if vapor transport is limiting. Each kg of
boiling and condensing nitrogen transports 193 kJ of heat.

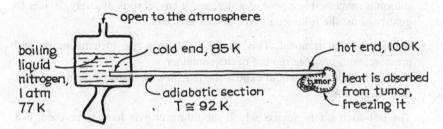

3.33. *Heat pipes for a solar-heated home.* In one design a set of ammonia heat pipes is used to transport heat from the solar-heated reservoir located in the basement of the home to its living quarters. One such pipe consists of an evaporator (ammonia at 1,200 kPa) immersed in the heat reservoir, a long adiabatic section (10 m), and a condenser (1,100 kPa). Find the heat transfer rate of this heat pipe if the vapor transport section has a diameter of 16 mm, a surface roughness $\epsilon = 0.096$ mm, and if vapor transport is limiting. The heat pipe is at about 30 °C; thus, each kilogram of ammonia going from the hot end to the cold end transports 1,155 kJ of heat.

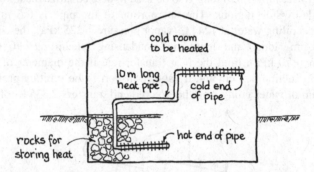

3.34. *Space satellite heat pipe.* An ammonia heat pipe is to transfer heat from the hot side to the cold side of a small space satellite. The large evaporator and condenser sections of this unit are connected by an adiabatic vapor transport section 5 mm i.d., surface roughness of 0.075 mm, and 0.84 m long. At the hot end of the heat pipe, ammonia boils at 25 °C ($\pi = 1$ MPa, $H_{vapor} = 1,465$ kJ/kg), and at the cold end ammonia condenses at -2 °C ($\pi = 400$ kPa, $H_{liquid} = 172$ kJ/kg). Find the heat transfer rate of this unit if vapor transport is limiting. Consider the average temperature of the adiabatic section to be 12 °C. [Problem modified from P. D. Dunn and D. A. Reay, *Heat Pipes*, second ed., p. 133, Pergamon Press, New York (1978).]

3.35. *Accidents at nuclear power stations.* The reactor core of a nuclear power station is immersed in a pool of water, and it has all sorts of safety devices to guard against the following two types of accidents:

• Inadequate heat removal. This would result in a rise in temperature and pressure, with possible rupture of the container.
• Loss of coolant. This would expose the reactor core, cause a meltdown, and destroy the unit.

The last-ditch safety device, which should never ever have to be used, is a water standpipe 10 m high, 1 m i.d., having a safety vent 0.35 m in diameter on top, and emergency water input from below, as shown below.

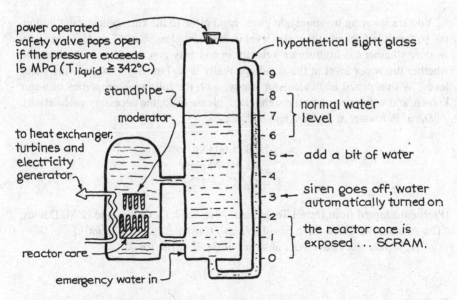

power operated
safety valve pops open
if the pressure exceeds
15 MPa ($T_{liquid} \geqq 342°C$)

hypothetical sight glass

standpipe

moderator

to heat exchanger,
turbines and
electricity
generator

reactor core

emergency water in

9
8
7
6
5
4
3
2
1
0

normal water level

add a bit of water

siren goes off, water automatically turned on

the reactor core is exposed ... SCRAM.

The inconceivable has happened. Elsewhere in the plant something went wrong, safety devices didn't work, the net result being that the heat exchanger fluid stopped removing heat from the reactor. Within 8 s the moderator rods fell into place, shutting down the reactor; however, the residual heat release caused the temperature and pressure to build up and thereby blow open the safety valve on the standpipe. Water boils madly and steam screams out the top of the standpipe, as shown here.

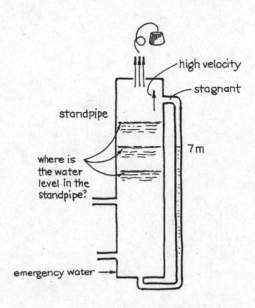

high velocity

stagnant

standpipe

where is
the water
level in the
standpipe?

7 m

emergency water

You are standing by your sight glass, hand close to the emergency water button, but so far all is fine, and the water level in the sight glass shows 7 m. However, you wonder whether the high steam velocity in any way can falsify your reading and whether the water level in the standpipe really is at 7 m. If not, what is the water level? With a pencil and calculator handy, with the following properties of water known, and with one eye on the sight glass, please make the necessary calculations.

Data: For water at 15 MPa and 342 °C,

$$\rho_{\text{liquid}}H_2O = 603 \text{ kg/m}^3$$
$$\rho_{\text{steam}} = 96.7 \text{ kg/m}^3$$
$$k = C_p/C_v = 1.26$$

[Problem adapted from Darrel Teegarden. Also see J. G. Collier and L. M. Davies, "The Accident at Three Mile Island," *Heat Transfer Eng.* **1**, 56 (1980).]

NOTE: If you don't have your answer within 82 s, run for it!

References

O. Levenspiel, The discharge of gases from a reservoir through a pipe. AIChE J. **23**, 402 (1977)

A.H. Shapiro, *The Dynamics and Thermodynamics of Compressible Fluid Flow*, vol. 1, Chapter 6, (Ronald, New York, 1953)

A.H. Streeter, *Fluid Mechanics*, 4th ed., Chapter 6, (McGraw-Hill, New York, 1966)

R. Turton, A new approach to non choking adiabatic compressible flow of an ideal gas in pipes with friction. Chem. Eng. J. **30**, 159 (1985)

Chapter 4
Molecular Flow

The mean free path (mfp) of molecules increases when the gas pressure is reduced, and at low enough pressure the (mfp) is so large that the molecules begin to bounce from wall to wall of the flow channel rather than collide with each other. When this happens the character of the flow changes. Thus, different flow regimes are encountered depending on the value of the ratio

$$Kn = \frac{(mfp)}{d} = \frac{\text{mean free path of molecules}}{\text{diameter of flow channel}}$$

where Kn is called the Knudsen number. These flow regimes are as follows:

- *Ordinary laminar flow* (Kn ≪ 1): Here Poiseuille's law applies. Flow in this regime is based on the following two assumptions:

   (a) $\tau = (\mu)(du/dy)$ with μ = const.
   (b)   Velocity at the wall is zero.

- *Intermediate or slip flow regime* (Kn ≅ 1): Here assumption (b) begins to break down.
- *Molecular flow regime* (Kn ≫ 1): Here there are very few collisions between molecules. Most collisions are with the wall. So the concept of viscosity has no meaning and assumption (a) also breaks down.

The velocity profiles in these three regimes are shown in Fig. 4.1.

Now the mean free path of gas molecules varies with pressure and from kinetic theory of gases is found to be about as follows:

$$\text{at 1 atm}: \quad (mfp) = 6.8 \times 10^{-8} m$$
$$\text{at 1 Pa}: \quad (mfp) = 6.8 \times 10^{-3} m$$

© Springer Science+Business Media New York 2014
O. Levenspiel, *Engineering Flow and Heat Exchange*,
DOI 10.1007/978-1-4899-7454-9_4

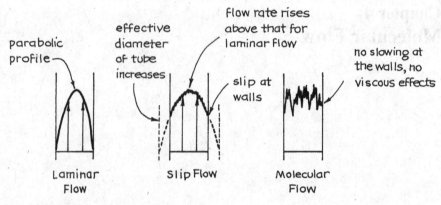

Fig. 4.1 Velocity profiles in various flow regimes

From this we have

$$
\left.
\begin{array}{ll}
\text{Laminar flow :} & \text{when } pd \;>\; 0.8 \,\text{Pa m} \\[4pt]
\text{Molecular flow :} & \text{when } pd \;<\; 0.01 \,\text{Pa m}
\end{array}
\right\}
\qquad (4.1)
$$

In finding how fluids flow in high-vacuum systems, we may have to consider all three flow regimes. In just about all cases, our concern reduces to handling the tank–line–pump problem, such as shown in Fig. 4.2.

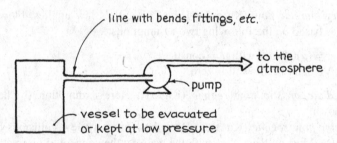

Fig. 4.2 Typical problem for vacuum systems

At one end of the system, flow may be in one regime, at the other end in another. We look at such problems. Also, in evacuating a system the pressure decreases with time, consequently the mass velocity decreases, and the Reynolds number becomes very small. Thus, the transition normally is from laminar to molecular flow, rarely from turbulent to molecular flow.

4.1 Equations for Flow, Conductance, and Pumping Speed

4.1.1 Notation

Molecular flow has its own particular and convenient notation. Let us introduce three essential terms.

1. *Flow rate*. This is measured by

$$Q = \left(\frac{\begin{array}{c} m^3 \text{ of gas flowing if the pressure is corrected} \\ \text{to unit pressure, or 1 Pa} \end{array}}{time} \right) \tag{4.2}$$

$$= p\dot{v} = \dot{n}RT = \frac{\dot{m}RT}{(mw)} = p\frac{\pi d^2}{4}u = \frac{\pi d^2}{4}\frac{GRT}{(mw)}$$

$$= \frac{\pi d\, RT\mu}{4\,(mw)}(\mathrm{Re}) \quad \left[\frac{\mathrm{Pa\ m^3}}{\mathrm{s}} = \frac{\mathrm{N\ m}}{\mathrm{s}} = \mathrm{W} \right]$$

2. *Conductance*. In a flow channel such as sketched in Fig. 4.3, the flow rate is proportional to the driving force, Δp. Thus,

$$Q = -C_{12}\,\Delta p = C_{12}\,(p_1 - p_2) \tag{4.3}$$

$$\mathrm{m^3/s}$$

where C_{12} is called the conductance between points 1 and 2 and is inversely proportional to the resistance to flow in that section of flow channel, or

$$C_{12} \propto \frac{1}{\text{resistance}}$$

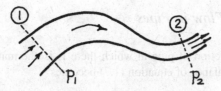

Fig. 4.3 Flow from 1 to 2 due to $p_1 > p_2$

3. *Pumping speed*. The volumetric flow rate of material across a plane normal to flow is called the pumping speed S. Thus, at planes A and B of Fig. 4.4, we have

$$Q = S_A p_A = S_B p_B \tag{4.4}$$

$$\mathrm{m^3\,/\,s}$$

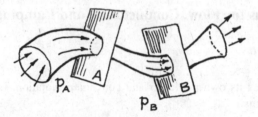

Fig. 4.4 Flow through plane A or B

Note the distinction between pumping speed and conductance. Although they have the same dimensions (m³/s), they are different measures and should not be confused. C refers to a section of flow system, while S measures what passes across a plane normal to flow. Thus, in Fig. 4.5, C_{12} refers to the section between points 1 and 2 and S_A refers to plane A. The following sections will present equations for conductances, pumping speed, and flow rates for various sorts of equipment: pipes, orifices, pumps, and fittings.

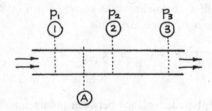

Fig. 4.5 Flow channel showing points 1, 2 and 3 and plane A

4.1.2 Laminar Flow in Pipes

In any differential section of pipe in which there is isothermal laminar flow, the mechanical energy balance of equation (1.7) becomes

$$g\,dz + u\,du + \frac{dp}{\rho} + W_s + d(\textstyle\sum F) = 0$$

$$\underbrace{}_{\text{Ignore}} \quad \underbrace{}_{\text{Ignore}} \quad \underbrace{\frac{dp}{\rho}}_{\substack{=\,0.\\ \text{no pump}}} \quad \underbrace{\frac{32u\mu}{d^2\rho}\,dL}_{\text{[from equation (2.5)]}}$$

Integrating and combining with equation (4.3) gives, between points 1 and 2,

$$Q_{lam} = C_{lam}\ p_1 - p_2 \quad \left[\frac{Pa\ m^3}{s}\right]$$

where

$$\bar{p} = \frac{p_1 + p_2}{2}$$

$$C_{lam} = \frac{\pi d^4 \bar{p}}{128\,\mu L} \frac{air}{20\,°C} 1{,}364 \frac{d^4\bar{p}}{L}$$

$$\frac{H_2O\ vapor}{20\,°C} 2{,}584 \frac{d^4\bar{p}}{L} \quad \left[\frac{m^3}{s}\right]$$

(4.5)

Equation (4.5) represents laminar flow in the "language" of molecular flow. Note that it looks different from the corresponding equation of Chap. 2.

4.1.3   Molecular Flow in Pipes

In this regime we assume no collision between molecules; they simply float from wall to wall of the pipe. But how do molecules leave the wall? Do they bounce off the wall (elastic collision) as shown in Fig. 4.6a, or do they hesitate for a long enough time on the surface to forget the direction they originally came from (diffuse reflection) as shown in Fig. 4.6b?

Let f = fraction of molecules diffusely reflected. For these Knudsen showed that

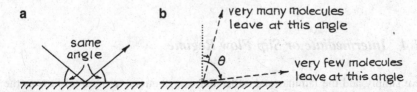

Fig. 4.6 Two types of collisions of molecules with the pipe wall: (**a**) elastic collisions (**b**) diffuse reflections

the number leaving at any particular angle is given by

$$n = k \cos \theta$$

Then $1-f$ = fraction reflected or bouncing off the wall.
Very little information is available on the value of f, but roughly

$$f \cong 0.77 \text{ for copper and glass tubing}$$
$$f \cong 0.90 \text{ for iron pipe}$$

Also, f values are suspected to vary with flow regime; for example, see Fig. 4.7.

Because of the uncertainty in f value and because it is close to unity, we will assume throughout that $f = 1$. Then, on applying the kinetic theory of gases with this assumption, it can be shown that

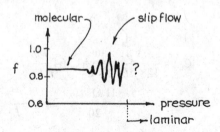

Fig. 4.7 Fraction of molecules diffusely rejected

$$Q_{mol} = C_{mol}(p_1 - p_2) \qquad \left[\frac{\text{Pa m}^3}{\text{s}}\right]$$

where

$$\left. C_{mol} = \frac{d^3}{L}\left[\frac{\pi RT}{18(mw)}\right]^{1/2} \underbrace{\text{air}}_{20\,°C} \left[\frac{\pi(8.314)293}{18(0.0289)}\right]^{\frac{1}{2}}\frac{d^3}{L} = 121.3\frac{d^3}{L} \\ \underbrace{\text{H}_2\text{O vapor}}_{20\,°C} \; 153.7\frac{d^3}{L} \quad \left[\frac{\text{m}^3}{\text{s}}\right] \right\} \qquad (4.6)$$

4.1.4 Intermediate or Slip Flow Regime

If we simply add the laminar and molecular contributions to the total flow as the pressure shifts from one flow regime to the other, we find the behavior shown in Figs. 4.8 and 4.9. Actually, the observed flow in the slip flow regime is somewhat lower (at most 20 %) than the sum of the individual contributions. Since the more exact treatment of this situation would lead to complications, we will assume simply that

$$\underset{\text{slip flow}}{Q_{\text{total in}}} = Q_{mol} + Q_{lam} \qquad \left[\frac{\text{Pa m}^3}{\text{s}} = \frac{J}{\text{s}} = W\right] \qquad (4.7)$$

More precise equations for this flow regime are found in Dushman (1949).

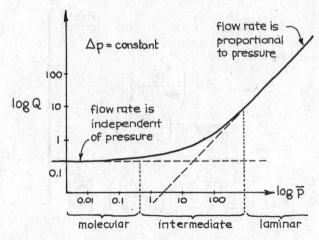

Fig. 4.8 Flow rate of a gas in a pipe for a fixed Δp between the two ends

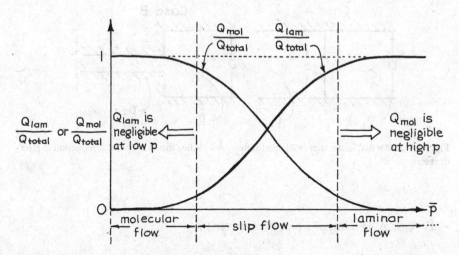

Fig. 4.9 Relative contribution of laminar and molecular mechanisms to the flow of gases in pipes

4.1.5 Orifice, Contraction, or Entrance Effect in the Molecular Flow Regime

As shown in Fig. 4.10, we have two situations here: an orifice or obstruction in a length of pipe (case A) and a smaller pipe leading from a larger pipe or a tank (case B). For both cases the kinetic theory of gases gives

$$Q_{or,\,mol} = C_{or,mol}(p_1 - p_2) \quad \left[\frac{Pa\,m^3}{s}\right]$$

where

$$C_{or,\,mol} = d^2 \left(\frac{D^2}{D^2 - d^2}\right) \left[\frac{\pi RT}{32(mw)}\right]^{1/2}$$

$$\frac{air}{20\,°C} 91d^2 \left(\frac{D^2}{D^2 - d^2}\right)$$

$$\frac{water}{vapor}{20\,°C} 115d^2 \left(\frac{D^2}{D^2 - d^2}\right) \quad \left[\frac{m^3}{s}\right]$$

(4.8)

1. *Equivalent length of pipe.* Comparing conductances with flow in a tube, or equation (4.8) versus equation (4.6), shows that the length of pipe which has the same resistance as the orifice or contraction is:

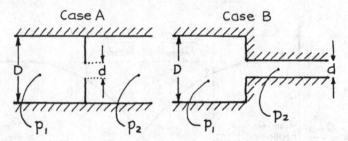

Fig. 4.10 Reduction in the area available for flow. A = orifice inside a pipe, B = reduction in pipe diameter

• For case A, in terms of a pipe of diameter D,

$$\frac{L_{eq}}{d} = \frac{4}{3}\left(\frac{D^2}{d^2} - 1\right)$$

(4.9)

This value of L_{eq} can be quite large. Thus, if $d = 0.1D$, then the resistance of the orifice is equivalent to the resistance of a pipe which is 132 pipe diameters long.

- For case B, in terms of the smaller following pipe of diameter d,

$$\frac{L_{eq}}{d} = \frac{4}{3}\left(1 - \frac{d^2}{D^2}\right) \tag{4.10}$$

This expression shows that the resistance contributed by the contraction is equivalent to a length of about one diameter of small pipe. This is often negligible compared to the other resistances in the vacuum system.

4.1.6 Contraction in the Laminar Flow Regime

Consider laminar flow of gases, not molecular flow, at not too high velocities (not near critical flow) through a contraction going from D to d. From the values of Table 2.2, we can show that

$$\left. \begin{array}{l} Q_{or,\ lam} = C_{or,\ lam}(p_1 - p_2) \\[2mm] \text{where} \qquad C_{or,\ lam} = \dfrac{\pi \overline{p}}{\rho u}\left(\dfrac{d^2 D^2}{D^2 - 0.8d^2}\right) \end{array} \right\} \tag{4.11}$$

The equivalent length of this contraction, in terms of the leaving pipe d, is then found to be

$$\frac{L_{eq}}{d} = \frac{Re}{160}\left(1.25 - \frac{d^2}{D^2}\right) \tag{4.12}$$

Here the equivalent length can be as much as 18 diameters of small pipe.

4.1.7 Critical Flow Through a Contraction

When the pressure ratio across a contraction is ≥ 2, the contraction behaves as a critical flow orifice. For this situation equation (3.27) can be written as

$$Q = \frac{\pi}{4}d^2 p_{\text{upstream}}\left[\frac{kRT}{(mw)}\left(\frac{2}{1+k}\right)^{(k+1)/(k-1)}\right]^{1/2} \tag{4.13}$$

4.1.8   Small Leak in a Vacuum System

Suppose we have a tiny leak in a vacuum system. One may look at this in one of a number of ways, for example, as a narrow channel or as a pinch point. These two extremes are shown in Fig. 4.11. Let us estimate the leak rate for these two extremes, remembering that $p_{\text{sytem}} \ll p_{\text{surroundings}}$.

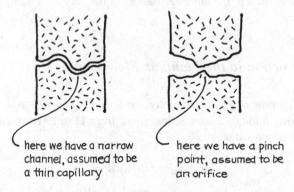

here we have a narrow   here we have a pinch
channel, assumed to be   point, assumed to be
a thin capillary   an orifice

Fig. 4.11 Two ways of looking at a leak in a vacuum system

1. *Assume a capillary.* If the diameter of the capillary is small compared to the mean free path of molecules at 1 atm, then we have molecular flow of the leaking gas, and from equation (4.6),

$$Q_{\text{leak}} = Q_{\text{mol}} = \frac{d^3 p_{\text{upstream}}}{L}\left[\frac{\pi RT}{18(mw)}\right]^{1/2} \tag{4.14}$$

However, if the diameter of the capillary is large compared to the mean free path of the molecules at 1 atm, then we have laminar flow of the leaking gas most of the way through the capillary (see Problem 4.3), in which case equation (4.5) applies. This gives

$$Q_{\text{leak}} = Q_{\text{lam}} = \frac{\pi d^4 \left(p_{\text{upstream}}^2 - p_{\text{downstream}}^2\right)}{256 \mu L} \tag{4.15}$$

2. *Assume an orifice.* Since the pressure ratio across the orifice is many times greater than 2, we should use the critical orifice expressions for compressible flow. Thus, from equation (3.27) we have

$$Q_{\text{leak}} = Q_{\text{crit}} = \frac{\dot{m}RT}{(mw)} = \frac{G^*ART}{(mw)}$$

$$= \frac{\pi}{4}d^2 p_{\text{upstream}} \left[\frac{kRT}{(mw)} \left(\frac{2}{1+k} \right)^{(k+1)/(k-1)} \right]^{1/2} \tag{4.16}$$

Which equation you use, (4.14), (4.15), or (4.16), depends on what you know of the leak and how you view it. However, if you know nothing of the nature of the leak, assume the critical orifice. Chances are that this extreme more closely represents the leak.

4.1.9 Elbows and Valves

In the molecular flow regime and Re < 100, the resistance of elbows and valves which have no flow restrictions is negligible. So just take the mean flow length of the fitting, bend, open valve, etc. However, if the pipe fitting or valve has a restriction, find the smallest cross section and apply equation (4.9).

4.1.10 Pumps

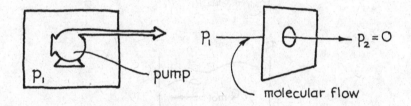

We define pumping speed S_p as follows

$$S_{p,\,\text{at}\,p_1} = \left(\begin{array}{c} \text{volume of gas removed,} \\ \text{measured at}\,p_1 \end{array} \middle/ \text{time} \right)$$

$$= \left(\begin{array}{c} \text{volume of gas entering the throat} \\ \text{of the pump, measured at the} \\ \text{entrance of the pump} \end{array} \middle/ \text{time} \right) \tag{4.17}$$

$$= \frac{Q}{p_1} \left[\frac{\text{m}^3}{\text{s}} \right]$$

The maximum theoretical pumping speed can be looked upon as the flow rate into an orifice which has no back pressure or with equation (4.16)

$$S_{p,\max} = \frac{Q_{or}}{p_1}\frac{\text{air}}{20\,^{\circ}\text{C}}91d^2 \tag{4.18}$$

The speed factor of a pump is defined as follows:

$$\text{Speed factor} = \left(\frac{\text{speed of actual pump}}{\text{speed of a perfect vacuum pump}}\right) \tag{4.19}$$

At pressures between 10^{-4} and 1 Pa, the speed factor is equal to $0.4 \sim 0.6$ for an oil diffusion pump and is equal to $0.1 \sim 0.2$ for a mercury vapor pump. The maximum practical speed factor of vacuum pumps $\cong 0.4$.

4.2   Calculation Method for Piping Systems

Suppose we have the piping system shown in Fig. 4.12. Its flow resistances, included in this figure, consist of a rather complex series–parallel combination. Let us see how to evaluate the overall resistance to flow for systems of this kind.

1. For *resistances in parallel*, an example being molecular and laminar flow, we write

$$Q_{\mathrm{mol}} = C_{\mathrm{mol}}(p_1 - p_2)$$
$$Q_{\mathrm{lam}} = C_{\mathrm{lam}}(p_1 - p_2)$$

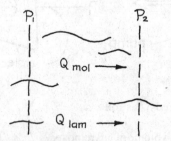

Adding these flow contributions gives

$$(Q_{\mathrm{mol}} + Q_{\mathrm{lam}}) = (C_{\mathrm{mol}} + C_{\mathrm{lam}})(p_1 - p_2)$$

or

$$Q_{\mathrm{tot}} = C_{\mathrm{tot}}(p_1 - p_2) \tag{4.20}$$

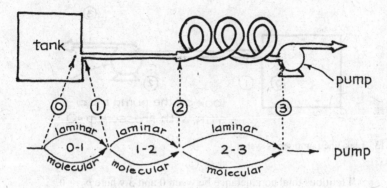

Fig. 4.12 A piping system and its corresponding series–parallel resistances

2. For *resistances in series*, noting that the flow rate is the same for each section, we can write

$$-Q = C_{12} \, \Delta \, p_{12} = C_{23} \, \Delta \, p_{23}$$

or

$$Q = C_{12} \, (p_1 - p_2) = C_{23} \, (p_2 - p_3)$$

Eliminating the intermediate pressure p_2, we find

$$Q = \frac{1}{(1/C_{12}) + (1/C_{23})}(p_1 - p_3) = C_{\text{tot}} \, (p_1 - p_3) \qquad (4.21)$$

Extending this procedure to any number of regions in series is direct, and generalization to any arrangement of series and parallel resistance is not too difficult. For example, for the tank–line–pump system of Fig. 4.13, a frequently met situation, we have

$$Q = C_{01}(p_0 - p_1) = C_{12}(p_1 - p_2) = S_2 p_2$$

Combining these expressions and eliminating intermediate partial pressures p_1 and p_2 gives

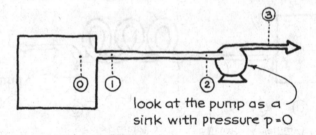

look at the pump as a
sink with pressure p = 0

Fig. 4.13 Tank-line-pump series system

Fictitious total conductance between 0 and 3 where $p_3 = 0$.
It is in fact the pumping speed at the tank, or S_0.

$$Q = C_{tot} \; p_0 - p_3 \; = S_0 p_0$$

$$= \cfrac{1}{\cfrac{1}{C_{or}} + \cfrac{1}{C_{line}} + \cfrac{1}{S_p}} p_0$$

$$= \cfrac{1}{\cfrac{1}{C_{01,lam} + C_{01,mol}} + \cfrac{1}{C_{12,lam} + C_{12,mol}} + \cfrac{1}{S_2}} p_0 \qquad (4.22)$$

$$\frac{air}{20\,^\circ C} \qquad \cfrac{1}{\cfrac{1}{\cfrac{\pi p d^2}{\rho u} + 91 d^2} + \cfrac{1}{1364 d^4 \bar{p} + 121.3 d^3} + \cfrac{1}{S_p}} p_0$$

Eq.(4.11) Eq.(4.8) Eq.(4.5) Eq.(4.6)

These two terms can
usually be ignored $= \dfrac{p_1 + p_2}{2}$,   and for anything
                                          but the shortest pipes
                                          take $\rho_1 \cong \rho_0$

4.3 Pumping Down a Vacuum System

Consider the changing conditions where gas is being pumped from the system, gas
leaks into the system, and the pressure within the system

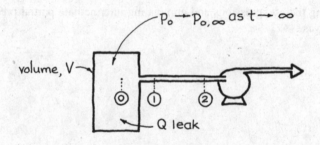

$P_0 \rightarrow P_{0,\infty}$ as $t \rightarrow \infty$

volume, V

Q leak

decreases with time. At any instant we can express the net flow rate of gas out of the system as

$$
\left.
\begin{aligned}
Q = -\frac{d}{dt}(p_0 V) = -V\frac{dp_0}{dt} &= \left(\begin{array}{c}\text{removal rate}\\\text{from tank}\end{array}\right) - \left(\begin{array}{c}\text{leak rate}\\\text{into tank}\end{array}\right)\\
&= S_2 p_2 - Q_{\text{leak}}\\
&= C_{\text{tot}} p_0 - Q_{\text{leak}}
\end{aligned}
\right\}
\tag{4.23}
$$

At $t = \infty$ the system reaches steady state, $Q = 0$, so the above expression reduces to

$$
Q_{\text{leak}} = C_{\text{tot}} p_{0,\infty}
\tag{4.24}
$$

Combining equations (4.23) and (4.24) at any time gives

$$
\left.
\begin{aligned}
-\frac{dp_0}{dt} &= \frac{C_{\text{tot}}}{V}(p_0 - p_{0,\infty})\\
&\qquad\qquad Q_{\text{leak}}/C_0\\
\frac{1}{C_{\text{tot}}} &= \frac{1}{C_{\text{or}}} + \frac{1}{C_{\text{line}}} + \frac{1}{S_p}
\end{aligned}
\right\}
\tag{4.25}
$$

In general C_{tot} varies with pressure.

Now if C_{tot} is constant and independent of p (this means that C_{line} and S_p do not change as the pressure is lowered), integration gives

$$
\ln\left(\frac{p_{0,\text{start}} - p_{0,\infty}}{p_{0,\text{finish}} - p_{0,\infty}}\right) = \frac{C_{\text{tot}}}{V}t
\tag{4.26}
$$

⎰ If unknown, start by ignoring this

If C_{tot} is not constant but varies with p (usually S_p varies drastically with pressure), integration gives

$$
-\int_{p_{0,\text{start}}}^{p_{0,\text{finish}}} \frac{dp}{C_{\text{tot}}(p_0 - p_{0,\infty})} = \frac{1}{V}\int_0^t dt
\tag{4.27}
$$

This integral can be solved either graphically (see Fig. 4.14) or numerically by taking small slices of p_0; thus, $p_{0,1}, p_{0,2}, \ldots, p_{0,i}, \ldots p_{0,n}$.

$p_{0,i+1} - p_{0,\infty}$	$p_{0,i} - p_{0,\infty}$	$\overline{C}_{tot}$	$\ln\left(\dfrac{p_{0,i+1} - p_{0,\infty}}{p_{0,i} - p_{0,\infty}}\right)$	t
$p_{0,2} - p_{0,\infty}$	$p_{0,1} - p_{0,\infty}$	—	—	—
$p_{0,3} - p_{0,\infty}$	$p_{0,2} - p_{0,\infty}$	—	—	—
$\vdots$	$\vdots$			
$p_{0,n} - p_{0,\infty}$	$p_{0,n-1} - p_{0,\infty}$	—	—	—

To save effort, take a constant ratio; for example, $\rho = 1, 2, 4, 8, \ldots$ In each interval assume a constant C_{tot} Total time $= \sum t$

These calculations simplify somewhat:

- When there is no leak; thus, when $p_{0,\infty} \to 0$
- When the pump is located right at the vessel to be evacuated, in which case $C_{tot} \to S_p$

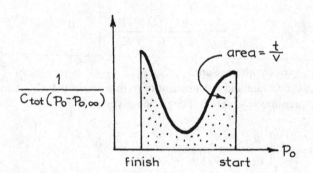

Fig. 4.14 Integrand for equation 4.27

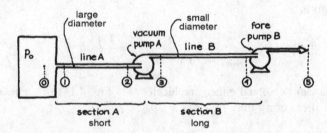

Fig. 4.15 Two-stage pumpdown system

4.4 More Complete Vacuum Systems

Often one employs a two-stage pumping system as shown in Fig. 4.15. Let us see how to treat this situation. For section A of Fig. 4.15 we write

$$Q_A = C_{tot,\,A}p_0 \ \text{ where } \frac{1}{C_{tot,\,A}} = \frac{1}{C_{or,\,A}} + \frac{1}{C_{line,\,A}} + \frac{1}{S_{p,\,A}}$$

Similarly for section B of Fig. 4.15

$$Q_B = C_{tot,\,B}p_3 \ \text{ where } \frac{1}{C_{tot,B}} = \frac{1}{C_{line,\,B}} + \frac{1}{S_{p,\,B}}$$

Since $Q_A = Q_B$ and $p_3 \gg p_0$ we must make $C_{A,tot} \gg C_{B,tot}$. This means that we should design the system so that most of the flow resistance is in line B, not A. Therefore:

- Use a big diameter pipe for line A.
- Keep pump A close to the vessel to be evacuated.
- A long, small-diameter tube can be used for line B with but little harm.

4.5 Comments

This chapter develops the language of molecular flow in the framework of the tank–line–pump system. The field is much broader than this. Here are some additional areas of study:

- The equations for the tank–line–pump situation only hold well if $L/d \geq 100$. For short pipes we may want to modify these expressions. Luckily these are second-order effects.
- Turbulent flow of gases. This situation occurs only very rarely—for high Δp in large pipes.
- Flow in conduits of other shapes: slits, rectangles, annuli, and triangles.
- Pumping speed of cold traps.
- Degassing problems—to remove adsorbed gases from metal and glass surfaces.
- Design of more complete vacuum systems.
- High-vacuum gauges and pumps.

> **Example 4.1. High-Vacuum Flow in a Pipe**
> Find the speeds S_1, S_2, S_3 and the conductances C_{12}, C_{13}, C_{23} for the pipeline below.

(continued)

(continued)

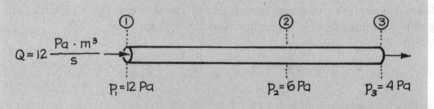

$$Q = 12 \frac{Pa \cdot m^3}{s}$$

$p_1 = 12\ Pa \qquad p_2 = 6\ Pa \qquad p_3 = 4\ Pa$

Solution

From the definition of pumping speed, we can write, for locations 1, 2, and 3,

$$Q = S_1 p_1 = S_2 p_2 = S_3 p_3$$

Therefore,

$$S_1 = \frac{Q}{p_1} = 1\ \mathrm{m^3/s}$$

$$S_2 = \frac{Q}{p_2} = 2\ \mathrm{m^3/s}$$

$$S_3 = \frac{Q}{p_3} = 3\ \mathrm{m^3/s}$$

Next consider the conductance of sections 1–2, 2–3, and 1–3 of the pipe. By definition

$$Q = C_{12}(p_1 - p_2) = C_{23}(p_2 - p_3) = C_{13}(p_1 - p_3)$$

Therefore,

$$C_{12} = \frac{Q}{p_1 - p_2} = \frac{12}{12 - 6} = 2\,\mathrm{m^3/s}$$

$$C_{23} = \frac{Q}{p_2 - p_3} = \frac{12}{6 - 4} = 6\,\mathrm{m^3/s}$$

$$C_{13} = \frac{Q}{p_1 - p_3} = \frac{12}{12 - 4} = 1.5\,\mathrm{m^3/s}$$

To check the results:

(continued)

(continued)

$$\frac{1}{C_{13}} \overset{?}{=} \frac{1}{C_{12}} + \frac{1}{C_{23}} \quad \text{or} \quad \frac{1}{1.5} \overset{?}{=} \frac{1}{2} + \frac{1}{6} = \frac{2}{3} \quad \text{(correct)}$$

Example 4.2. Conditions in a Steady-State Vacuum System
A vacuum pump ($S_p = 0.2 \text{ m}^3/\text{s}$) is connected by 10 m of 100-mm-i.d. pipe to a large vessel which is to be evacuated of air at room temperature. At a time when the pressure in the tank is 10 mPa:

(a) Calculate the pressure at the mouth of the pump (point 2).
(b) Determine the pumping speed at the tank (point 0). This is the rate at which air at 10 mPa is being removed from the tank.
(c) Locate the major resistance to the evacuation.

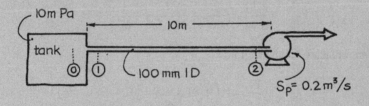

Solution
We can see from the figure above that $S_2 = S_p$ and $p_2 = p_p$. Then

$$Q = S_0 p_0 = C_{01}(p_0 - p_1) = C_{12}(p_1 - p_2) = S_2 p_2 = C_{\text{tot}}(p_0 - 0) \qquad \text{(i)}$$

Including the pump

Use underlined terms only

Imaginary region
beyond the pump

where, from equation (4.22)

(continued)

(continued)

$$\frac{1}{C_{tot}} = \frac{1}{C_{or,\ lam} + C_{or,\ mol}} + \frac{1}{C_{line,\ lam} + C_{line,\ mol}} + \frac{1}{S_2} \quad \text{(ii)}$$

$$\underbrace{\qquad\qquad}_{\substack{\text{From} \\ \text{Eq.}(4.11)}} \quad \underbrace{\qquad}_{\substack{\text{From} \\ \text{Eq.}(4.8)}} \qquad \underbrace{\qquad}_{\substack{\text{From} \\ \text{Eq.}(4.5)}} \quad \underbrace{\qquad}_{\substack{\text{From} \\ \text{Eq.}(4.6)}}$$

Before we proceed to evaluate all the terms, let us see whether we are completely in one or the other flow regime. If we are this would simplify matters. At the point of highest pressure, at the tank, we have

$$pd = (0.1)(0.01) = 10^{-3}\,\text{Pa m}$$

According to equation (4.1) this condition means that flow is completely in the molecular regime, so we can safely and happily drop the laminar terms in equation (ii). Evaluating the remaining terms gives

$$\frac{1}{C_{tot}} = \frac{1}{91d^2} + \frac{L}{121.3d^3} + \frac{1}{S_2}$$

and on replacing values we find

$$\frac{1}{C_{tot}} = 1.1 + 81.5 + 5 = 87.6 \quad \text{(iii)}$$

or

$$C_{tot} = 0.0114\,\text{m}^3/\text{s}$$

Replacing in equation (i) gives

$$p_2 = \frac{C_{tot}p_0}{S_2} = \frac{(0.0114)(0.01)}{0.2} \quad \text{(a)}$$
$$= 5.7 \times 10^{-4}\ \text{Pa} = 0.571\ \text{mPa}$$

The pumping speed at the tank is also given by equation (i). Thus,

$$S_0 = \frac{C_{tot}p_0}{p_0} = C_{tot} = 0.0114\ \text{m}^3/\text{s} \quad \text{(b)}$$

Equation (iii) shows that the relative resistances are

(continued)

(continued)

$$\text{Entry orifice} : \quad \frac{1.1}{87.6} = 1.3\,\%$$

$$\text{The line} : \quad \frac{81.5}{87.6} = 93\,\% \tag{c}$$

$$\text{The pump} : \quad \frac{5}{87.6} = 5.7\,\%$$

Thus, the line provides the major resistance $(\sim 93\,\%)$.

Note: To speed the evacuation either shorten the pipe or increase the pipe diameter. The latter change is better by far since the pumping speed varies as d^3. Using a bigger pump won't help much. For example, even with the biggest pump in the world, equation (iii) becomes

$$\frac{1}{C_{\text{tot}}} = 1.1 + 81.5 + \frac{1}{\infty} = 82.6 \quad \text{vs.} \quad 87.6$$

Thus, the conductance will only increase by about 6 %.

If we would have included the laminar pipe resistance term in our calculations, our answer would only have changed by about 1 %. This justifies our dropping this term.

Since the molecular orifice resistance is only about 1 % that of the pipe [see equation (iii)], and the laminar orifice resistance can be expected to be so much smaller than either, the latter can well be ignored.

Actually, to evaluate the laminar conductances of pipe and orifice is awkward and requires a trial-and-error procedure. The next example shows how this is done for pipe flow.

Example 4.3. Conditions in Another Vacuum System
Repeat Example 4.2 with just one change; let the pressure in the tank be 10 Pa.

Solution
At the pipe entry $pd = 1$ Pa m; thus, we are in the regime of laminar flow and should use equation (4.11). However, from Example 4.2 we find that the resistance of the entry is negligible (about 1 % of the total), so let us ignore it. Next, following the procedure of Example 4.2, we write

(continued)

(continued)

$$Q = S_0 p_0 = S_2 p_2 = C_{\text{tot}}(p_0 - 0) \tag{i}$$

where

$$\frac{1}{C_{\text{tot}}} = \cancel{\frac{1}{C_{\text{or}}}} + \frac{1}{\dfrac{1364 d^4 \bar{p}}{L} + \dfrac{121.3 d^3}{L}} + \frac{1}{S_2} \tag{ii}$$

Neglect

$$\bar{p} = \frac{p_1 + p_2}{2} \text{ where } p_2 \text{ is unknown}$$

Since p_2 is unknown guess that $p_2 = p_1 = 10$ Pa. Then replacing all values in equation (ii) gives

$$\frac{1}{C_{\text{tot}}} = \frac{1}{0.1364 + 0.0121} + \frac{1}{0.2} = 6.7326 + 5 = 11.7326$$

or

$$C_{\text{tot}} = 0.0852 \text{ m}^2/\text{s}$$

Then from equation (i)

$$p_2 = \frac{C_{\text{tot}} p_0}{S_2} = \frac{(0.0852)(10)}{0.2} = 4.2616 \text{ Pa}$$

This value for p_2 does not agree with our guess. So try again. With the help of an astrologer, let us guess that $p_2 = 3.4$ Pa. Then equation (ii) gives

$$\frac{1}{C_{\text{tot}}} = \frac{1}{\dfrac{1364(0.1)^4(10 + 3.4)/2}{10} + \dfrac{121.3(10^{-3})}{10}} + \frac{1}{0.2}$$

$$= \frac{1}{0.0914 + 0.01213} + 5 = 9.66 + 5 = 14.66$$

or

$$C_{\text{tot}} = 0.0682 \text{ m}^3/\text{s}$$

Then equation (4.1) gives

(continued)

(continued)

$$p_2 = \frac{C_{tot}p_0}{S_2} = \frac{(0.0682)(10)}{0.2} = 3.41 \, \text{Pa}$$

Our guess was right, so this is the right pressure for p_2, or

$$p_2 = 3.41 \, \text{Pa} \tag{a}$$

The pumping speed at the tank is also given by equation (i). Thus

$$S_0 = \frac{C_{tot}p_0}{p_0} = C_{tot} = 0.0682 \, \text{m}^3/\text{s} \tag{b}$$

Again the line provides the major resistance, or

$$\frac{9.66}{14.66} \cong 66\% \tag{c}$$

NOTE: Laminar flow is the main mechanism causing the movement of fluid in the line. In fact it accounts for

$$\frac{0.0914}{0.0914 + 0.01213} = 88\%$$

of the total flow. Molecular flow accounts for just 12 % in the conditions of this problem.

Example 4.4. Pumping Down a Leaky Vacuum System
A pump ($S_p = 1 \, \text{m}^3/\text{s}$) is placed within a vessel ($V = 10 \, \text{m}^3$) and is pumping it out. However, because of leaks into the vessel, the pressure within the vessel decreases to a limiting value $p_{0,\infty} = 1$ Pa. Find the leak rate into the vessel.

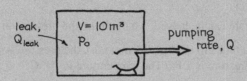

Solution
When the system reaches steady state,

$$Q_{\text{leak}} = Q = C_{\text{tot}} p_0$$

But since there is no line present,

$$\frac{1}{C_{\text{tot}}} = \frac{1}{C_{\text{or}}} + \frac{1}{C_{\text{line}}} + \frac{1}{S_p} = \frac{1}{S_p}.$$

Therefore,

$$Q_{\text{leak}} = S_p p_0 = \left(1\,\text{m}^3/\text{s}\right)(1\,\text{Pa}) = 1\,\text{Pa}\,\text{m}^3/\text{s}$$

Thus, 1 m^3 of air measured at 1 Pa, or 10 cm^3 of air measured at 1 atm, leaks into the vessel each second.

Problems on Vacuum Flow

4.1. A molecular still, to be kept at 0.01 Pa, is connected to an oil diffusion pump by a 0.1-m-i.d. line. Data on the pump indicates that it has a pumping speed of 250 lit/s at low pressure. Assume that air at 20 °C is being pumped.

(a) What is the speed factor of the pump?
(b) What length of this line can be used and still not reduce the pumping speed at the still to less than 50 lit/s?

4.2. A large vacuum system is connected by 1 m of 0.1-m-i.d. pipe to a 0.4 m^3/s vacuum pump. After pumpdown and with the pump working at full speed, the pressure in the system is 1 mPa.

(a) What is the pressure at the pump intake?
(b) What is the leak rate of room air into the system? Give this as lit/h of air at 1 atm.
(c) If the leak rate can be halved, what will be the pressure in the system?

4.3. At one end of a tube (10 m long, 0.1 mm i.d.), the pressure is 1 atm, and at the other end the pressure is 1 Pa. The temperature is 20 °C.

(a) Find the flow rate of air (measured at 1 atm) through this tube.
(b) Plot the pressure 1/4, 1/2, and 3/4 of the way along the tube.
(c) How will the flow rate change if the tube length is halved?
(d) How will the flow rate change if the tube diameter is doubled? Ignore the entrance (or orifice) effects.

4.4. An apparatus is connected by 1 m of 4-cm-i.d. glass tubing, free from leaks, to a combined mercury diffusion and mechanical pump (speed $= 40$ lit/s). Because of small unavoidable leaks in the apparatus itself, the minimum pressure attainable in the apparatus is 10 mPa.

   (a) What pressure can be maintained in the apparatus if the tubing connecting the pumps to apparatus were shortened to 0.1 m?

   (b) What pressure can be maintained in the original apparatus if the pumps were replaced by new ones 10 times as fast?

4.5. During the pumpdown of our vacuum system, I noted that it took 1 day for the pressure to drop from 0.3 to 0.2 Pa. Further pumping reduced the pressure down to 0.1 Pa, but no lower. Yesterday I put extra sealant on what I thought was a leaky joint. Sure enough the pressure in the system started to fall and after 24 h had gone from 0.1 to 0.06 Pa. Estimate the new limiting low pressure of the system. Since the pump characteristics are not given, assume a constant pumping speed at all pressures.

4.6. At present an apparatus is kept at 1 mPa by a vacuum pump and backup pump running at a speed of 10 lit/s and connected to the apparatus by 1 m of 20-mm-i.d. tubing. We wish to lower the pressure in the apparatus, and three choices come to mind.

   (i)   Double the tube diameter.

   (ii) Shorten the connecting pipe from 1 to 0.1 m.

   (iii) Replace the present pump with a larger one having a pumping speed of 30 lit/s.

Rate these alternatives and present calculations to back your rating.

4.7. A vacuum system consists of a vessel connected by pipe (2–3) to a vacuum pump, more pipe (4–5) to a forepump, and then to the atmosphere as shown.

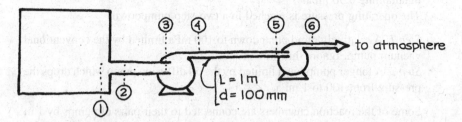

For the vacuum pump, $S_p = 0.1$ m^3/s, $p_3 = 0.2$ Pa, $p_4 = 100$ Pa. Determine the size of forepump which is being used.

4.8. Repeat the previous problem with just one change: the 1 m of 100-mm-i.d. line is to be replaced by 1 m of 1-mm-i.d. line.

4.9. The pumpdown of a 7.5-m^3 leak-free vacuum system from 1 to 0.1 Pa takes
      2 h. The pump is connected to the vessel by 1 m of 30-mm-i.d. pipe. Estimate
      the speed of the pump in this pressure range.

4.10. Our laboratory has a monstrous white elephant, a good-for-nothing three-
      story-high 12-m^3 distillation column. We tried to sell it, to give it away, and
      finally in desperation we offered $1,000 to anyone who'd remove it. Metal
      dealers who looked it over agreed that there was a good bit of metal there, but
      it was too bulky for them.

      Well, I think I can help with that. I will have our machinist connect the
      diffusion pump to the column and evacuate it. From my *Strength of Materials*
      text, I calculate that the column will collapse with a mighty bang into a
      compact easily moved ball of metal when its interior pressure just reaches
      0.1 Pa.

      Since our department head is due to make his annual whirlwind visit of our
      laboratory about 2 pm next Friday, wouldn't it be a nice surprise if he were
      there, may be even leaning on the tower when it collapsed? What a delicious
      thought. When should we start evacuating the column so that our honored
      chief will forever remember his visit?

      Data: The pump is connected directly to the tower, and from the manufac-
      turer's data sheet, the pump speed is as follows:

Pressure (Pa)	0.8	0.2	0.5	1	2	5	10	20	50	100	1 atm
Pump speed (l/s)	18	25.5	30.5	32.5	32	19	11	5.5	2.5	1.5	1

One of the key steps in a local company's manufacture of integrated circuits
is the low-pressure chemical vapor deposition (LPCVD) of exotic materials.
This operation takes place in a battery of special reaction chambers, or
furnaces, kept at 1 mPa by using an oil diffusion pump followed by a
conventional vacuum pump. Without reaction the chamber is capable of
maintaining 0.36 mPa.

The operating pressure is reached in a two-step pumpdown:

Step 1. A relatively rapid pumpdown to 100 mPa limited by the conventional
vacuum pump. Ignore this time.

Step 2. A longer pumpdown limited by the oil diffusion pump which drops the
pressure from 100 to 1 mPa.

Some of the reaction chambers are connected to their pairs of pumps by 1 m
of 5-cm line and have a pumpdown time of 42 min; others are connected with
2 m of 5-cm line and have a pumpdown time of 63 min.

Operations are expanding, the maintenance isle is too crowded, and so the
staff is thinking of relocating some of the pumps on the next floor. This would
require using 8-m lines between chamber and pump pairs.

4.11. How would this affect the pumpdown time?

4.12. What diameter of connecting line should be used to keep the pumpdown time at 42 min? [Problem prepared by Jim McDaniel]

4.13. A simple way of detecting small leaks in heat exchangers is as follows. Pressurize the unit with air, say to 2 atm absolute; immerse it in hot water containing a sprinkle of detergent to reduce the surface tension; and carefully look for bubbles. This technique is sufficiently sensitive to detect very small leak rates roughly equivalent to forming a 1-mm-diameter bubble each minute. How big a hole do you estimate this to represent? Consider the hole to be a pinch point or orifice.
NOTE: This problem involves material from both Chaps. 3 and 4

At the beginning of the week, my bicycle tire (wall thickness = 1.7 mm) contained 1 lit of air at 700 kPa and 20 °C. But after 5 days the pressure was down to 690 kPa, and I am sure that the air leaked out of just one hole—which was made by *you* when you kicked the tire. Yes, I saw you do it! Estimate the size of the hole.

4.14. Assuming that the hole is a "pinch point," an orifice

4.15. Assuming that the hole is tubular in form
Note: These problems involve material from both Chaps. 3 and 4.

References and Further Readings

A.S.D. Barrett, B.D. Power, in *Chemical Engineering Practice*, ed. by H.W. Cremer, T. Davies, vol. 5 (Academic, New York, 1958)

S. Dushman, *Scientific Foundation of Vacuum Technique* (Wiley, New York, 1949). Complete and comprehensive

M. Knudsen, Ann. Phys. **28**, 75, 999 (1909). These papers laid the foundations and developed the basics of the whole field

R. Loevinger, in *Vacuum Equipment and Technique*, ed. by A. Guthrie, R.K. Wakerling (McGraw-Hill, New York, 1949). A nice, simple treatment of pumping systems

Chapter 5
Non-Newtonian Fluids

5.1 Classification of Fluids

This chapter introduces and uses the following symbols to describe the viscous behavior of flowing fluids.

Absolute or dynamic viscosity $= \mu = kg/m \cdot s = Pa \cdot s$
Kinematic viscosity $= v = \mu/\rho = m^2/s$

5.1.1 Newtonian Fluids

For the viscous flow of Newtonians, the resistance of adjacent layers of fluid as they slide past each other, or when they flow through tubes.

In equation form we have

dynamic viscosity

Shear stress $\tau = \mu/s = Pa = kg/m \cdot s^2 = N/m^2$

Shear rate $du/dy = (m/s)/m = 1/s$

Newtonians are the simplest of fluids and are characterized by the fact that the stress rate at a point in a flow channel is proportional to the shear stress at that point,

© Springer Science+Business Media New York 2014
O. Levenspiel, *Engineering Flow and Heat Exchange*,
DOI 10.1007/978-1-4899-7454-9_5

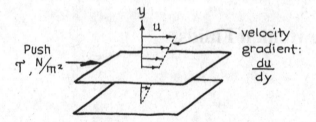

Air, water, all gases, and all fluids consisting of simple molecules are Newtonians.

5.1.2   Non-Newtonian Fluids

Non-Newtonians (NNs) can be divided into three broad classes of materials.

1. *Time-independent NNs* for which at a point in the flow stream

$$\begin{pmatrix}\text{Shear}\\\text{rate}\end{pmatrix} = f\begin{pmatrix}\text{shear stress}\\\text{alone}\end{pmatrix} \quad \text{or} \quad \frac{du}{dy} = f(\tau\,\text{alone})$$
$$\underset{\text{Pa}}{\uparrow}$$

There are a number of types here depending on the form of the τ vs. du/dy relationship. These are shown and named in Fig. 5.1.

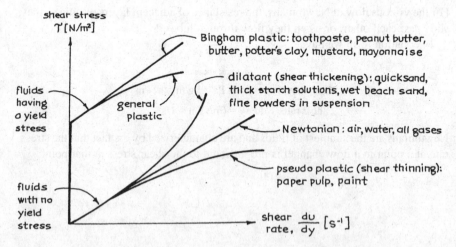

Fig. 5.1 Stress vs. shear rate for various kinds of time-independent non-Newtonians

2. *Time-dependent but nonelastic NNs* are fluids whose present behavior is influenced by what has happened to them in the recent past. For example, tomato ketchup, which has been resting unmoved for a while, will not pour; however, a recently shaken bottle of ketchup pours easily. These fluids seem to have a memory which fades with time; thus, we can write

$$\begin{pmatrix} \text{Shear} \\ \text{rate} \end{pmatrix} = f \begin{pmatrix} \text{shear stress,} \\ \text{past history of stress} \end{pmatrix}$$

This behavior is shown in Fig. 5.2.

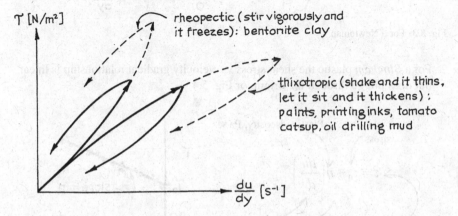

Fig. 5.2 Stress vs. shear rate for two classes of time-dependent, but nonelastic, non-Newtonians

3. *Viscoelastic NNs* are materials which combine the elastic properties of solids with the flow behavior of fluids, and as examples we have saliva and nearly all biological fluids, concentrated tomato soup, bread dough, and many polymeric solutions. With viscoelastics the τ vs. du/dy diagram only tells part of the story; transient experiments (give the can of tomato soup a quick twist and watch the fluid swish left and right) are needed to characterize their elastic properties.

This chapter develops the flow equations for time independent NNs. For other types of NNs the flow equations, if they can be developed at all, are much more complicated. Fortunately, however, in steady state flow without acceleration (flow in straight pipes without nozzles, bends, orifices, etc.), these fluids can often be treated as time independent, too.

5.2 Shear Stress and Viscosity of a Flowing Fluid

1. For a *Newtonian* the velocity gradient is proportional to the imposed shear stress on the fluid; see Fig. 5.3.

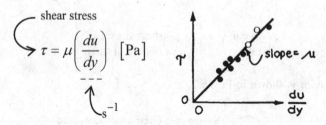

Fig. 5.3 For a Newtonian

2. For a *Bingham* plastic the shear stress vs. velocity gradient relationship is linear but does not go through the origin, or Fig. 5.4

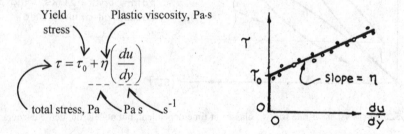

Fig. 5.4 For a Bingham plastic

3. For *pseudoplastics and dilatants* which follow power law behavior, called power law fluids, the relationship between shear stress and velocity gradient is not linear; thus,

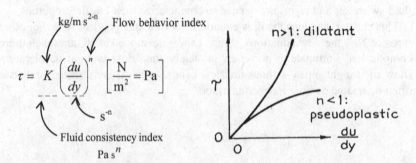

Fig. 5.5 For pseudo plastics and dilatents which follow power law behavior called power-law fluids

A log–log plot will give K and n, as shown in Fig. 5.6.

Fig. 5.6 Method of finding the flow parameters of a power law fluid

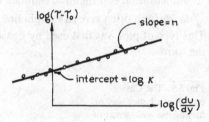

There are many other ways of characterizing fluids with no yield stress; however, the power law is a simple representation which reasonably well fits all these fluids.

The general plastic has characteristics of both Bingham plastics and power law fluids, and it represents a very broad class of fluids, including all previously mentioned time-independent NNs. To determine the three parameters of this type of fluid, τ_0, K, and n, first determine the yield stress τ_0 from Fig. 5.7, and then make the log–log plot of Fig. 5.8 to find K and n.

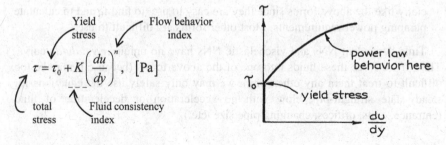

$$\tau = \tau_0 + K\left(\frac{du}{dy}\right)^n, \quad [\text{Pa}]$$

Fig. 5.7 Finding the yield stress of a general plastic

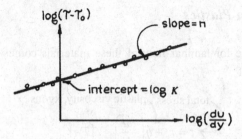

Fig. 5.8 Finding K and n for a general plastic once τ_0 is known

4. *For general plastics* we have

Sometimes it is difficult to estimate τ_0 reliably. One useful help is to plot $\sqrt{\tau}$ vs. $\sqrt{du/dy}$ for it often gives a straight line near the zero abscissa, as shown in Fig. 5.9. This type of plot was first used by Casson for printer's ink. Thus, the expression of the form

Fig. 5.9 The Casson equation: a better way of finding the yield stress of a general plastic

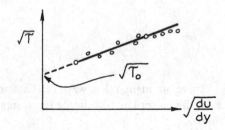

$$\sqrt{\tau} = \sqrt{\tau_0} + K'\sqrt{du/dy} \tag{5.1}$$

is called the Casson equation. It has only a vague association with theory but is useful for finding τ_0.

5. *Comments.* There are many other equation forms for NNs. However, we especially like the above forms since they are easy to use, to find f_F and to calculate pumping power requirements. Most other forms are difficult to use.

Time-dependent NNs and viscoelastic NNs have no unique τ vs. *du/dy* curve. Thus, if we force these fluids into one of the above forms (because it may be too difficult to treat them any other way), we may only safely use the equations for steady-state straight-pipe flow with no acceleration or deceleration of fluid (entrance, exits, orifices, changing pipe size, etc.).

5.3   Flow in Pipes

5.3.1   Bingham Plastics

The velocity profile for laminar flow of these materials comes from the shear–velocity relationship

$$\underset{\text{yield stress}}{\overset{\text{total stress}}{\tau}} = \tau_0 + \eta \overset{\text{plastic viscosity, kg/ms}}{\left(\underset{\text{1/s}}{\frac{du}{dy}}\right)} \quad [Pa] \tag{5.2}$$

and always shows a pluglike core region of diameter

$$d_c = \frac{4\tau_0 L}{\rho \sum F} = \frac{4\tau_0 L}{-\Delta p} \tag{5.3}$$

as shown in Fig. 5.10. Proper integration of the velocity profile across the flow tube gives the mean velocity of flow as

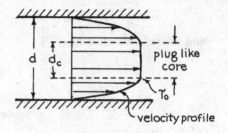

Fig. 5.10 Velocity profile of Bingham plastics in laminar flow

$$u = \frac{d^2 \rho \sum F}{32 \eta L}\left(1 - \frac{4}{3}m + \frac{1}{3}m^4\right)$$

Newtonians have this term alone since $\tau_0 = 0$ $\tag{5.4}$

where

$$m = \frac{4\tau_0 L}{\rho d \sum F} = \frac{\tau_0}{\tau_w}, \quad \text{and} \quad m \leq 1$$

This expression shows that flow is "frozen" when

$$\tau_w \leqslant \tau_0 \quad \text{or} \quad \rho \sum F \leqslant \frac{4\tau_0 L}{d} \quad \text{or} \quad m \geqslant 1 \tag{5.5}$$

Looking at it in a different way, when $m = 1$, the term in brackets of equation (5.4) becomes zero, meaning that the yield stress is nowhere exceeded. Hence, there will be no flow. Thus, the above expression only has meaning when $m < 1$; so when $m \geq 1$, then $u = 0$.

For turbulent flow there is but little information; hence, the best we can do today is to use the f_F vs. Re relationship for Newtonians (see Chap. 2).

For Bingham plastics the strategy for using the mechanical energy balance is the same as for Newtonians. Thus, between any two points of a pipe, we write equation (1.5) as

this whole term is nearly
always negligible

$$g\Delta z + \Delta\left(\frac{u^2}{\alpha}\right) + \int_1^2 \frac{dp}{\rho} + W_s + \sum F = 0$$

somewhere between 1 and 2

where

$$\sum F = \frac{2f_F Lu^2}{d}$$

and where, with equation (5.7),

$$f_F = f\left[\frac{du\rho}{\eta}, \frac{\tau_0 d^2\rho}{\eta^2}\right] = f\left(\text{Re}, \text{He}\right)$$

Reynolds
number

Hedstrom number.
He = 0 for Newtonians

(5.6)

The relationship between friction factor and Reynolds and Hedstrom numbers is shown in Fig. 5.11. Use it to determine pumping requirements or flow rates in a piece of equipment.

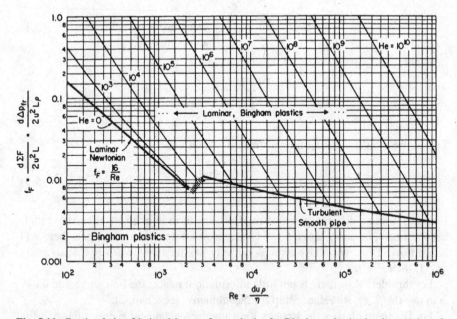

Fig. 5.11 Graph relating frictional loss to flow velocity for Bingham plastics in pipes (Adapted from B. O. A. Hedstrom, *Ind. Eng. Chem.* **44**, 651 (1952))

5.3.2 Power Law Fluids

For laminar flow the velocity profile, shown in Fig. 5.12, comes from the defining shear expression

$$\tau = K\left(\frac{du}{dy}\right)^n \qquad (5.7)$$

Integration across the flow tube gives the mean velocity of flow as

$$u = \frac{d^2\rho\sum F}{32KL}\frac{4n}{1+3n}\left(\frac{d\rho\sum F}{4KL}\right)^{(1-n)/n} \qquad (5.8)$$

⌃ This term alone for Newtonians

Not much is known about the velocity profile in turbulent flow.

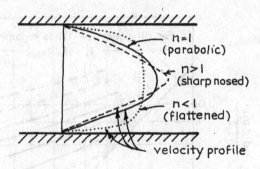

Fig. 5.12 Velocity profile for power law fluids in laminar flow

For power law fluids the mechanical energy balance for flow between any two points of a pipe, equation (1.5), becomes

$$g\Delta z + \Delta\left(\frac{u^2}{2}\right) + \int_1^2 \frac{dp}{\rho} + W_s + \sum F = 0$$

where

$$\sum F = \frac{2f_F L u^2}{d}$$

$$f_F = f\left(\begin{array}{l}\text{generalized Reynolds number } \mathrm{Re_{gen}},\\ \text{and probably roughness as well}\end{array}\right) \left[\frac{J}{kg}\right]\left[\frac{m^2}{s^2}\right]$$

and where, with equation (5.9),

$$\mathrm{Re_{gen}} = \frac{d^n u^{2-n}\rho}{8^{n-1}K}\left(\frac{4n}{1+3n}\right)^n$$

$$(5.9)$$

Figure 5.13 shows the relationship between f_F and $\mathrm{Re_{gen}}$ as known today. Note that experiments are not far enough along to include the effect of roughness in this chart.

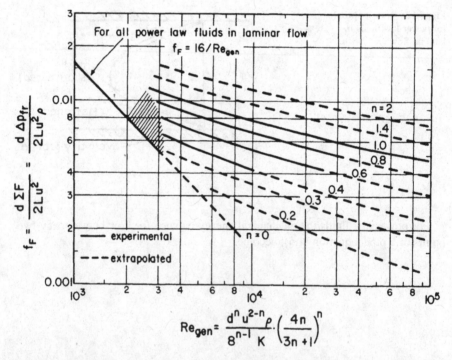

$$\mathrm{Re_{gen}} = \frac{d^n u^{2-n}\rho}{8^{n-1}K}\cdot\left(\frac{4n}{3n+1}\right)^n$$

Fig. 5.13 Graph relating frictional loss to flow velocity for power law fluids in pipes (Adapted from D. S. Dodge and A. B. Metzner, *AIChE J.* **5**, 189 (1959))

For the flow of power law fluids in circular pipes, we use this generalized definition of the Reynolds number because it brings together all the laminar flow lines on the f_F vs. Re chart. It should be noted however that this definition is not useful for flow in other shapes of flow channel or for other than power law fluids.

5.3.3 General Plastics

The velocity profile is complex and comes from the shear–velocity relationship:

$$\tau = \tau_0 + K \left(\frac{du}{dy}\right)^n \tag{5.10}$$

A plug flow core section is always present here as shown in Fig. 5.14.

Fig. 5.14 Velocity profile for general plastics

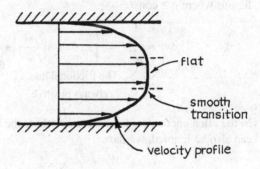

For laminar flow integration of the velocity profile expression gives the mean velocity of flow as

$$u = \frac{d^2 \rho \sum F}{32KL} \left(\frac{d\rho \sum F}{4KL}\right)^{(1-n)/n} 4n(1-m)^{(n+1)/n}$$

$u = 0$ when $m \geq 1$ $\longleftarrow$ This term alone for Newtonians

$$\times \left[\frac{(1-m)^2}{1+3n} + \frac{2m(1-m)}{1+2n} + \frac{m^2}{1+n}\right]$$

where

$$m = \frac{4\tau_0 L}{\rho d \sum F} = \frac{\tau_0}{\tau_{\text{wall}}} \quad \text{and} \quad m \leq 1$$

$$\tag{5.11}$$

This laminar flow expression reduces to the corresponding expressions for Bingham plastics (when $n = 1$), for power law fluids (when $\tau_0 = 0$), and for Newtonians (when both $n = 1$ and $\tau_0 = 0$).

The f_F vs. Re chart for general plastics has not yet been developed. So to evaluate the frictional loss in flow, we must use equation (5.11) for laminar flow, and we must somehow interpolate between Bingham plastics and power law fluids for turbulent flow.

5.3.4 Comments on Flow in Pipes

Many NNs are rather viscous liquids whose flows are in the laminar regime. Consequently the flow equations for this regime are of particular interest.

The frictional pressure drop Δp_{fr} is sometimes used in place of the frictional loss ΣF. These terms are related by the mechanical energy balance of Chap. 1. For liquids where $\rho \cong \text{const}$,

$$\Delta p_{fr} = \rho \sum F = -\left[g\rho\Delta z + \Delta\left(\frac{\rho u^2}{2} \right) + \Delta p \right]$$

The frictional loss is always positive $(p_2 - p_1)_{measured}$

$$(5.12)$$

In the often encountered special case where the kinetic and potential energy terms can also be ignored, we have

$$\Delta p_{fr} = \rho \sum F = -\Delta p_{measured} \qquad (5.13)$$

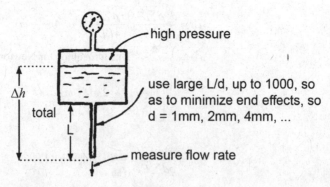

Fig. 5.15 Tube viscometer or extrusion rheometer: the simplest of devices to measure the flow properties of non-Newtonians

Fig. 5.16 Concentric
cylinder rotary viscometer:
especially useful are the
narrow gap viscometer
where $r_1/r_2 \rightarrow 1$ and the
infinite medium viscometer
where $r_2 \rightarrow \infty$

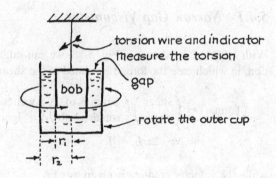

Fig. 5.17 Cone and plate
viscometer

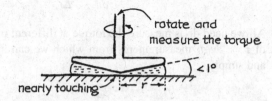

5.4 Determining Flow Properties of Fluids

Three classes of devices are commonly used to measure the flow properties of
non-Newtonians. These are shown in Figs. 5.15, 5.16, and 5.17.

Let us see how to evaluate the flow properties of a fluid from experiments made
in two variations of the concentric cylinder viscometer and in the tube viscometer.

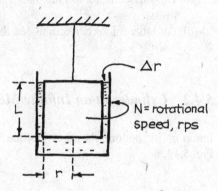

Fig. 5.18 Narrow gap
viscometer

5.4.1   Narrow Gap Viscometer

With $r_1/r_2 \to 1$, as shown in Fig. 5.18, we can safely use the flat-plate approximation, in which case the torque is related to the shear by

$$(\text{Torque}) = \begin{pmatrix} \text{shear} \\ \text{stress} \end{pmatrix} \begin{pmatrix} \text{radius of} \\ \text{rotating bob} \end{pmatrix} \begin{pmatrix} \text{wetted surface of the bob} \\ \text{ignoring its bottom} \end{pmatrix}$$

$$= Nr \cdot 2\pi rL, \quad [\text{J}] \tag{5.14}$$

while the velocity gradient is given by

$$\frac{du}{dr} = \frac{2\pi rN}{\Delta r} \qquad [\text{s}^{-1}] \tag{5.15}$$

All we need do is measure the torque at different rotation rates. This gives a series of τ vs. du/dr measurements from which we can find the flow parameters directly and simply.

Remarks

(a)   This device has severe frictional effects which heat the fluid, changing its properties. Thus, cooling is often necessary.

(b)   End effects such as friction at the bottom of the bob are accounted for by measuring the torque for a Newtonian of known viscosity and then finding a value for effective length L_e such that the following shear–torque equation is satisfied

$$(\text{Torque}) = 2\pi r^2 L_e \, \mu \left(-\frac{du}{dr} \right) \quad [\text{N m}] \tag{5.16}$$

(c)   This device is not too flexible in its ability to treat a large range in shear stress.

(d)   For other minor corrections see Skelland (1967).

5.4.2   Cylinder in an Infinite Medium

Ignoring its bottom the shear–torque relationship for this device, shown in Fig. 5.19, is

Fig. 5.19 Infinite medium
viscometer

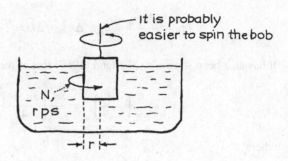

$$\tau = \frac{\text{torque}}{2\pi r^2 L} \quad [\text{N/m}^2 = \text{J/m}^3] \tag{5.17}$$

while the velocity gradient at the surface of the spinning cylinder is calculated to be

$$\frac{du}{dr} = 4\pi N \left[\frac{d(\log N)}{d(\log \text{torque})} \right] \tag{5.18}$$

Experimentally all we need do is measure the torque at a number of N values, plot on a log–log scale, and evaluate the slope at various N values. With equations (5.17) and (5.18), this gives the τ vs. du/dr curve directly. Thus, this device can be used to find the flow parameters for all time-independent NNs.

Remarks

(a) This device is quick and easy to use.

(b) To correct for end effects, use an equivalent length L_e, as with the narrow gap viscometer.

(c) To insure that the outer cylinder is large enough for the infinite medium approximation to apply, use two different sizes of outer cylinders and compare results.

(d) When the gap is somewhere between the above two extremes, the equations become complicated.

(e) For additional minor corrections see Skelland (1967).

5.4.3 Tube Viscometer

For laminar flow a force balance on a section of tube of length L gives

$$\tau_w = \frac{d\Delta p_{\text{fr}}}{4L} = d\rho g \ \Delta h_{total}/\Delta L = \frac{d\rho \sum F}{4L} \quad [\text{Pa}] \qquad (5.19)$$

It has also been shown by Skelland (1967) that at the wall of the tube

$$\left(-\frac{du}{dr}\right)_w = \frac{8u}{d}\left(\frac{3n' + 1}{4n'}\right) \qquad (5.20)$$

where

$$n' = \frac{d[\log(d\rho \sum F/4L)]}{d[(\log(8u/d)]} \qquad (5.21)$$

The procedure for finding the flow characteristics of a power law fluid is then as follows:

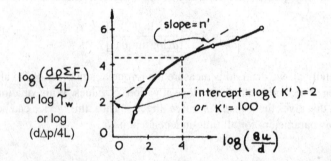

Fig. 5.20 Evaluation of flow parameters at a given flow rate in a pipe

1. Make a series of runs using different tube diameter, tube length, or pressure drop. For each run measure Δp and u, and then evaluate both $d\rho\Sigma F/4\,L$ and $8u/d$.

2. Make a log–log plot of $d\rho\Sigma F/4\,L$ vs. $8u/d$ and determine its slope n' and intercept K', as shown in Fig. 5.20.

   • If $n' = 1$ then the fluid is a Newtonian with a viscosity $\mu = K'$.
   • If n' is constant, but different from unity, then the fluid is a power law fluid with flow parameters

$$n = n'$$
$$K = K'\left(\frac{4n}{3n + 1}\right)^n$$

- If n' is not constant, calculate τ_w and $(-du/dr)_w$ for each point from equations (5.19), (5.20), and (5.21), and construct the τ_w vs. $(-du/dr_w)$ chart. From this find the type of fluid at hand and evaluate its flow parameters.

Remarks

(a) This device is very flexible in that it can explore a wide range of shear values. For slurries one should have an agitator in the reservoir.

(b) The above equations only apply to the laminar flow regime. Be sure to verify that flow is in this regime by evaluating the generalized Reynolds number

$$\text{Re}_{\text{gen}} = \frac{d^n u^{2-n} \rho}{8^{n-1} K} \left(\frac{4n}{1+3n}\right)^n \qquad (5.22)$$

for the run at the highest flow rate, after having found the flow parameters.

(c) In the above expressions the term $\rho\Sigma F$ is evaluated with equations (5.12) and (5.13). To account for the kinetic energy and potential energy effects in these expressions, plus slip effects, once again see Skelland (1967).

5.5 Discussion on Non-Newtonians

5.5.1 Materials Having a Yield Stress, Such as Bingham Plastics

These materials show both solid-like (elastic) and liquid-like (plastic) behavior depending on the intensity of shear acting on them. Many substances, normally treated as solids, exhibit this elastic–plastic behavior, metals, for example. With sufficient shear they give, they flow, and they can be extruded, shaped, punched, etc. On the other hand glass and many rocks do not have a yield stress on man's time scale. But even they do on the geological scale.

On the microscopic level, this elastic–plastic behavior can occur either in *metals* by atomic displacement, or the sliding of molecules over each other, or in *two-phase systems*, one finely dispersed in the other with large surface forces acting between phases. Examples of such two-phase systems are:

- Mayonnaise—oil dispersed in aqueous solution
- Whipped cream—air dispersed in protein solution
- Margarine—fat crystals dispersed in oil
- Chocolate—sugar and cocoa dispersed in cocoa butter (35 %)
- Mashed potatoes—water (90 %) dispersed in food material
- Toothpaste—chalk dispersed in water

Table 5.1 Flow parameters of some familiar Bingham plastics, $\tau = \tau_0 + \eta \left(\dfrac{du}{dy} \right)$

Material	Yield stress, τ_0	Plastic viscosity, η
Ketchup (30 °C)	14	0.08
Mustard (30 °C)	38	0.25
Oleomargarine (30 °C)	51	0.72
Mayonnaise (30 °C)	85	0.63
Butter, on a warm day, very soft and nearly melted	10–20	
Butter, just out of refrigerator, hard, but just spreadable	100–150	
Lead (20 °C)	1.3×10^7	
Solder (20 °C)	2.7×10^7	
Copper (20 °C)	7×10^7	
Various iron and steel (20 °C)	$20–50 \times 10^7$	
Titanium alloy (20 °C)	120×10^7	

The Bingham plastic is the simplest representation for materials having a yield stress, and Table 5.1 gives values of the flow parameters for some commonly met materials of this type.

Table 5.2 Flow parameters of some familiar power law fluids (All are shear thinning or pseudoplastics)

Material	n (−)	K (kg/m s^{2-n})
Applesauce, different recipes (24 °C)	0.41	0.66
(24 °C)	0.65	0.50
Banana puree, different samples (24 °C)	0.46	6.5
(24 °C)	0.33	10.7
Human blood	0.89	0.00384
Soups and sauces	0.51	3.6–5.6
Tomato juice (5.8 % solids, 32 °C)	0.59	0.22
(30 % solids, 32 °C)	0.40	18.7
4 % paper pulp in water (20–30 cm pipe)	0.575	20.7
33 % lime in water (2–5 cm pipe)	0.171	7.16
15 % carboxy-methyl-cellulose in water (2–4 cm pipe)	0.554	3.13

5.5.2 Power Law Fluids

The power law is a convenient representation for many fluids, and Table 5.2 shows values of the flow parameters for some familiar substances which can reasonably be represented by this model.

5.5.3 Thoughts on the Classification of Materials

1. The observation that the measured viscosity of a fluid changes with shear rate or velocity is a sure sign of non-Newtonian behavior (see Figs. 5.1 and 5.2).

2. Power law fluids and Bingham plastics are the simplest models for non-Newtonians. Many more complicated ones are available. Luckily these two simple approximations are often quite satisfactory for engineering purposes. Even time-dependent and other more complex materials flowing at steady state in pipes can often be treated by these simple models.

3. Sometimes the rate of displacement, du/dy, will determine whether a material behaves as fluid or solid. High rates will cause the material to break, while low rates will result in flow. "Silly Putty" and concentrated starch–cold water mixtures are familiar examples. Even liquid water breaks at high shear. On the other hand, even glass will give and flow at room temperature if we wait long enough. For example, window panes from medieval times are thinner at the top than at the bottom.

4. Most biological fluids are NNs and must be treated as such.

5. Most NNs can be classified in more than one way depending on how they are being processed.

6. We have only presented a simple approach to NNs. The whole question of time dependency (shake vigorously and ketchup thins, let it stand and it thickens) and viscoelastic behavior is something we do not touch.

Example 5.1 Flow of a Bingham Plastic from a Tank
A Bingham plastic ($\tau_0 = 20$ Pa, $\eta = 0.02$ kg/m s, $\rho = 2{,}000$ kg/m^3) discharges from the bottom of a storage tank through a horizontal 100-mm-i.d. pipe of

(continued)

(continued)

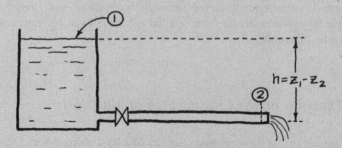

equivalent length of 19.6 m. What head of fluid h will give an outflow velocity of 1 m/s?

Solution

We can find the head either by using the design charts for Bingham plastics (Fig. 5.11) or by using the flow equation (5.4). Let us find the head both ways.

Method A, using the design chart of Fig. 5.11. For this determine

$$\text{He} = \frac{\tau_0 d^2 \rho}{\eta^2} = \frac{(1)(20)(0.1)^2(2{,}000)}{(0.02)^2} = 10^6$$

$$\text{Re} = \frac{du\rho}{\eta} = \frac{(0.1)(1)(2{,}000)}{0.02} = 10^4$$

$$f_F = 0.025 \quad (\text{from} \quad \text{Fig. 8})$$

To find the head needed, apply the mechanical energy balance between points 1 and 2 to give

$$\Delta z + \overset{\approx 0}{\cancel{\frac{\Delta u^2}{2g}}} + \overset{= 0}{\cancel{\frac{1}{g}\frac{\Delta p}{\rho}}} + \overset{= 0}{\cancel{\frac{1}{g}W_s}} + \frac{1}{g}\Sigma F = 0$$

Or

$$h = z_1 - z_2 = \frac{2f_F u^2 L}{gd}$$

$$= \frac{2(0.025)(1)^2(19.6)}{0.1(9.8)} = 1\,\text{m}$$

NOTE: In this solution we ignored the kinetic energy term because it was

(continued)

(continued)

felt that 1 m/s represented a minor energy effect. If this term were included, we'd have

$$h = \frac{2f_F L u_2^2}{gd} + \frac{u_2^2}{2g} = 1 + \frac{1^2}{2(9.8)} = 1.05\,\text{m}$$

or a 5 % correction.

Method B, using the flow equation (5.4). First we need to evaluate a number of terms. Between points 1 and 2, the mechanical energy balance gives

$$\sum F = -\left[g\Delta z + \overset{=0}{\cancel{\frac{\Delta u^2}{2}}} + \overset{=0}{\cancel{\frac{\Delta p}{\rho}}} + \overset{=0}{\cancel{W_s}} \right]$$

$$= -9.8(\Delta z) = 9.8h\,\frac{\text{J}}{\text{kg}}$$

then

$$m = \frac{4\tau_0 L}{\rho d \sum F} = \frac{4(20)(19.6)}{(2{,}000)(0.1)9.8h} = \frac{0.8}{h}$$

Replacing all the known values in equation (5.4) gives

$$1 = \frac{(0.1)^2 (2{,}000)9.8h}{32(0.02)(19.6)} \left[1 - \frac{40.8}{3} \frac{1}{h} + \frac{1}{3} \frac{(0.8)^4}{h^4} \right]$$

or

$$h^4 - 1.1306 \dot{h}^3 + 0.13653 = 0$$

or

$$h = 0.99\,\text{m}$$

which is close to the answer in method A. Actually the answer to method A is a bit off because of chart reading error.

Example 5.2. Transporting Coal by Pipeline

In northern Arizona the Peabody Coal Co. transports coal ($\rho = 1{,}500$ kg/m^3) by slurry pipeline. For this purpose coal is crushed and pulverized to minus 8 mesh and is then pumped 440 km as a 50 %-by-weight slurry in a 0.45-m-i.d. pipeline having four pumping stations. Transit time is 3 days.

(a) What pumping power is needed if the pump and motor are 70 % efficient?
(b) What is the pumping cost/ton to transport the coal from the mine to its destination?

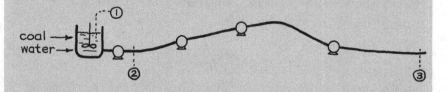

Data:

- A 50 % by weight coal slurry behaves as a power law fluid with $n = 0.2$ and $K = 0.58$, in SI units.
- Electricity costs 3¢/kW h.

[See *Chemical and Engineering News*, p. 17 (April 15, 1974) for more information on this operation.]

Solution

From the data we have

$$u = \frac{440 \text{ km}}{3 \text{ days}} \left(\frac{1 \text{ day}}{24 \times 3{,}600 \text{ s}} \right) \left(\frac{1{,}000 \text{ m}}{1 \text{ km}} \right) = 1.7 \frac{\text{m}}{\text{s}}$$

Density of a 50 % by weight mixture, after much fiddling about, is found to be

$$\bar{\rho} = \frac{2\rho_1\rho_2}{\rho_1 + \rho_2} = \frac{2(1{,}000)(1{,}500)}{1{,}000 \;+\; 1{,}500} = 1{,}200 \frac{\text{kg}}{\text{m}^3}$$

Mass flow rate of slurry transferred, which is 50 % coal

(continued)

(continued)

$$\dot{m} = uA\rho = (1.7)\left[\frac{\pi}{4}(0.45)^2\right](1,200) = 324\frac{kg}{s}$$

Since the slurry is a power law fluid,

$$
\begin{aligned}
Re_{gen} &= \frac{d^n u^{2-n}\rho}{8^{n-1}K}\left(\frac{4n}{1+3n}\right)^n \\
&= \frac{(0.45)^{0.2}(1.7)^{1.8}(1,200)}{8^{-0.8}(0.58)}\left[\frac{4(0.2)}{1+3(0.2)}\right]^{0.2} = 21,060
\end{aligned}
$$

So from Fig. 5.13 we find

$$f_F = 0.002$$

Now to the mechanical energy balance. Across the whole pipeline (point 1 to 3) we may write

$$g\underbrace{\Delta z}_{=0} + \underbrace{\frac{\Delta u^2}{2}}_{=0} + \underbrace{\frac{\Delta p}{\rho}}_{=0} + W_s + \Sigma F = 0$$

Let us explain the canceled terms. In the third term p_1 and p_3 are both at 1 atm so $\Delta p = 0$. Next, because of the great length of the pipeline, the frictional losses therein should dominate and overwhelm the other losses. Thus, the kinetic energy losses should be negligible, and entrance effects and elevation of the storage tank can safely be ignored. Finally, the difference in elevation, since it is not given, will be ignored. This leaves us with

$$
\begin{aligned}
-W_s = \Sigma F &= \frac{2f_F u^2 L}{d} \\
&= \frac{2(0.002)(1.7)^2(440,000)}{(0.45)} = 11,303\frac{J}{kg}
\end{aligned}
$$

and the actual total power requirement

$$
\left.
\begin{aligned}
-\dot{W}_s &= (11.3 \text{ kJ/kg})(324 \text{ kg/s})(1/0.7) = 5,230 \text{ kW} \\
-\dot{W}_s &= 1,310 \text{ kW/pumping station}
\end{aligned}
\right\} \quad (a)
$$

or

The cost of transporting coal by pipeline is then

(continued)

(continued)

$$(5{,}239\,\mathrm{kW})\left(0.03\,\frac{\$}{\mathrm{kWh}}\right)\left(\frac{\mathrm{h}}{3{,}600\,\mathrm{s}}\right)\frac{1\,\mathrm{s}}{(324/2)\mathrm{kg\,coal}}\left(1{,}000\,\frac{\mathrm{kg}}{\mathrm{t}}\right) \qquad \text{(b)}$$
$$= 27\mathrm{c/t}\ \text{of coal}$$

Problems on Non-Newtonians

5.1. Consider a Bingham plastic flowing in a horizontal pipe. If the pressure drop from end to end of the pipe is lowered, then flow slows naturally. Eventually, if Δp is lowered far enough, a critical point is reached where flow stops and material "freezes" in the pipe. Determine this critical Δp for tomato ketchup flowing in a horizontal pipe ($L = 10$ m, $d = 10$ cm).

5.2. If flow of a Bingham plastic just "freezes" in a 10-m length of horizontal 10-cm-i.d. pipe, what length of 20-cm-i.d. pipe would cause the fluid to just freeze for the same overall pressure drop across the pipe?

5.3. What diameter of vertical tube would allow mayonnaise ($\rho = 1{,}200$ kg/m^3) to flow under its own weight?

5.4. A 3-mm-i.d. tube 100 mm long is connected to the bottom of a vat of mustard and is pointing straight down. When the vat is full (depth of 1 m), mustard flows out the tube, but when the depth in the tank falls to 0.4 m, flow stops. From the above information find the yield stress of mustard, a Bingham plastic of density $\rho = 1{,}200$ kg/m^3.

A Bingham plastic ($\tau_0 = 20$ Pa, $\eta = 0.2$ kg/m s, $\rho = 2{,}000$ kg/m^3) discharges from the bottom of a storage tank through a horizontal 0.1-m-i.d. pipe. Determine the outflow velocity from the pipe if the pipe is located 10 m below the liquid level in the tank and has an equivalent length of:
5.5. 4.9 m
5.6. 19.6 m

5.7. Leer toothpaste is to be pumped through a 50-mm-i.d. stainless steel tube from the ingredient-blending machine to the toothpaste tube-filling machine. The equivalent length of line, including bends, fittings, and entrance and exit losses, is 10 m, and the mean velocity of flow is to be 1 m/s.
(a) What pressure difference (in atm) will give this flow rate?
(b) What size motor will do the job for a pump–motor efficiency of 30 %?
   Data: Leer can be considered to be a Bingham plastic with the following properties: $\rho = 1{,}600$ kg/m^3; $\tau_0 = 200$ Pa; $\eta = 10$ kg/m s.

5.8. We wish to pump homogenized soy bean butter ($\rho = 1{,}250$ kg/m^3; $\tau_0 = 80$ Pa; $\eta = 1$ kg/m s) from a storage tank on the upper floor of our little co-op factory to the packaging department below. What size of pump and motor, at 50 % efficiency, should we build into the line to guarantee that the flow rate never gets below 0.8 m/s. See drawing below for additional data.

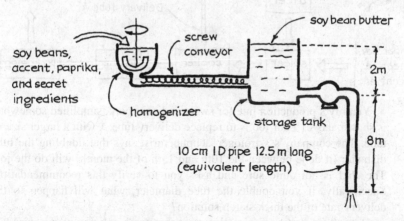

5.9. At the time of the year when honey is harvested, we plan to use the equipment of Problem 5.8 to pump spiced, blended honey at a velocity never lower than 0.8 m/s from the storage tank to the packaging department. The pump and motor are expected to be 50 % efficient for this operation. What size of pump and motor should we order?

　　　　Data: Flavored honey is a power law fluid with: $n = 2$, $K = 5/98$ kg/m, and $\rho = 1,250$ kg/m^3.

Paint is to be pumped at 1 m/s through a horizontal 1-cm-i.d. pipe 25 m long. Find the size of pump–motor, at 40 % efficiency, needed.

5.10. Solve using the design charts given in this chapter.

5.11. Solve using the flow equation given in this chapter.

　　　　Data: This paint follows power law fluid behavior with $n = 0.5$, $K = 2.53$ kg/m s$^{3/2}$, and $\rho = 2,000$ kg/m^3.

5.12. Yummy Oriental Delicacies, Inc., has completely changed American home eating habits, and of the 23 Chinese dishes that it prepares and markets, sweet and sour pork dinners are the overwhelming favorite. Grocery stores just can't meet the demand, and a black market is developing for this item. Yummy will have to speed production of this item as soon as possible, and management has decided on 10 times the present production rate.

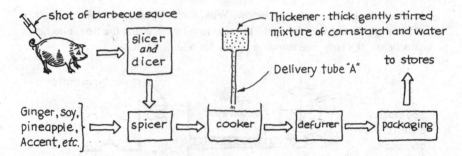

Yummy's production line for sweet and sour pork, simplified somewhat, is shown above. Your job is to replace delivery tube A with a larger size of tube. The company's astrologer–acupuncturist says that doubling the tube diameter (if done between the 10th and 15th of the month) will do the job. The chef is not quite sure and asks you to verify this recommendation. Specifically, if you double the tube diameter, what will happen to the delivery rate of the thick starch solution?

Data: As a first approximation treat the starch solution as a power law dilatant with $n = 2$.

5.13. A 4 % paper pulp slurry is to be pumped from a well-mixed storage tank through a 20-cm-i.d. pipe to a processing tank. The liquid level in the processing tank is 10 m higher than in the storage tank, and the equivalent length of connecting pipe is 40 m. The pump and motor on the line are rated at 25 kW and are 50 % efficient overall. With this arrangement, as shown below, calculate the expected flow rate of slurry in m^3/s.

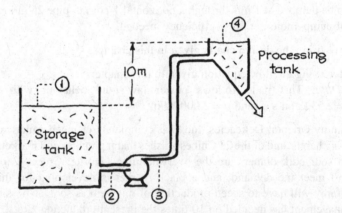

5.14. The flow rate of the previous problem is too low. We wish to raise it to 0.2 m^3/s. What size of pump and motor (at 50 % efficiency) will do the job?

5.15. *Banana pipeline*. Why ship bananas by boat from Honduras to the United States? Look at all the hauling, loading onto ships, squashing, spoilage, worry about rats and cockroaches, unloading, trucking, handling, labor, etc.

Why not peel the bananas in Honduras and pump them as a puree by pipeline straight to Chicago, just 5,000 km away, and then add a touch of vegetable glue to reform the bananas into any desired shape. What an exciting opportunity for creative banana reshaping. However, before plunging into banana reformation, let us see if this proposal is economical. So ignoring capital costs in building the pipeline, estimate the pumping cost for such an operation. Give costs as $/year and ¢/kg of shipped bananas.

Data: Assume: Pump and motor are 50 % efficient overall.
Electricity costs 3.6¢/kW h, or 1¢/MJ.
Pipe size = 10-cm-i.d.
Mean banana velocity = 1 m/s.

Flow characteristic of banana puree is given as

$$\tau = 6.3(du/dy)^{1/3} \text{(all in SI units.)}$$

5.16. Chicago and the Midwest have really taken to our new reformed bananas (see above problem), so we contemplate installing more pumping stations and raising the flow rate in our pipeline to 6 m/s. With this change what does it cost per kilogram to pump the bananas from Honduras to Chicago?

5.17. Consider the following facts.

- West Asian countries presently ship vast quantities of oil all over the world in giant tankers. These return empty—what a waste.
- These countries also have lots of natural gas which they don't know what to do with.
- Some of the oil-importing countries have much iron ore. The fines are not much liked because they cannot be used directly in blast furnaces because of clogging.
- In recent years a number of firms have developed processes for reducing pelletized iron ore fines with natural gas.

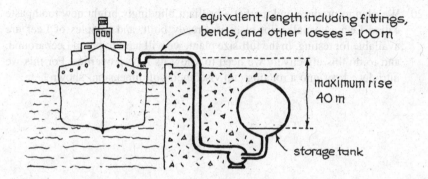

An obvious thought—why not fill the empty returning tankers with a thick iron ore slurry and make steel right there in West Asia. It would at one fell swoop solve the problems of empty tankers, wasted ore fines, and wasted natural gas and at the same time would produce steel for these growing economies.

Let us explore one tiny aspect of the overall process, the pumping of the dense slurry of iron ore into the hold of the tankers from buried storage tanks vented to the atmosphere. What size of pump and motor is needed (at 33 % efficiency) for a flow velocity of 2 m/s of dense slurry in the 0.3-m-i.d. pipe?

Data: Density of slurry: $\rho = 3{,}000$ kg/m^3. The slurry is definitely a non-Newtonian, represented as a power law fluid with $K = 3$ kg/m s^{2-n} and $n = 0.15$.

5.18. Dutch Masters paint pigment is made of TiO_2 powder in water plus dispersant (to keep the solid from settling), thickener (a cellulose derivative), formaldehyde (to keep bugs from eating the thickener), plus this and that. The final result, after much difficult mixing and stirring, is a beautiful white Bingham pseudoplastic whose flow characteristics are given by

$$\tau = 20 + 2(du/dy)^{0.5}$$

The pigment ($\rho = 1{,}700$ kg/m^3) flows from vat to tank car through 25 m of 20-cm pipe. How long would it take to fill a tank car (36 m^3) by gravity alone if the level of pigment in the vat is about 6 m above the level in the tank car?

5.19. *Coal to Texas.* Texas Eastern Corp. is planning to pump coal ($\rho = 1{,}500$ kg/m^3) in a 0.96-m-i.d. slurry pipeline from the coal mining regions of Montana (elevation $= 1{,}400$ m) to the Texas Gulf coast (elevation $= 30$ m), a distance of 3,000 km. Water from the Little Bighorn River (25×10^6 m^3/year) will be used to make the 50 % by volume slurry, which is a power law fluid ($n = 0.2$; $K = 0.65$ kg/m s$^{1.8}$).

What will be the cost per ton of coal transported this way if power costs 3¢/kW h and if the whole pumping system is 50 % efficient? [Information from *Chem. Eng. News,* p. 20 (March 12, 1979).]

5.20. We plan to produce and market a brilliant blindingly bright new toothpaste called Leer. A small pilot plant is already built, and samples of Leer are available for testing. In the full-size plant, we will have to pump Leer around, and to do this effectively we need to know its flow properties. For this we introduce Leer into a rotating cup viscometer of dimensions shown here:

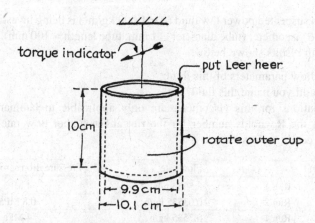

We find that the cup is able to rotate only when the torque exceeds $\pi/10$ N m, and the cup rotates at 3.8 rpm when the torque is $\pi/5$ N m. What kind of fluid is Leer, and what are the values of its flow parameters?

5.21. Find the flow properties of a scrumptious 5-t batch of warm chocolate, after 72 h of blending, from the following data obtained in a narrow gap viscometer ($r_1 = 25$ mm; $r_2 = 28$ mm; $L_e = 76.4$ mm)

Torque (N m)	0.0051	0.0077	0.0158	0.0414
Rotational rate (min^{-1})	Just begins to rotate	0.39	2.62	14.81

[Data inspired by Charm, p. 63 (1971).]

5.22. A cylinder ($r = 0.95$ cm, $L_e = 4$ cm) is lowered into a vat of concentrated orange juice at 0 °C, rotated, and the torque is measured, with the following results

Rotational rate (s^{-1})	0.1	0.2	0.5	1.0
Torque (N m)	42×10^{-6}	63×10^{-6}	107×10^{-6}	152×10^{-6}

Find the flow characteristics of this sample of orange juice. [Data from Charm, p. 64 (1971).]

5.23. Find the flow characteristics of pureed missionary soup (Niugini variety) from the following data taken in a tube viscometer.

d (mm)	L (cm)	Δp (MPa)	$\dot{v}$(cm^3/s)
0.8	10	1.6	0.05
0.8	10	5	0.5
8	200	1	5

[Data from Port Moresby.]

The flow of a suspected power law fluid ($\rho = 1{,}000$ kg/m^2) is being investigated in a capillary tube viscometer (tube diameter $= 1$ mm; tube length $= 100$ mm). Two runs are made with results shown below.

(a) Find the flow parameters of this fluid.
(b) What would you name this fluid?
(c) The equations for this viscometer are only applicable to laminar flow, so calculate the Reynolds number for the run at the higher flow rate to verify that this condition is satisfied.

		Flow rate	Pressure drop across the tube
5.24.	Run 1	3.535 kg/h	4 MPa
	Run 2	0.03535 kg/h	0.8 MPa
5.25.	Run 1	0.3535 kg/h	4 MPa
	Run 2	0.03535 kg/h	0.8 MPa

5.26. Find the flow characteristics of tomato paste from the following data, taken in a tube viscometer: $L = 1.22$ m, $d = 12.7$ mm, height of tomato paste in the reservoir $= 0.11$ m, and $\rho = 1{,}120$ kg/m^3.

$\dot{v}$ (cm^3/s)	0.1	0.5	1.3	4.3
$-\Delta p$ (Pa)	19,600	27,500	34,800	43,800

[This data comes from Charm, p. 62 (1971).]

5.27. Observation shows that tomato paste has a yield stress, so develop an equation which includes this factor to represent the flow data of the previous problem.

5.28. The strained and cooled (10 °C) juice of Rocky Mountain oyster stew is reported to have flow properties displayed below. How would you describe the viscous properties of this regional delicacy? Give an equation if you can.

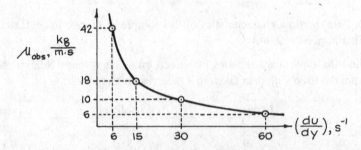

5.29. We have already determined the flow properties of Leer toothpaste (Bingham plastic: $\tau_0 = 200$ Pa; $\eta = 10$ kg/m s), and we have pumped it from the blender to the toothpaste tube-filling machine. Now consider the design of the toothpaste tube itself.

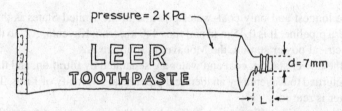

For a reasonable ribbon of toothpaste, it is well accepted that the nozzle of the tube should be 7 mm i.d. Also, the "finger-force" needed to push the toothpaste out of the tube should not be too large or too small. How long should the tube nozzle be so that toothpaste will just squeeze out when the pressure on the toothpaste is 2 kPa?

5.30. Consider the following four piston-cylinder arrangements containing a Bingham plastic. When a low pressure is exerted on the pistons, none of the fluids will flow. However, as the pressure is increased first, one of these units will flow, then another, and so on. Indicate the order in which this will occur as the pressure is increased.

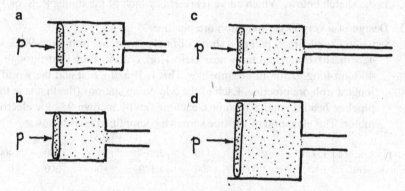

5.31. Yucky stuff flows out of its storage tank if the horizontal drain pipe is shorter than 8 m; it freezes if the drain pipe is longer than 8 m.

   If it flows at a velocity of 20 mm/s in a 6-m-long drain pipe, estimate the flow velocity in a 4-m-long drain pipe. Ignore entrance effects.

5.32. Mustard ($\rho = 1{,}300$ kg/m^3; also see Table 5.1) occupies a 1-m length of smooth horizontal 25.4-mm-i.d. tube. If the tube is slowly tipped up, at what angle will the mustard start to slide down the tube?

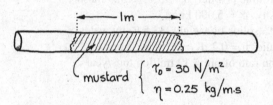

5.33. The longest and only coal–water pipeline in the United States is the Black
Mesa pipeline. It is 0.45 m id and runs 440 km from Kayenta, AZ, to the giant
electrical power station, the Mohave, in southern AZ

At the power station, coal and water are separated by filtration, and the water
is returned to Kayenta by an identical pipeline at a velocity of 1 m/s. Then the
water is reused.

With a pumping efficiency is 70 %, what is the pumping cost per m^3 of water
(roundtrip) if electricity costs $.03/kW/h?

5.34. Consider two parallel flat plates one above the other, immersed in a large
tank of fluid. Plate A is stationary, while plate B is above plate A and moving
to the right at v m/s. The fluid itself is flowing right in laminar flow at v m/s
past plate A and is:

R) A Newtonian fluid
S) A pseudoplastic fluid
T) A power law fluid

The velocity distribution of the 3 fluids flowing between the plates are shown
in the sketch below. Which curve represents which of the fluids, R, S, or T?

5.35. Design of a very large, long iron ore pipeline

Samarco Co. of Brazil pumps a slurry of finely powdered iron ore (~90 % <
425 mesh) from the mines near Belo Horizonte to Port Ubu through a
400-km-long, 0.508-m-i.d. pipeline. This is Brazil's first and the world's
longest iron ore pipeline. Each of the two pump stations (the first is at the
pipeline head) contains 6 pumps, each driven by its own 932 kw electric
motor. The elevation of the line somewhat simplified is as follows:

From pipeline head, L(km)	0	140	200	270	280	400
Elevation, z(m)	1,000	400	1,120	800	300	0

(a) Determine the efficiency of the pumping system.
(b) Within 20 km where should the second pumping station be located?
(c) With no pressure reducing valves in the system, where in the line would the
pressure be highest, and what would it be at that point?
(d) Since the slurry is flowing downhill, it may be possible to recover some of the
potential energy of the flossing slurry. If such a recovery system were
installed and if it were 75 % efficient, what power, if any, could be covered?

Data: Iron ore fine powder: 85–95%, < 325 mesh
Iron ore density: $\rho_s = 5,000$ kg/m^3
Slurry: 66.5 % solids by weight, 33.5 % by water
Power law fluid: n = 0.2, K = 0.75, in SI units
Transportation rate of ore = 12 million tons/year

Suggestion: It may be useful to prepare an accurate elevation–distance diagram, calculate the lost head gradient, dh_L/dL, (see equation (2.4)), show this on the diagram, and then use it to help solve the problem.

Sources: From Chemical Engineering, April 12, 1975, pg 39, and from the booklet "Samarco Industrial Complex" by Samarco Mineracao S.A., second edition, 1981. Thanks to Carlos.

NOTE: Most of the data in this chapter on the flow properties of non-Newtonians are gathered from Skelland, Charm, and Muller and from the *Handbook of Chemistry and Physics*, 47th ed., Chemical Rubber, Boca Raton, FL (1966).

References and Related Readings

R.B. Bird, G.C. Dai, B.J. Yarusso, A good, well-organized review of theory and experiment on flow and heat transfer of non-Newtonians having a yield stress. Reviews in Chemical Engineering **1**, 1 (1982)

S.E. Charm, *The Fundamentals of Food Engineering*, 2nd ed., Ch. 3, (Avi, Westport, 1971). Hurriedly written and difficult to follow but has much useful material.

G.W. Govier, K. Aziz, *The Flow of Complex Mixtures in Pipes* (Von Nostrand Reinhold, New York, 1972). Afraid to leave anything out, has everything; an omni-book

H.G. Muller, *An Introduction to Food Rheology* (Crane Russak, New York, 1973). Delightful gem, even recommended for bedtime reading

A.H.P. Skelland, *Non-Newtonian Flow and Heat Transfer* (Wiley, New York, 1967). Excellent primary reference; go here first

Chapter 6
Flow Through Packed Beds

There are two quite different types of porous media:

- *Packed beds:* These include rock piles, sand filters, soil in flower pots, cigarettes, and absorption columns. For good gas–liquid contacting, absorption columns are usually packed with specially shaped ceramic, plastic, or metal objects such as rings and saddles which have both a large surface area and high voidage; thus, they have a low resistance to flow.
- *Porous solids*: These include both natural porous materials such as oil-bearing rock and pumice and prepared structures such as fused alumina particles, polyurethane foam sponges, foam rubber mattresses, etc.

This chapter concentrates on flow in packed beds.

6.1 Characterization of a Packed Bed

6.1.1 Sphericity φ of a Particle

The sphericity is the most useful single measure for characterizing the shape of irregular and other nonspherical particles. It is defined as

$$\phi = \left(\frac{\text{surface of sphere}}{\text{surface of particle}}\right)_{\substack{\text{same}\\\text{volume}}} = \frac{\bullet}{\clubsuit} \leq 1 \qquad (6.1)$$

Table 6.1 shows values of sphericity for various familiar shaped particles.

© Springer Science+Business Media New York 2014
O. Levenspiel, *Engineering Flow and Heat Exchange*,
DOI 10.1007/978-1-4899-7454-9_6

Table 6.1 Sphericity of particles[a]

Particle shape	Sphericity ϕ
Sphere	1.00
Cube	0.81
Cylinder	
$h = d$	0.87
$h = 5d$	0.70
$h = 10d$	0.58
Disks	
$h = d/3$	0.76
$h = d/6$	0.60
$h = d/10$	0.47
Old beach sand	As high as 0.86
Young river sand	As low as 0.53
Average for various types of sand	0.75
Crushed solids	0.5–0.7
Granular particles	0.7–0.8
Wheat	0.85
Raschig rings	0.26–0.53
Berl saddles	0.30–0.37
Nickel saddles	0.14

[a]Data from Brown (1950) and from geometrical considerations

6.1.2   Particle Size, d_p

When we measure the size of a spherical particle, with a ruler, by screen analysis, or whatever, we know what the measurement means. But with nonspherical particles we have difficulties. When we come up with a size for the particle of Fig. 6.1, do we mean dimension "a" or "b" or something else?

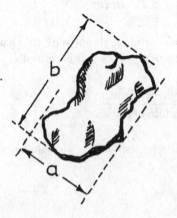

Fig. 6.1 An irregular-shaped particle

Let us define the particle size d_p in such a way as to be useful for flow and pressure drop purposes. With this in mind, how we evaluate d_p depends on the type of instrument available to measure size.

1. For *large particles* (>1 mm) we determine the size:
   - By weighing a known number of particles, if their density is known
   - By fluid displacement of a known number of particles, if the particles are nonporous
   - By calipers or micrometer, if the particles are regular in shape

From these measurements first calculate the equivalent spherical diameter, defined as follows:

$$d_{sph} = \left(\begin{array}{c} \text{diameter of sphere} \\ \text{having the same volume} \\ \text{as the particle, } V \end{array} \right) = \left(\frac{6V}{\pi} \right)^{1/3} \tag{6.2}$$

from which the particle size d_p is given as

$$d_p = \phi \cdot d_{sph} \tag{6.3}$$

where ϕ is measured directly or is estimated from Table 6.1.

Table 6.2 Tyler standard screen sizes

Mesh number (number of wires/in)	Aperture, μm (opening between adjacent wires)	Mesh number (number of wires/in)	Aperture, μm (opening between adjacent wires)
3	6,680	35	417
4	4,699	48	295
6	3,327	65	208
8	2,362	100	147
10	1,651	150	104
14	1,168	200	74
20	833	325	53
28	589	400	38

2. For *intermediate sizes* screen analysis is the most convenient way of measuring the size. Each manufacturer has his own designated sizes, so be sure you know whose screens you are using. As an example, the Tyler standard screens, widely used, have sizes shown in Table 6.2. The screen size d_{scr} is then the mean of the aperture of the screen which just lets the particle through and the screen on which they are resting. For example, particles passing through a 150-mesh screen but resting on a 200-mesh screen would be called $-150+200$ mesh particles and would have a screen size

$$d_{\text{scr}} = \frac{104 + 74}{2} = 89\,\mu\text{m}$$

Unfortunately there is no general relationship between d_{scr} and d_p. The best we can say for pressure drop considerations in packed beds is:

- For irregular particles with no seeming long or shorter dimension take

$$d_p \cong \phi d_{\text{scr}} = \phi d_{\text{sph}} \tag{6.4}$$

- For irregular particles with one somewhat longer dimension, but with length ratio not greater than 2:1, eggs, for example,

$$d_p \cong d_{\text{scr}} = \phi d_{\text{sph}} \tag{6.5}$$

- For irregular particles with one somewhat shorter dimension, but with length ratio not less than 1:2

$$d_p \cong \phi^2 d_{\text{scr}} = \phi d_{\text{sph}} \tag{6.6}$$

- For extreme needle or fiber-like particles

$$d_p > d_{\text{scr}}$$

- For very flat pancake-shaped particles

$$d_p < \phi^2 d_{\text{scr}}$$

- For these flat-plate or needlelike particles, it may be better to approximate the shape with idealized cylinders or disks.

3. For *very small particles* ($<40\,\mu$m) we rely on indirect methods such as sedimentation or Brownian motion to measure the particle size.

4. For *a size distribution of particles*, such as shown in Fig. 6.2, there are many ways of defining an average size. Since it is the surface of the particle which gives frictional resistance to flow, we want a surface-average particle size. Thus, we define this mean size as

$$\bar{d}_p = \left(\begin{array}{c}\text{the single size of particle which would have the same total}\\ \text{surface area as the size mixture in question—same total}\\ \text{bed volume and same bed voidage in both cases}\end{array}\right)$$

This definition leads to the following simple expression:

$$\bar{d}_p = \frac{1}{\displaystyle\sum_{\text{all size cuts}} \left(\frac{x_i}{d_{pi}}\right)} \tag{6.7}$$

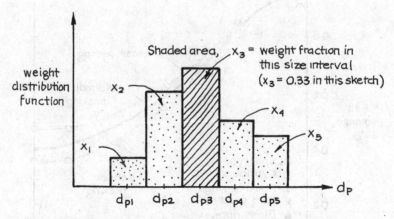

Fig. 6.2 Histogram representing the size distribution of particles in a packed bed

6.1.3   Determination of the Effective Sphericity ϕ_{eff} from Experiment

There are a number of serious problems in using ϕ to relate d_{scr} to d_p for irregularly shaped solids. First of all, all sorts of particle shapes can have the same sphericity, for example, pencils, donuts, and coils. Next, how does one quantify the "egg-shapedness" of an irregular particle? And how to account for particle roughness? Finally, and most important of all, it is very difficult and tedious to properly evaluate the sphericity of irregular particles.

Because of these reasons, we recommend the following experimental procedure for relating d_p to d_{scr}. Carefully and accurately determine the bed voidage ϵ_m. Then measure the frictional loss of this bed, ΣF, at a number of flow velocities. Finally insert ϵ_m, ΣF, and all of the system properties and extract the value of d_p which best fits the data. This relationship of d_p and d_{scr} is called the Ergun equation.

$$d_p = \phi_{eff} d_{scr}$$

This value of effective sphericity ϕ_{eff} can then be used to predict frictional losses in beds of this material, but of any size and also of any wide distribution.

In general, this is the most reliable procedure for relating d_{scr} to d_p.

6.1.4   Bed Voidage, ε

Typical voidage data for packed beds are shown in Figs. 6.3 and 6.4. Also, many equations and correlations have been proposed to give bed voidage. However, if it is important to get a good value for design, it is best to measure the voidage of the packed bed directly. This is not that hard to do.

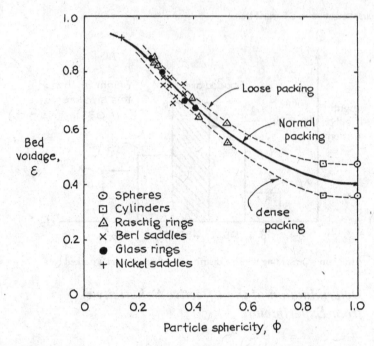

Fig. 6.3 The voidage increases as the sphericity decreases for randomly packed beds of uniform particles (Adapted from Brown et al. (1950))

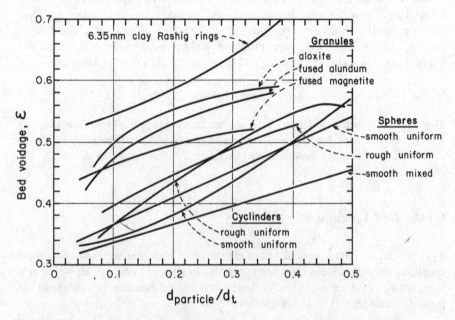

Fig. 6.4 The wall region of a packed bed has an increased voidage (From Leva (1957))

6.2 Frictional Loss for Packed Beds

For fluid flowing through a packed bed of solids as shown in Fig. 6.5, the characteristic Reynolds number is defined as

$$\mathrm{Re}_p = \frac{d_p u_0 \rho}{\mu} \tag{6.8}$$

where ρ is the density of the fluid and u_0 is called the superficial velocity of the fluid. This is the velocity that the fluid would have if the vessel contained no solids.

The frictional loss for flow through packed beds can be expressed as

$$\Sigma F = \underbrace{\frac{150(1-\varepsilon)^2 \mu u_0 L}{\varepsilon^3 d_p^2 \rho}}_{\substack{\text{Viscous} \\ \text{losses}}} + \underbrace{\frac{1.75(1-\varepsilon)u_0^2 L}{\varepsilon^3 d_p}}_{\substack{\text{Turbulent or inertial} \\ \text{losses}}} \quad \left[\frac{\mathrm{J}}{\mathrm{kg}}\right] \tag{6.9}$$

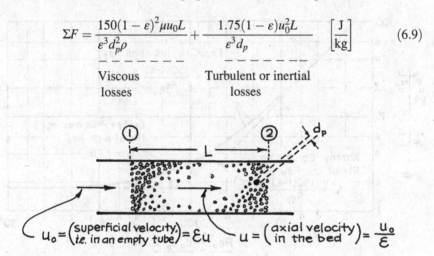

Fig. 6.5 Flow through a packed bed

Equation (6.9) is an expression proposed by Ergun (1952) based on the clever combination of the Kozeny–Carman equation for flow in the viscous region and the Burke–Plummer equation for the turbulent region. This two-term expression fits the data well, is widely used, and is called, not surprisingly, the Ergun equation. When $\mathrm{Re}_p < 20$ the viscous loss term dominates and can be used alone with negligible error. On the other hand, when $\mathrm{Re}_p > 1{,}000$ only the turbulent loss term need be used.

Let us represent this correlation on a friction factor versus the Reynolds number plot, somewhat as in pipe flow. For this define the friction factor as

$$f_f = \left(\frac{\dfrac{\text{frictional}}{\text{energy loss}} \big/ \text{kg of fluid}}{\text{kinetic energy}/\text{kg of fluid}} \right) = \frac{\varepsilon}{1-\varepsilon} \cdot \frac{(\Sigma F)d_p}{u_0^2 L} \quad [\text{-}] \tag{6.10}$$

or

$$\Sigma F = \frac{1 - \varepsilon}{\varepsilon^3} \cdot \frac{f_f u_0^2 L}{d_p}, \; \ldots \text{compare with equation (2.2)}$$

The Ergun equation then becomes

$$f_f = 150\frac{1 - \varepsilon}{Re_p} + 1.75 \quad [\text{-}] \tag{6.11}$$

which is displayed in Fig. 6.6.

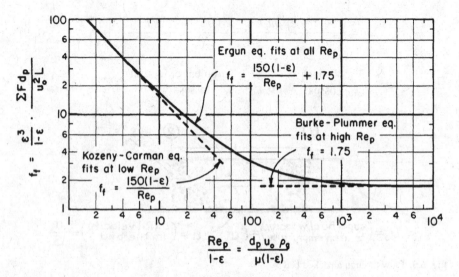

Fig. 6.6 Friction factor vs. Reynolds number for flow through packed beds

Either use equation (6.11) with equation (6.10), or use equation (6.9) directly to determine the frictional losses in packed beds.

6.3 Mechanical Energy Balance for Packed Beds

Consider the flow system of Fig. 6.7. A mechanical energy balance between points 1 and 5 of the system gives

$$g\Delta z + \Delta\left(\frac{u^2}{2}\right) + \int\frac{dp}{\rho} + W_s + \Sigma F = 0 \quad \left[\frac{J}{kg}\right] \tag{6.12}$$

see equation (6.9) or (6.10)

Consider the individual terms in this expression.

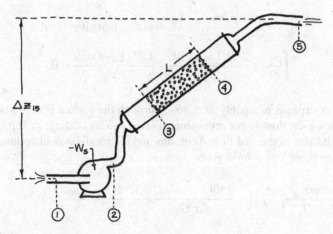

Fig. 6.7 Flow system which includes a packed bed

1. *Friction term, ΣF.* Because the packed bed section has a very large interfacial area, its frictional loss is usually much greater than that of the rest of the piping system. Thus, we often can consider its frictional loss alone, or

$$\Sigma F_{\text{total}} \cong \Sigma F_{\text{packed section}}$$

2. *The potential energy term, $g\Delta z$.* For gases this is usually negligible, but for liquids this can be an important term in the mechanical energy balance.

3. *The work flow term in systems with small density changes, $\int dp/\rho$.* When the density of the fluid does not vary much as it passes through the packed bed, one can use an average fluid density in the system. Thus,

$$\int \frac{dp}{\rho} \cong \frac{\Delta p}{\bar{\rho}} \qquad (6.13)$$

This condition is satisfied for all liquids and for gases where the relative pressure variation is less than 10 %, or where

$$\Delta p < 0.1\bar{p}$$

4. *The flow work term for gases experiencing large density changes, $\int dp/\rho$.* When the frictional pressure drop is large, meaning when $\Delta p > 0.1\,\bar{p}$, one should properly account for the change of density with pressure. Thus, combining equation (9) with the mechanical energy balance of equation (6.12), dropping the kinetic and potential energy terms, dividing by u_0^2, and introducing the superficial mass velocity

$$G_0 = \frac{\dot{n}(mw)}{A} = \frac{kg}{m^2} = u_0\rho = u\varepsilon\rho \text{ give}$$

$$\int \frac{\rho dp}{G_0^2} + \frac{150(1-\varepsilon)^2 \mu L}{\varepsilon^3 d_p^2 G_0} + \frac{1.75(1-\varepsilon)u_0^2 L}{\varepsilon^3 d_p} = 0 \qquad (6.14)$$

If the gas expands reversibly as it flows through the packed bed, it would cool. Here, however, flow is not reversible, friction causes heating, so it probably is best to assume isothermal flow. With this and the ideal gas assumption, integration from point 3 to point 4 gives

$$\frac{(mw)}{2G_0^2 RT}\left(p_4^2 - p_3^2\right) + \frac{150(1-\varepsilon)^2 \mu L}{\varepsilon^3 d_p^2 G_0} + \frac{1.75(1-\varepsilon)u_0 L}{\varepsilon^3 d_p} = 0 \qquad (6.15)$$

Note how this expression parallels equation (3.21) without the kinetic energy term.

This isothermal packed bed expression is the same one used in gas flow systems experiencing large pressure changes.

5. *The kinetic energy term*, $(\Delta u^2/2)$. This is usually negligible for both liquids and gases because one rarely gets to very high velocities in packed beds. However, if one has to include this factor, one should use the approach of Chap. 3 [see equations (3.12) and (3.21)] and add the correct term to equation (6.15).

6. *The work term*, W_s. Taking an energy balance about the whole system from point 1 to point 5 gives the shaft work directly. Alternatively, knowing p_1 and p_2, a mechanical energy balance about the ideal compressor alone (KE, PE, and ΣF all ignored) gives the energy actually received by the flowing fluid as

$$-W_s = \int_1^2 \frac{dp}{\rho} \qquad \left[\frac{J}{kg}\right] \qquad (6.16)$$

For liquids and for gases experiencing a small fractional change in pressure (or $\Delta p < 0.1 \bar{p}$), this expression simplifies to give

$$-W_s = \frac{\Delta p}{\rho} \qquad (6.17)$$

For gases experiencing a high fractional change in pressures (or $\Delta p > 0.1\bar{p}$), equation (6.16) is integrated to give equations (1.9), (1.10), (1.11), (1.12), (1.13) and (1.14).

7. *Comments*. Although most workers will agree that the form of the Ergun equation reasonably represents the frictional loss in packed beds, some have suggested that we do not trust too strongly the values of the Ergun constants, 150 and 1.75.

For example, in blast furnace work with very large particles, Standish and Williams (1975) find that the Ergun constants should both be doubled or tripled. Again, Macdonald et al. (1979) recently suggested that the 150 be replaced by 180 and that the 1.75 be replaced by 1.8 for smooth particles and 4.0 for very rough particles.

However, keeping in mind that most of the data points vary by a factor of two from the mean, we prefer, for the time being, to stay with the original Ergun equation with its considerable backing of experimental verification.

Example 6.1   A Laboratory Packed Bed Experiment
In a second floor laboratory of the East China University of Science and Technology, in Shanghai, is a 0.22-m-i.d. glass tube packed to a depth of 1 m with 10-mm spheres as shown below. What will be the superficial velocity of water at 20 °C through the bed if the water level is kept 3 m above the exit of the bed?

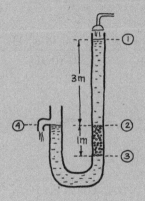

Solution
We can choose to take the mechanical energy balance between various pairs of points in the system, thus, between points 1 and 3, 1 and 4, 2 and 3, or 2 and 4. However, in all cases the packed bed section should be included, because that is where most of the frictional loss occurs. Let us take the balance between points 1 and 4, in which case $\Delta p/\rho$ and the kinetic energy terms drop out. Thus we obtain

(continued)

(continued)

$$g\Delta z + \cancel{\frac{\Delta u^2}{2}}^{=0} + \cancel{\frac{\Delta p}{\rho}}^{=0} + \cancel{W_s}^{=0} + \Sigma F = 0 \qquad \text{(i)}$$

With no pump or turbine in the system $W_s = 0$, so we are left with

$$\Sigma F + g(z_4 - z_1) = 0$$

Now the frictional loss from points 1 to 4 consists of the resistance of the packed bed section 2–3 and the empty pipe sections 1–2 and 3–4. However, it is reasonable to assume that the resistance of the empty pipe sections is negligible compared to that of the packed section. Thus, applying the Ergun equation (6.9), and estimating the bed voidage to be 0.38 from Figs. 6.8 and 6.9, we obtain

$$\frac{150(1-0.38)^2(10^{-3})u_0}{(0.38)^3(0.01)^2(1{,}000)} + \frac{1.75(1-0.38)u_0^2(1)}{(0.38)^3(0.01)} + (9.8)(0-3) = 0 \quad \text{(ii)}$$

or

$$10.51\,u_0 + 1{,}977\,u_0^2 - 29.4 = 0 \qquad \text{(iii)}$$

Solving the quadratic gives

$$u_0 = \boxed{0.119 \text{ m/s}}$$

NOTE: The above solution uses the complete Ergun equation. Had we guessed that the turbulent losses dominated, we would have dropped the linear term in the above quadratic, equation (iii), and this would give

$$u_0 = \left(\frac{29.4}{1{,}977}\right)^{1/2} = 0.122 \text{ m/s } (2.5\,\% \text{ high})$$

A Reynolds number check would then show

$$\text{Re}_p = \frac{d_p u_0 \rho}{\mu} = \frac{(0.01)(0.122)(1{,}000)}{10^{-3}} = 1222$$

which justifies the assumption that turbulent losses dominate in this situation.

The solution to this problem is quite sensitive to the value chosen for the bed voidage because the voidage, in effect, appears to the fourth and fifth

(continued)

(continued)

powers in equation (ii). So if we had chosen $\varepsilon = 0.42$, instead of $\varepsilon = 0.38$, we would have found for equation (i)

$$6.81\, u_0 + 1{,}370\, u_0^2 - 29.4 = 0$$

or

$$u_0 = \boxed{0.144 \text{ m/s}}\ (21\% \text{ high})$$

This large uncertainty in reading ε from Figs. 6.8 and 6.9 is a reflection of the large variation of voidage obtainable in a packed bed. How you pack the bed, whether you shake and tap as you pour the solids in, and so on can give widely different values for voidages.

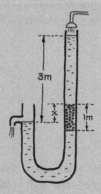

Fig. 6.8 Packed bed partially above exit in a U tube

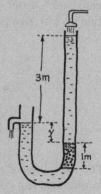

Fig. 6.9 Packed bed below exit in a U tube

In the setups above, with any value for x and y, the driving force remains at 3 m, the resistance is unchanged—still at 1 m of packed bed—so the solution is the same as presented above.

Problems on Packed Beds

6.1. I predict that "Happy Bananas" will soon sweep the country. They will be produced by passing readily absorbed nitrous oxide (laughing gas) through a packed bed of green but full-grown Central American bananas. In developing this process we will need to know the pressure drop in these beds of bananas. To achieve this, estimate the effective banana size d_p from the geometric considerations shown here:

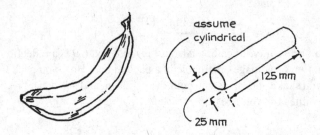

6.2. We plan to pack a tower with Raschig rings of dimensions shown below. Determine the effective particle size d_p of this packing material.

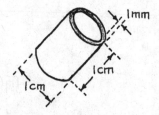

NOTE: The advantage of these specially designed packings is that they give small particle behavior coupled with large voidage—and high voidage gives a low pressure drop.

6.3. For pressure drop purposes a packed bed of spheres of what size would behave like a randomly packed mixture of equal weights of 1 and 2 mm spheres? Assume the same voidage in both beds.

6.4. Air at about 20 °C and 1 atm passes upward through a fixed bed ($L = 0.8$ m, $d_{bed} = 0.1$ m, $\varepsilon = 0.4$) of spherical particles ($\rho_s = 3{,}000$ kg/m^3, $d_p = 10$ mm) at a superficial velocity of 1.5 m/s. Find the pressure drop across the bed.

6.5. Water flows downward through a tube inclined 30° from the horizontal and packed 10 m in length with metal spheres ($d_p = 1$ mm, $\rho_s = 5{,}200$ kg/m^3, $\varepsilon = 0.34$). At a particular flow velocity, the pressure is 3 atm at both ends of the bed.

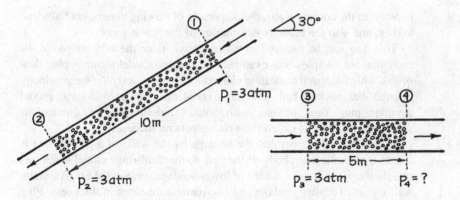

The tube is now tipped horizontally, the packed length is reduced to 5 m, and the water flows at the same rate through the bed. If the pressure at the entrance of the bed is 3 atm, what is it at the exit of the bed?

6.6. Water flows downward through a vertical tube packed 10 m deep with metal spheres ($d_p = 1$ mm, $\rho_s = 5{,}200$ kg/m^3, $\varepsilon = 0.34$). At a particular flow velocity, the pressure just above the bed is 3 atm. Just below the bed it is also 3 atm.

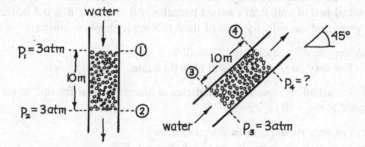

The tube is now tipped 45° from the vertical, and water at the same flow rate flows upward through the bed. If the pressure just at the entrance is 3 atm, what is it at the exit of the bed?

6.7. For the regenerator of Example 15.1, air is compressed, cooled back to 20 °C, and then passed at a mass velocity $G_0 = 4.8$ kg/m^2 s through a vessel 54.5 m high, 1 m^2 in cross section, and filled with close to spherical packing ($d_p = 0.05$ m, $\varepsilon = 0.4$). Air leaves the packed bed at 20 °C and 1 atm. What size of ideal compressor will provide this airflow rate?

6.8. *Gas flow in packed bed catalytic reactors.* Packed beds of catalyst with wall cooling are widely used in the process industry to effect strongly exothermic reactions; for design purposes die engineer must be able to develop a model which reasonably approximates what is happening in these reactors. Such a model should consider a number of phenomena: the flow of generated heat outward to the cooling walls, the dispersion of flowing material radially and axially, and also the velocity distribution of the flowing gas.

This flow can be pictured in various ways, from the very simple to the more real but complex. For example, the simplest model assumes plug flow of gas, which means that all fluid elements move at exactly the same velocity through the packed bed with no overtaking. The second-stage model assumes plug flow of gas with small random velocity fluctuations superimposed. This is called the axial dispersion model.

Now it is well known that the voidage by the walls of a packed bed is higher than in the main body of the bed, so the third-stage model visualizes two flow regions, a central core of lower voidage surrounded by an annular wall region of higher voidage and one-particle diameter in thickness. Plug flow is assumed to occur in each region but with a higher velocity in the high-voidage region.

More precise models may try to incorporate the true velocity profile in the packed bed, a profile which in fact deviates rather significantly from plug flow; however, this profile is not reliably known today. Schlünder (1978) gives a good accounting of the state of knowledge of the factors involved in the proper modeling of fixed bed reactors.

To go back to the third-stage model, if the voidage in the main body of the packed bed of spherical catalyst particles is 0.36 and if it is 0.5 in the wall region, find the velocity ratio of fluid flowing in these regions for:

(a) Very small particles and slow flow
(b) For very large particles and high flow rate

If the packed bed reactor is 8 particles in diameter, what fraction of the fluid flows up the wall region:

(c) For very small particles and slow flow?
(d) For very large particles and high flow rate?

6.9. In one design of a direct-contact crossflow gas–solid heat exchanger, the solids are conveyed horizontally on a screen conveyor while gas percolates upward through the solid. Suppose that hot solids $(\overline{d}_p = 10 \text{ mm}, \varepsilon = 0.4)$ are conveyed at 0.2 m/s in a layer 0.2 m thick while cold air flows upward through the solid mass from a high-pressure chamber (pressure is 2 kPa above atmospheric) to atmospheric pressure, the whole exchanger being at an average temperature of 100 °C. Find the direction of airflow through the layer of solid, and give this as the angle from the vertical θ, as shown below.

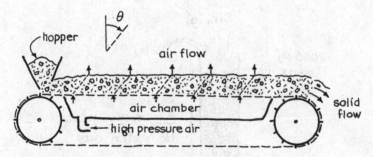

6.10. The critical step in Motorola's proposed process for producing ultra pure solar-cell grade silicon is the reaction

$$Si\left(\begin{array}{c}\text{commercial grade,}\\ \text{solid, 98\% pure}\end{array}\right) + SiF_4\left(\begin{array}{c}\text{gas,}\\ \text{pure}\end{array}\right) \rightarrow 2\,SiF_2\left(\begin{array}{c}\text{gas,}\\ \text{pure}\end{array}\right) \ldots \text{exothermic}$$

The SiF_2 is then decomposed elsewhere to produce pure silicon. Thermodynamics says that at the chosen operating temperature of 1,350 K, the reaction should only become appreciable when the pressure falls below 100 Pa. The reactor will be a vertical 0.4-m-i.d. tube packed with 6-mm silicon particles ($\varepsilon = 0.5$). Reactant gas (pure SiF_4) will enter the reactor at 2,000 Pa and 1,350 K, and 1.2 m downstream we expect the pressure to have dropped to 100 Pa. From here on reaction takes off and conversion to product is soon complete. Estimate the production rate of silicon (kg/h) obtainable from this reactor.

Data:

$$\mu_{SiF_4} = 4 \times 10^{-5}\,\text{kg/m s (estimated)}$$
$$(mw)_{Si} = 0.028$$
$$(mw)_{SiF_4} = 0.104$$
$$(C_p/C_v)_{SiF_4} = 1.15 \text{ (estimated)}$$

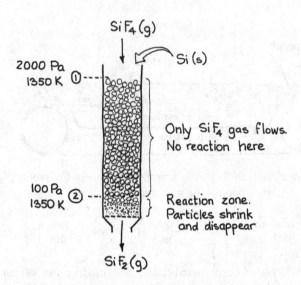

6.11. *Nitrogen fixation by the Haber–Bosch process* (1909–1914). This remarkable
hydrogenation discovered or invented by Prof. Haber and commercialized by
Dr. Bosch makes ammonia (a key ingredient in fertilizers and in explosives)
from nitrogen and hydrogen by the very high-pressure reaction

$$N_2 + 3H_2 \xrightarrow[\text{atmospheres}]{\text{hundreds of}} 2\,NH_3$$

Its use in agriculture is directly responsible for saving millions of people from
starvation these last 70 years. On the other hand, it also allowed the First World
War to drag on for three extra years, resulting in untold casualties and misery.

As part of this process, suppose a mixture of 3 parts H_2 and 1 part N_2 at
0 °C and 2.5 MPa flows at a superficial velocity of $u_0 = 1$ m/s downward
through a packed bed of spherical particles which rest on a holey plate which
in turn rests on 8 L-shaped supports, evenly spaced around the tube wall.

Find the downward force (in newtons and in kilograms) exerted on each of
the L-shaped supports by the gas flowing through the packed bed of catalyst.
Ignore the resistance of the holey plate.

Data: Particles: spherical, $d_p = 2$ mm.

Gas: At these conditions the feed mixture behaves as an ideal gas with
$\mu = 2 \times 10^{-5}$ kg/m · s.

Bed: diameter = 1 m, height = 300 mm, voidage = 0.4, total weight of
packing and with its support plate = 700 kg.

6.12. *Rotating disk heat exchanger.* Our firm is having difficulty in designing a
rotating disk heat exchanger. Heat exchanger test books do not seem to help,
so I am turning to you for help.

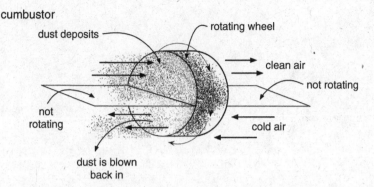

cumbustor

dust deposits

rotating wheel

clean air

not rotating

not rotating

cold air

dust is blown
back in

This unit is to transfer heat from dusty combustion gas leaving a fluidized combustor (60 mol/s, 800 °C, $C_p = 60$ J/mol · K,) to the incoming air entering the combustor (60 mol/s, 0 °C, $C_p = 60$ J/mol · K). The big advantage to this device is that dusty gas is returned to the combustor by the rotating wheel which is to be made of metal honeycomb ($C_p = 1,000$ J/kg · K, mass $= 1,000$ kg):

(a) What should be the rotation rate of the wheel in RPM?
(b) Would you design the wheel to be a large flat pancake unit or a longer smaller diameter unit? Can you explain why?

References

G.G. Brown et al., *Unit Operations,* Chapter 16, (Wiley, New York, 1950)

S. Ergun, Fluid flow through packed columns. Chem. Eng. Prog. **48**, 89 (1952)

M. Leva, Articles on fixed, moving and fluidized systems. Chem. Eng. Prog. **48**, 89 (1952)
    Flow thru packings and moving beds. Chem. Eng. 204, February, pp. 263–268, March, pp. 261–265, (1957)
    Gas-liquid flow in stacked towers. Chem. Eng. 267–270 (1957)
    Pressure drop in stacked towers. Chem. Eng. 269–272 (1957)
    Fluidized heat transfer nomograph. Chem. Eng. 254–257 (1957)
    Equipment for fixed and moving beds. Chem. Eng. 258–262 (1957). Also see Fig. 6.4
    Variables in fixed bed systems. Chem. Eng. 263–266 (1957)
    Correlations in fixed bed systems. Chem. Eng. 245–248 (1957)
    Flow behavior in fluidized systems. Chem. Eng. 289–293 (1957)
    Liquid flow in pack towers. Chem. Eng. 267–272 (1957)

I.F. Macdonald, M.S. El-Sayed, K. Mow, F.A.L. Dullien, Flow through porous media—the Ergun equation revisited. Ind. Eng. Chem. Fundam. **18**, 199 (1979)

N. Standish, I.D. Williams, *Proceedings of a Blast Furnace Aerodynamics Symposium*, (Aus. I.M. M., Wollongong, 1975)

X. Schlünder, *Chemical Reaction Engineering Reviews—Houston*, ACS Symposium Series No. 72, p. 110 (1978)

Chapter 7
Flow in Fluidized Beds

7.1 The Fluidized State

Suppose we progressively increase the velocity of fluid which is flowing upward through a batch of solids. The frictional resistance increases, and eventually a point is reached where the resistance just equals the weight of solids. At this point the solids become suspended—in other words, they

$$A_t L_m (1 - \varepsilon_m) = A_t L_{mf} (1 - \varepsilon_{mf}) = A_t L_f (1 - \varepsilon_f)$$

become "fluidized"—and the superficial velocity of fluid needed to just fluidize the solids is called the minimum fluidizing velocity u_{mf}. As the flow rate of fluid is increased beyond this point, a liquid fluidized bed keeps expanding, while a gas bed expands only slightly. This progression is shown in Fig. 7.1. The fluidized state has many desirable liquid-like properties. For example, we can easily move solids about just by pumping or by gravity flow.

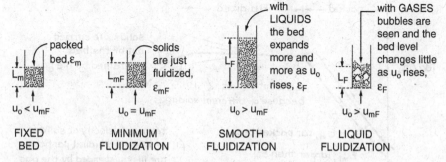

Fig. 7.1 Transition from a packed to a fluidized bed as the velocity of upflowing fluid is raised. Note that the volume of solids in the vessel is

© Springer Science+Business Media New York 2014
O. Levenspiel, *Engineering Flow and Heat Exchange*,
DOI 10.1007/978-1-4899-7454-9_7

The progression from packed bed to fluidized bed is best followed on a pressure drop versus velocity graph, as shown in Figs. 7.2 and 7.3. The packed bed behavior, shown on the left of these figures, is reasonably represented by the Ergun equation (6.9)

$$\Delta p_{fr} = \rho_g \Sigma F = 150 \frac{(1-\varepsilon)^2}{\varepsilon^3} \frac{\mu u_0 L}{d_p^2} + 1.75 \frac{1-\varepsilon}{\varepsilon^3} \frac{\rho_g u_0^2 L}{d_p} \quad \left[Pa = \frac{N}{m^2} = \frac{kg}{ms^2} \right] \quad (7.1)$$

where the characteristic particle size d_p is given by equations (6.3), (6.4), (6.5), (6.6), and (6.7). From this equation we see that

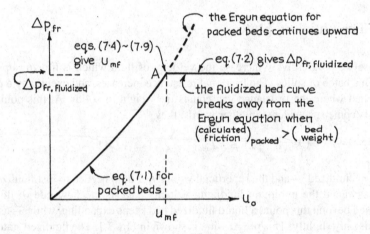

Fig. 7.2 Frictional loss in the packed and in the fluidized state

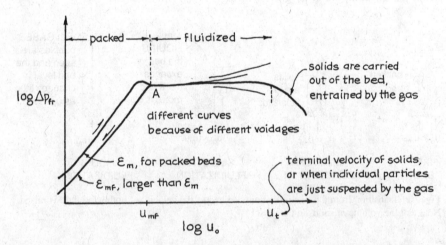

Fig. 7.3 The pressure drop vs. velocity curve in more detail and on log–log coordinates

$$\Delta p_{fr} \propto u_0 \quad \text{(at low velocity)}$$
$$\Delta p_{fr} \propto u_0^2 \quad \text{(at high velocity)}$$

Point A in Figs. 7.2 and 7.3 represents the just fluidized bed; thus, it is at the minimum fluidizing velocity. To the right of point A, the bed is well fluidized. In the following sections we will find

- The pressure drop and pumping power needed to fluidize a bed of solids
- The minimum fluidizing velocity, given by point A of Figs. 7.2 and 7.3

7.2   Frictional Loss and Pumping Requirement Needed to Fluidize a Bed of Solids

Consider the setup of Fig. 7.4. Solids will just fluidize ($u_0 = u_{mf}$, $\varepsilon = \varepsilon_{mf}$) when

$$\left| \left(\begin{array}{c} \text{total drag force} \\ \text{by fluid on} \\ \text{particles} \end{array} \right) \right| = \left| \left(\begin{array}{c} \text{weight of} \\ \text{solids in} \\ \text{the bed} \end{array} \right) \right|$$

or

$$\left(\begin{array}{c} \text{frictional} \\ \text{pressure drop} \end{array} \right) \left(\begin{array}{c} \text{cross-sectional} \\ \text{area of bed} \end{array} \right) = \left(\begin{array}{c} \text{volume} \\ \text{of bed} \end{array} \right) \left(\begin{array}{c} \text{fraction} \\ \text{consisting} \\ \text{of solids} \end{array} \right) \left(\begin{array}{c} \text{specific} \\ \text{weight} \\ \text{of solids} \end{array} \right)$$

or, in symbols,

$$\Delta p_{fr} \cdot A_t = A_t L_{mf} \left(1 - \varepsilon_{mf} \right) \left(|\rho_s - \rho_g| \right) (g)$$

Writing the mechanical energy balance of equation (6.12) between points 1 and 2 of Fig. 7.4 and ignoring kinetic energy effects and the resistance of the distributor give for unit cross-sectional area of bed

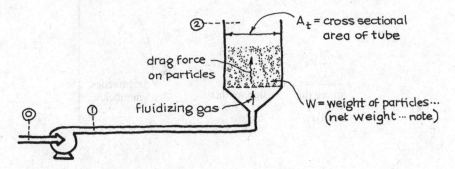

Fig. 7.4 Fluidized bed of solids

$$\Delta p_{fr} = \rho_g \Sigma F = -\left(\Delta p + \rho_g g \Delta z\right) = L_{mf}\left(1 - \varepsilon_{mf}\right)\left(|\rho_s - \rho_g|\right)g$$

$\underset{\substack{\text{Frictional}\\\text{pressure drop}}}{\Big\uparrow}\qquad\qquad \underset{\substack{\text{Ignore}\\\text{for gases}}}{\Big\uparrow}\ \underset{}{\text{For gases}}\ \overset{\Big\uparrow}{=W/A_t}$

(7.2)

The pumping requirement is found by writing the mechanical energy balance between points 0 and 2. This gives

$$-W_s = \Sigma F_{\text{for the bed}} = \frac{\Delta p_{fr}}{\overline{\rho}_g} \qquad \left[\frac{\text{J}}{\text{kg}}\right]$$

(7.3)

Equations (1.9), (1.10), (1.11), (1.12), (1.13), and (1.14) deal further with pumping requirements.

In more realistic situations the frictional loss of the air distribution plate holding up the solids, the cyclone separator or bag filters above the bed, and the other equipment in the line between points 1 and 2 must be accounted for. Thus, in place of equation (7.3), we should more properly write

$$-W_s = \Sigma F_{\text{for the bed}} + \Sigma F_{\substack{\text{distributor}\\\text{plate}}} + \Sigma F_{\substack{\text{cyclone}\\\text{separator}}} + \text{etc.}$$

(7.3a)

At flow velocities higher than u_{mf}, the frictional loss per unit of fluid passing through the bed remains practically unchanged, as shown in Fig. 7.2 or 7.3; however, the pumping power needed will change in proportion to the fluidizing velocity u_0.

7.3   Minimum Fluidizing Velocity, u_{mf}

The intersection of equations (7.1) and (7.2) represents the condition where the solids just fluidize (point A of Fig. 7.2). So combining these two equations gives the following expression for finding the minimum fluidizing velocity:

$$\frac{1.75}{\varepsilon_{mf}^3}\left(\frac{d_p u_{mf}\rho_g}{\mu}\right) + \frac{150\left(1-\varepsilon_{mf}\right)}{\varepsilon_{mf}^3}\left(\frac{d_p u_{mf}\rho_g}{\mu}\right) = \frac{d_p^3 \rho_g\left(|\rho_s - \rho_g|\right)g}{\mu^2}$$

or

$$\underset{\substack{\text{Ignore this}\\\text{term for}\\\text{small } d_p}}{\underbrace{\frac{1.75}{\varepsilon_{mf}^3}\text{Re}_{p,mf}^2}} + \underset{\substack{\text{Ignore this}\\\text{term for}\\\text{large } d_p}}{\underbrace{\frac{150\left(1-\varepsilon_{mf}\right)}{\varepsilon_{mf}^3}\text{Re}_{p,mf}}} = \underset{\substack{\text{Archimedes}\\\text{number}}}{\text{Ar}}$$

(7.4)

In the special case of very small or very large particles, the above expression simplifies as follows: for *very small particles*

$$u_{mf} = \frac{d_p^2(|\rho_s - \rho_g|)}{150\mu} \cdot \frac{ge_{mf}^3}{1 - \varepsilon_{mf}} \qquad Re_{p,\,mf} < 20 \qquad (7.5)$$

and for *very large particles*

$$u_{mf}^2 = \frac{d_p(|\rho_s - \rho_g|)}{1.75\rho_g} \cdot ge_{mf}^3 \quad Re_{p,\,mf} > 1{,}000 \qquad (7.6)$$

If ε_{mf} and/or sphericity ϕ (which enters in d_p) is not known, we can use the observation of Wen and Yu (1966) that $1/\phi\varepsilon_{mf}^3$ and $(1 - \varepsilon_{mf})/\phi^2\varepsilon_{mf}^3$ are both close to constant for all systems. Chitester et al. (1984) summarized the reported studies and recommended using

$$\frac{1}{\phi\varepsilon_{mf}^3} \cong 11.57 \quad \text{and} \quad \frac{1 - \varepsilon_{mf}}{\phi^2\varepsilon_{mf}^3} \cong 7.74$$

Replacing these expressions in equation (7.4) and combining with equation (6.3) then give

$$\frac{d_p u_{mf} \rho_g}{\mu} = \left[(28.7)^2 + \frac{0.0494\, d_p^3 \rho_g \left(|\rho_s - \rho_g|\right) g}{\mu^2} \right]^{1/2} - 28.7 \qquad (7.7)$$

For more on the flow characteristics of fluidized beds, see Kunii and Levenspiel (1991).

Example 7.1. Power to Run a Fluidized Incinerator for Municipal Garbage

Room temperature air (20 °C) is to be compressed and fed to a fluidized incinerator which operates at high temperature and uses sand as a carrier solid. Find the power requirement of the compressor if the unit is to operate at ten times the minimum fluidizing velocity.

(continued)

(continued)

Data:

- Incinerator: 3-m-i.d., slumped bed height $= 0.56$ m

$$\varepsilon_m = 0.36, \quad \varepsilon_{mf} = 0.44, \quad \varepsilon_f = 0.54$$

- Solids: $-28 + 35$ mesh sand, $\rho_s = 2{,}500$ kg/m^3, $\phi = 0.875$.
- Temperature of the bed $= 850$ °C. Because of its remarkably good heat distribution characteristics, a well-fluidized bed is close to uniform in temperature, so assume that the air reaches 850 °C as soon as it enters the bed.
- Take $\mu_{\text{air, 850 °C}} = 4.5 \times 10^{-5}$ kg/m s.
- Efficiency of the compressor and motor $= 70$ %.
- Ignore the pressure drop of the distributor and of the cyclones.

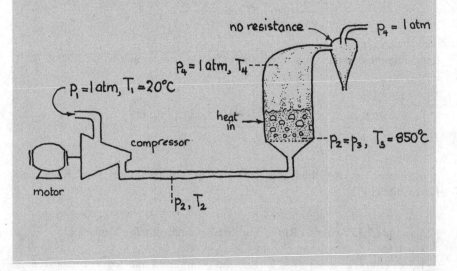

Solution

Our strategy will be to determine, in turn, the pressure drop through the bed, the gas velocity through the bed, and finally, the power requirement. But let us first evaluate a few needed quantities.

$$L_{mf} = \frac{L_m(1 - \varepsilon_m)}{1 - \varepsilon_{mf}} = \frac{0.56(1 - 0.36)}{1 - 0.44} = 0.64 \text{ m}$$

$$d_{scr} = \frac{417 + 589}{2} = 503 \ \mu m, \text{ from Table 6.2}$$

$$d_p = \phi d_{scr} = 0.875(503) = 440 \ \mu m, \text{ from equation (6.4)}$$

Calculate Δp_{fr} from equation (7.2)

$$\Delta p_{fr} = \rho_g \sum F = L_m(1 - \varepsilon_m)\left(\rho_s - \rho_g\right)g$$

$$= 0.56(1 - 0.36)(2,500)9.8 = 8,781 \text{ Pa}$$

$$p_2 = 101,325 + 8,781 = 110,106 \text{ Pa}$$

and

$$\bar{\rho}_{g,\text{in bed}} = \frac{(mw)\bar{p}}{RT_3} = \frac{0.0289\left(\dfrac{110,106 + 101,325}{2}\right)}{(8.314)(1,123)}$$

$$= 0.327 \text{ kg/m}^3$$

Calculate u_{mf} from equation (7.1)

$$\Delta p_{fr} = \frac{150\left(1 - \varepsilon_{mf}\right)^2 \mu u_{mf} L_{mf}}{\varepsilon_{mf}^3 d_p^2} + \frac{1.75\left(1 - \varepsilon_{mf}\right)\bar{\rho}_g u_{mf}^2 L_{mf}}{\varepsilon_{mf}^3 d_p}$$

Replacing known values gives

$$8,781 = \frac{150(1 - 0.44)^2\left(4.5 \times 10^{-5}\right)u_{mf}(0.64)}{(0.44)^3(0.00044)^2}$$

$$+ \frac{1.75(1 - 0.44)(0.327)u_{mf}^2(0.64)}{(0.44)^3(0.00044)}$$

or

(continued)

(continued)

$$8,781 = 82,148\, u_{mf} + 5,472\, u_{mf}^2$$

and solving gives

$$u_{mf} = 0.106\ \text{m/s}$$

Hence, the velocity of gas entering the bed at 850 °C and 110 kPa is

$$u_0 = 10\, u_{mf} = 1.06\ \text{m/s}$$

NOTE 1: Instead of finding Δp_{fr} and the u_{mf}, we could have found u_{mf} directly from equation (7.4). It will give the same answer.

NOTE 2: At the start we could have guessed that this was a fine particle system, in which case equation (i) becomes

$$8,781 = 82,147\, u_{mf}, \quad \text{or } u_{mf} = 0.107\ \text{m/s}$$

A check of the Reynolds number then shows that $\text{Re}_{p,mf} < 20$, which satisfies our original guess.

NOTE 3: If we were not given ϕ and/or ε_{mf}, we would have to use equation (7.7) to evaluate u_{mf}. This gives

$$\frac{(503 \times 10^{-6})u_{mf}(0.327)}{4.5 \times 10^{-5}} = \left[(28.7)^2 + \frac{(0.0494)(503 \times 10^{-6})^3(0.327)(2,500)(9.8)}{(4.5 \times 10^{-5})^2} \right]^{1/2} - 28.7$$

from which

$$u_{mf} = \overline{0.118\,\text{m/s}}, \quad \text{which is about } 10\,\% \text{ high}$$

Calculate the power requirement from equations (1.12) *and* (1.14) but first determine the molar flow rate of gas from the conditions of the gas just entering the bed. Thus,

$$\dot{n} = \frac{p_3 \dot{v}_3}{RT_3} = \frac{(110,106)(1.06)(\pi/4)(3)^2}{(8.314)(1,123)} = 88.5\ \text{mol/s}$$

and now an energy balance about the compressor gives

$$-\dot{W}_s = \frac{k}{k-1} \cdot \frac{\dot{n}RT_1}{\eta} \left[\left(\frac{p_2}{p_1} \right)^{(k-1)/k} - 1 \right]$$

$$= \frac{1.4}{1.4 - 1}\, \frac{(88.5)(8.314)(293)}{0.7} \left[\left(\frac{110,106}{101,325} \right)^{(1.4-1)/1.4} - 1 \right]$$

$$= 25,902\ \text{J/s} = 26\ \text{kW, to gas}$$

Problems on Fluidized Beds

7.1. Air at 20 °C passes upward through a bed (2 m high, 0.5 m i.d., $\varepsilon_m = 0.4$, $\varepsilon_{mf} = 0.44$) of limestone ($d_p = 2$ mm, $\rho_s = 2{,}900$ kg/m³). What inlet air pressure is needed to fluidize the solids (outlet pressure = 1 atm)?

7.2. Air at 1.1 atm and 20 °C filters upward through a 10-m high bed of 1-mm solid silver spheres ($\varepsilon_{mf} = 0.4$) and discharges to the atmosphere. Ten meters!! What a waste! Silver is valuable, so let's sell some of it. No one will know as long as the remaining solids remain as a fixed bed. What height of silver should we leave behind so that the bed won't fluidize?

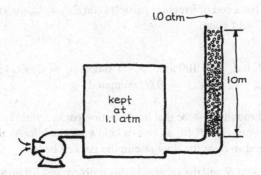

7.3. We plan to pass air upward through a bed of solids resting on a screen. Will the solids fluidize?

Data:

$$\text{Solids}: \ \rho_s = 3{,}000 \ \text{kg/m}^3, \ d_p = 2 \ \text{mm}, \ \varepsilon_m = 0.36$$
$$\text{Bed}: \ 2 \ \text{m high, } 0.5 \ \text{m i.d.}$$
$$\text{Air}: \ p_{\text{inlet}} = 130 \ \text{kPa}, \ p_{\text{outlet}} = 100 \ \text{kPa}, \ T = 100 \ °\text{C}$$

7.4. Air at 150 kPa and 20 °C will flow upward at a superficial velocity of 1 m/s through a bed of solids.

(a) Will the solids fluidize?
(b) Find the outlet air pressure for this flow rate.

Data:

$$\text{Solids}: \ \rho_s = 4{,}500 \ \text{kg/m}^3, \ d_p = 1 \ \text{mm}$$
$$\text{Bed}: \ \varepsilon_m = 0.36, 1 \ \text{m high, } 0.3 \ \text{m i.d.}$$

7.5. Air enters ($u_0 = 1$ m/s, $p = 0.2$ MPa, $T = 293$ K) and passes upward through a bed of packed solid ($d_p = 1$ mm, $\varepsilon_m = 0.4$, $\rho_s = 9{,}500$ kg/m³, $L_m = 1$ m) sandwiched and kept in place between two screens having negligible resistance to airflow.

(a) What is the outlet pressure?

(b) The upper screen is removed. Now what is the outlet pressure?

7.6. Air at about 20 °C and 1 atm passes upward through a fixed bed of solids.

(a) At what superficial air velocity will the particles just fluidize?

(b) At minimum fluidizing velocity what will be the pressure drop across the bed?

Data:

$$\text{Solids}: d_{sph} = 15 \text{ mm}, \ \phi = 0.67, \ \rho_s = 3{,}000 \text{ kg/m}^3$$
$$\text{Bed}: \varepsilon_m = \varepsilon_{mf} = 0.4, \ \text{height} = 0.5 \text{ m}, \ \text{diameter} = 0.1 \text{ m}$$

7.7. Calculate u_{mf} for a bed of irregular particles (take $d_p = d_{scr} \cdot \phi_s$) fluidized by air at about 20 °C and 1 atm.

Data:

$$\text{Solids}: d_{scr} = 100 \ \mu\text{m}, \ \phi_s = 0.63, \ \rho_s = 5{,}000 \text{ kg/m}^3$$
$$\text{Bed voidage}: \varepsilon_m = \varepsilon_{mf} = 0.6 \text{ (estimated)}$$

7.8. Water flows through a U-tube that has a section packed with 1-mm glass beads ($\rho_s = 2{,}200$ kg/m^3, $\varepsilon_m = 0.40$) as shown below on the left. As the upper water level is changed so does the flow rate in the packed bed.

(a) At what height H will the spheres in this problem just lift up and wash away?

(b) What is the velocity of the water when this occurs?

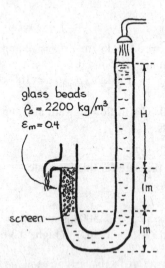

glass beads
$\rho_s = 2200$ kg/m^3
$\varepsilon_m = 0.4$

H

screen

1m

1m

7.9. A packed bed flow experiment is sketched below on the right. As the water flow rate is increased, the water level at point 1 rises and H increases. Eventually, a point is reached when the plastic balls just fluidize downward. Determine the height H and the liquid flow rate when this should occur.

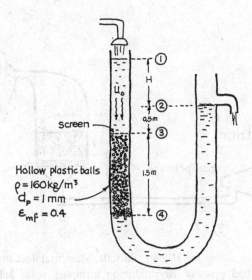

7.10. Here is a proposed ride for the Tokyo Disneyland. Crowd a group of 75 children into a Plexiglas cylinder 3 m i.d. and fluidize them. Hands and feet flying, what fun it would be. And we'd call it "Typhoon", the Chinese word for "big wind."

(a) What air velocity would be needed to just fluidize the children?
(b) What size of compressor at 50 % efficiency (to account for inefficiencies, distributor resistance, screen above the bed to catch the occasional blown-out child) is needed if $u_0 = 1.1\ u_{mf}$?

   Data: Children weigh from 25 to 50 kg with 40 kg as a mean. For suitably dressed screaming children (crash helmets, padding, and all) $\phi = 0.22$, $\rho = 800$ kg/m³, and I suppose that $\varepsilon_m \cong 0.4$, $\varepsilon_{mf} = \varepsilon_f = 0.5$ are fair estimates of the void fraction between children.

7.11. Example 7.1 made the simplifying assumption that the pressure drop of the distributor plate and of the cyclone are relatively small and can be ignored. Relax this assumption and repeat Example 7.1 with the more realistic assumption that $\Delta p_{\text{distributor}} = 20\ \%\ \Delta p_{\text{bed}}$ and $\Delta p_{\text{cyclones}} = 10\ \%\ \Delta p_{\text{bed}}$.

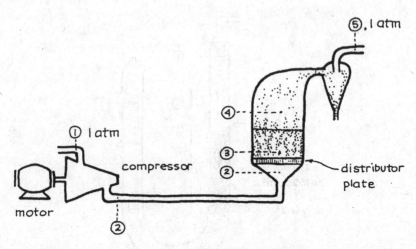

7.12. *Production of ultrapure silicon.* Battelle Memorial Institute is developing a fluidized bed process for producing ultrapure solar cell grade silicon. The reaction is to proceed at 1,200 K as follows:

$$2Zn(g) + SiCl_4(g) \rightarrow 2ZnCl_2(g) + Si \downarrow$$

and the process looks somewhat as shown on the following page. If the discharge tube plugs up, the process is ruined. To avoid this plugging, sticking, or bridging of solids, we pass a bit of $SiCl_4$ gas *up* the tube to fluidize the solids. This will insure smooth downflow of solids.

(a) How would you be able to tell whether the solids in the discharge tube are fluidized or not?
(b) What gas flow up the tube would you need to keep the downflowing solids fluidized?

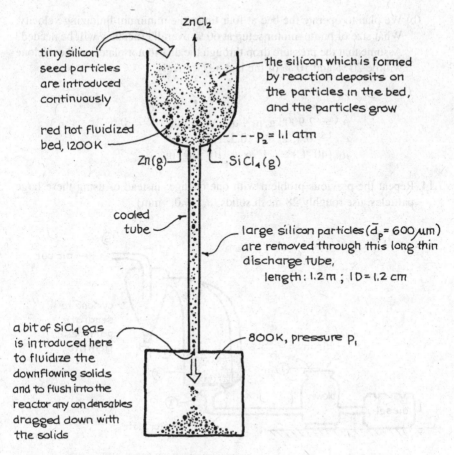

$ZnCl_2$

tiny silicon seed particles are introduced continuously

the silicon which is formed by reaction deposits on the particles in the bed, and the particles grow

red hot fluidized bed, 1200 K

$P_2 = 1.1$ atm

$Zn(g)$ $SiCl_4(g)$

cooled tube

large silicon particles ($\bar{d}_p = 600\,\mu m$) are removed through this long thin discharge tube,
length: 1.2 m ; ID = 1.2 cm

a bit of $SiCl_4$ gas is introduced here to fluidize the downflowing solids and to flush into the reactor any condensables dragged down with the solids

800 K, pressure P_1

Data: Silicon particles in discharge tube:

$$d_p = 600 \ \mu m, \ \rho_s = 2{,}200 \ \text{kg/m}^3, \ \varepsilon_{mf} = 0.4$$

$SiCl_4$ gas at average conditions (1,000 K) in the discharge tube:

$$(mw) = 0.170 \ \text{kg/mol}, \ \mu = 4 \times 10^{-5} \text{kg/m s}$$

7.13. *Cold model of an FBC (fluidized bed combustor).* Our laboratory has a cold model of a tube-filled fluidized furnace for the generation of steam from the combustion of coal. The unit consists of a vessel 1 m^2 in cross section, which fluidizes crushed rock, roughly 4 mesh in size, with a settled bed height of 0.5 m. Room temperature air (20 °C) enters the compressor, but it is estimated that the bed itself will be hotter, roughly 40 °C.

(a) Find the superficial gas velocity which will just fluidize the solids.

(b) We plan to operate the bed at four times the minimum fluidizing velocity. What size of pump–motor setup at 60 % overall efficiency will be needed? Assume that the pressure drop through the distributor plate and the cyclone will be 20 % and 10 % of the bed, respectively.

Data:

$$\rho_s = 2{,}900 \text{ kg/m}^3, \ d_{scr} = 5 \text{ mm}, \ \phi = 0.6$$
$$\varepsilon_m = 0.4, \ \varepsilon_{mf} = 0.5, \ \varepsilon_f = 0.6$$
$$\mu_{air}(40\,^\circ\text{C}) = 1.96 \times 10^{-5} \text{ kg/m s}$$

7.14. Repeat the previous problem with one change: instead of using these large particles, use roughly 28 mesh solids ($d_{scr} = 0.5$ mm).

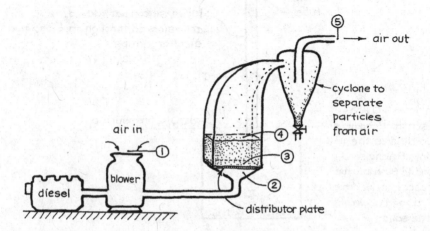

7.15. Our university laboratory has a large fluidized test unit $1 \text{ m} \times 1$ m in cross section and 7 m high. In this unit we plan to fluidize limestone ($\rho_s = 2{,}930 \text{ kg/m}^3$, $d_p = 0.65$ mm, $\varepsilon_{bed} = 0.5$) at 1.5 u_{mf}. The fluidizing air comes from a compressor which delivers just about any flow rate but with a maximum outlet pressure of 122 kPa. What height of solids can we fluidize with this unit?

Data: Room air is at 20 °C and the pressure here in Corvallis is about 100 kPa.

References

D.C. Chitester et al., Characteristics of fluidization at high pressure. Chem. Eng. Sci. **39**(253) (1984)

D. Kunii, O. Levenspiel, Fluidization Engineering, 2nd edn, Chap. 3, (Butterworth, Boston, 1991)

C.Y. Wen, Y.H. Yu, A generalized method for predicting the minimum fluidization velocity. AIChE J. **12**(610) (1966)

Chapter 8
Solid Particles Falling Through Fluids

8.1 Drag Coefficient of Falling Particles

8.1.1 The Small Sphere

The forces acting on a sphere falling through a fluid (call it gas for convenience) are
as follows:

$$\left| \begin{pmatrix} \text{Force causing} \\ \text{particle to} \\ \text{accelerate} \end{pmatrix} \right| = \left| \begin{pmatrix} \text{net weight} \\ \text{of particle} \end{pmatrix} \right| - \left| \begin{pmatrix} \text{drag} \\ \text{force} \end{pmatrix} \right| \qquad \left[\frac{\text{kg} \cdot \text{m}}{\text{s}^2} \right]$$

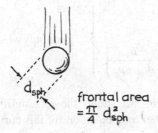

$$\text{frontal area} = \frac{\pi}{4} d_{\text{sph}}^2$$

In symbols

$$|F| = m\frac{du}{dt} = \left(\frac{\pi}{6}d_{\text{sph}}^3\right)\left(|\rho_s - \rho_g|\right)g - |F_d| \tag{8.1}$$

© Springer Science+Business Media New York 2014
O. Levenspiel, *Engineering Flow and Heat Exchange*,
DOI 10.1007/978-1-4899-7454-9_8

where the drag force

$$|F_d| = C_D \cdot \frac{\pi d_{\text{sph}}^2}{4} \frac{\rho_g u^2}{2} \tag{8.2}$$

Drag coefficient

At the *terminal velocity* $du/dt = 0$, in which case equations (8.1) and (8.2) give

$$u_t = \left(\frac{4 g d_{\text{sph}} \left(|\rho_s - \rho_g| \right)}{3 \rho_g C_D} \right)^{1/2} \quad \text{or} \quad C_D = \frac{4 g d_{\text{sph}} \left(|\rho_s - \rho_g| \right)}{3 \rho_g u_t^2} \tag{8.3}$$

The drag coefficient in equation (8.3) has been found by experiment to be related to the particle Reynolds number at terminal velocity, defined as

$$\text{Re}_{\text{sph},\,t} = \frac{d_{\text{sph}} u_t \rho_g}{\mu} \tag{8.4}$$

For the special case of *viscous flow of a sphere*, occurring when $\text{Re}_t < 1$, Stokes developed the following theoretical expression for the drag force:

$$F_d = 3 \pi d_{\text{sph}} \mu u \quad \left[\frac{\text{kg} \cdot \text{m}}{\text{s}^2} \right] \tag{8.5}$$

Replacing in equation (8.4) then gives

$$\left. \begin{array}{l} u_t = \dfrac{\left(|\rho_s - \rho_g| \right) g d_{\text{sph}}^2}{18 \mu} \\[3mm] C_D = \dfrac{24}{\text{Re}_{\text{sph},\,t}} = 24 \left(\dfrac{\mu}{d_{\text{sph}} u_t \rho_g} \right) \end{array} \right\} \quad \text{good only when } \text{Re}_{\text{sph},\,t} < 1 \tag{8.6}$$

The lowest curve on Fig. 8.1 shows the relationship of C_D vs. $\text{Re}_{\text{sph},\,t}$ for spherical particles. The rather long equation for this curve is given by Haider and Levenspiel (1989).

8.1.2 Nonspherical Particles

The drag coefficient for these irregular particles is also shown on Fig. 8.1. Again, see Haider and Levenspiel (1989) for the ugly equation which represents these curves.

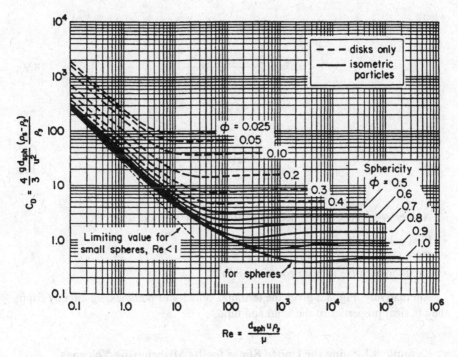

Fig. 8.1 Design chart for drag coefficients of single free-falling particles

8.1.3   Terminal Velocity of Any Shape of Irregular Particles

Larger spheres and other shaped particles generate and are followed by a wake as they fall at their terminal velocity, at $Re_t > 1$. Here no satisfactory theoretical drag force expression has been developed. Consequently, the frictional loss and terminal velocity have to be found by experiment. These findings, by Achenbach (1972), Pettijohn and Christiansen (1948), Schiller (1932), Schlichting (1979), Schmiedel (1928), and Wadell (1934), have been well correlated by Haider and Levenspiel (1989) by the following expressions:

For spheres, $\phi=1$

$$u_t^* = \left[\frac{18}{(d^*)^2} + \frac{0.591}{(d^*)^{1/2}} \right]^{-1} \tag{8.7}$$

For nonspherical particles, $0.5 < \phi < 1$

$$u_t^* = \left[\frac{18}{(d^*)^2} + \frac{2.335 - 1.745\,\phi}{(d^*)^{1/2}} \right]^{-1} \tag{8.8}$$

where

$$u_t^* = u_t \left[\frac{\rho_g^2}{g\mu\left(\left|\rho_s - \rho_g\right|\right)} \right]^{1/3} \tag{8.9}$$

and

$$d^* = d_{sph} \left[\frac{g\rho_g\left(\left|\rho_s - \rho_g\right|\right)}{\mu^2} \right]^{1/3} \tag{8.10}$$

A three-step procedure is needed to evaluate u_t, given d_{sph} and ϕ.

- First calculate d^* from equation (8.10).
- Then find u_t^* from equation (8.7) or (8.8).
- Finally determine u_t from equation (8.9).

Alternatively, Fig. 8.2 gives the terminal velocity of particles, u_t, directly from the physical properties of the solid and fluid.

Example 8.1. Suing the United States for Its Misbehaving Volcanos

On May 18, 1980, Mount St. Helens on the West Coast of the United States erupted catastrophically, spewing an ash plume to an altitude of 20 km. The winds then carried these millions of tons of particles, consisting mainly of silica (~70 %), across the United States, depositing a 2-cm layer on my campsite high (2 km) in the Rockies and 1,000 km away. It started raining ash just 50 h after the eruption, and although I left for home soon after, I had to breathe this contaminated air.

I am worried because I read on page 19 of the June 9, 1980, issue of *Chemical and Engineering News* that silica particles smaller than $10\,\mu m$ are respirable and can cause silicosis. No one told me this, I've since developed a cough, and being a normal American, I'm ready to sue the government for gross negligence in not warning me of this danger. But, of course, I will only do this if the particles are in the dangerous size range.

Please estimate the size of particles which settled on me at the start of this ash rain.

Data: Assume that ash particles consist of pure silica for which

$$\rho_s = 2{,}650 \ \text{kg/m}^3 \quad \text{and} \quad \phi = 0.6$$

The average atmospheric conditions from 2 to 20 km are $T = -30$ °C, $p = 40$ kPa, at which $\mu_{air} = 1.5 \times 10^{-5}$ kg/m s.

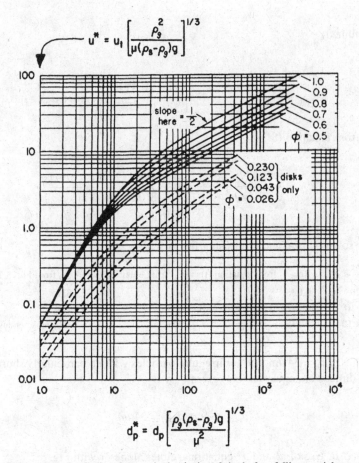

Fig. 8.2 Design chart for finding the terminal velocity of single free-falling particles

Solution

Method A. Use the C_D and Re_p equations with Fig. 8.1.

The problem statement does not specify what is meant by the words "particle size"—silica is quite nonspherical—so let us determine both the screen size and the equivalent spherical size. First we find

$$\rho_g = \frac{(mw)p}{RT} = \frac{(0.0289)(40,000)}{(8.314)(243)} = 0.5722 \text{ kg/m}^3$$

$$u_t = \frac{\text{distance fallen}}{\text{time}} = \frac{20,000 - 2,000}{50 \times 3,600} = 0.1 \text{ m/s}$$

(continued)

(continued)

Then,

$$\mathrm{Re}_{\mathrm{sph},\,t} = \frac{d_{\mathrm{sph}}u_t\rho_g}{\mu} = \frac{d_{\mathrm{sph}}(0.1) \times (0.5722)}{1.5 \times 10^{-5}} = 3{,}815\,d_{\mathrm{sph}} \qquad (i)$$

and from equation (8.3)

$$C_D = \frac{4gd_{\mathrm{sph}}\left(\rho_s - \rho_g\right)}{3\rho_g u_t^2} = \frac{4(9.8)d_{\mathrm{sph}}(2{,}650 - 0.57)}{3(0.5722)(0.1)^2} \qquad (ii)$$
$$= 6.05 \times 10^6 d_{\mathrm{sph}}$$

Now solve by trial and error using Fig. 8.1.

Guess d_{sph}	$\mathrm{Re}_{\mathrm{sph},\,t}$ from equation (i)	C_D from equation (ii)	C_D from Fig. 8.1
1×20^{-5} m	0.038	60	632
10×10^{-5}	0.38	600	67
3.4×10^{-5}	0.13	206	190 (close enough)

Thus, $d_{\mathrm{sph}} = 34\,\mu$m, and for irregular particles with no particularly short or long dimension, equation (6.4) gives

$$d_{\mathrm{scr}} = \frac{d_p}{\phi} = d_{\mathrm{sph}} = 34\mu m$$

Method B. Use of u_t^* and d^* equations, either alone or with Fig. 8.2
   In this problem u_t^* is known, and d^* is to be found. So from equation (8.9)

$$u_t^* = 0.1\left[\frac{(0.5722)^2}{9.8\left(1.5 \times 10^{-5}\right)(2{,}650)}\right]^{1/3} = 0.084\,05 \qquad (iii)$$

We next find d^* either directly from Fig. 8.2 or from equation (8.8).
   From Fig. 8.2 we find

$$d^* = 0.13$$

Alternatively, from equation (8.8)

$$u_t^* = 0.084\,05 = \left[\frac{18}{(d^*)^2} + \frac{2.335 - 1.745(0.6)}{d^{1/2}}\right]^{-1}$$

(continued)

(continued)

Rearranging gives

$$\frac{18}{(d^*)^2} + \frac{1.288}{(d^*)^{1/2}} = 11.8977 \qquad \text{(iv)}$$

Now solve equation (iv) by trial and error.

Guess d^*	LHS of equation (ii)
1.0	19.28
1.3	11.78
1.2931	11.8975 (close enough to the RHS of equation (iv))

So from equation (8.10)

$$d_{sph} = d_{scr} = 1.293 \left[\frac{(1.5 \times 10^{-5})^2}{9.8(0.5722)(2,650)} \right]^{1/3} = 32 \times 10^{-6}\text{m} = 32 \ \mu m$$

Comment Method B does not require pulling a value from a chart and overall is simpler to use than Method A.

Conclusion. Either way you define particle size, you'd better not sue.

Problems on Falling Objects

8.1. *Skydiving.* Indoor skydiving has come to Saint Simon, near Montreal, in the form of the "Aerodium," a squat vertical cylinder 12 m high and 6 m i.d., with safety nets at top and bottom. A DC-3 propeller which is driven by a 300-kW diesel engine blasts air upward through the Aerodium at close to 150 km/h, while the "jumper" dressed in an air inflated jumpsuit floats, tumbles, and enjoys artificial free fall in this rush of air without the danger of the real thing.

   If an 80-kg, suited adult (density = 500 kg/m^3) in the spread-eagled position can hang suspended when the air velocity is 130 km/h, what is his sphericity in this position? (information from *Parachutist*, pg. 17, August 1981)

8.2. The free-fall velocity of a very tiny spherical copper particle in 20 °C water is measured by a microscope and found to be 1 mm/s. What is the size of the particle?

8.3. Who else could it happen to but "Bad Luck" Joe? He goes hunting, gets lost, fires three quick shots into the air, and gets hit squarely on the head by all three bullets as they come down. How fast were the bullets going when they hit him?

Data: Each bullet has a mass of 180 grains or 0.0117 kg, $\phi = 0.806$, and $\rho_{bullet} = 9{,}500$ kg/m^3.

8.4. Water at 20 °C flows downward through a 1-m deep packed bed ($\varepsilon = 0.4$) of 1-mm plastic spheres ($\rho_s = 500$ kg/m^3). What head of water is needed to keep the spheres from floating upward?

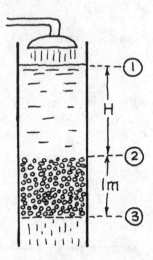

8.5. Rutherford Arlington, a freshman who weighs 80 kg stripped, yens to streak with a difference, and stepping out of a balloon at 3,000 m directly above Central Square during a noon revival meeting appeals to him. Imagine the impact—a real heavenly body entering their midst. At what speed would he join the faithful:

(a) If he curls up into a perfect sphere?
(b) If he is spread eagled? In this orientation, $\phi = 0.22$.

8.6. Find the upward velocity of air at 20 °C which will just float a ping pong ball.
   Data: Nittaku 3-Star ping pong balls, used for the 37th World Championships in Tokyo in 1983, have a diameter of 37.5 mm and a mass of 2.50 g.

8.7. Lower Slobovia recently entered the space race with its own innovative designs. For example, the touchdown parachute of their lunar space probe, the "Lunik," was ingeniously stored in the mouth of the braking rocket to save space. Unfortunately, instead of releasing the parachute, then firing the rocket, our intrepid spaceman, the "lunatic," first fired the rocket, which then used all its fuel to burn up the parachute—all this 150 km above the Earth. When the Slob finally returned to Earth, he had a somewhat rough landing. At what speed do you estimate that he hit the ground?
   Information on the rocket: Volume $= 5$ m^3, mass $= 2.5$ t, and surface area $= 20$ m^2.

8.8. Rutherford Arlington, famed streaker, plans to use a helium balloon for the ascent prior to his spectacular free fall (see Problem 8.5). To reach a height of 1,000 m in 10 min, what size of balloon would he need?

Data: The combined mass of Ruthy and his balloon = 120 kg, $\phi \cong 1$, $T = 20\ ^\circ C$, $\pi = 100$ kPa.

(problem prepared by Dan Griffith)

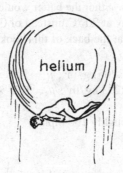

8.9. Referring to the proposed ride called "Typhoon" for the Tokyo Disneyland (Problem 7.10) in which children are fluidized in a large Plexiglas cylinder, the only worry is that some small child may rise above his screaming fellows to get away from it all. To see if this is likely to occur, calculate the terminal velocity of a little Japanese child. See Problem 7.10 for additional data.

8.10. Referring to the data of Example 8.1, how long would it take for 1-μm ash particles (equivalent spherical diameter) disgorged to an altitude of 20 km to settle out of the atmosphere down to sea level? How far around the world would particles of this size go in this time?

Data: Average westerly wind speed at 45° latitude (close to the location of the volcano) is 800 km/day. Ignore updrafts and downdrafts; in the long run they should cancel out.

8.11. The *Official Baseball Rules* states, in part:

"1.09 The ball should be a sphere formed by yarn wound round a small sphere of cork, rubber, or similar material covered with two strips of white horsehide or cowhide, tightly stitched together. It shall weigh not less than 5 nor more than 5¼ ounces avoirdupois and measure no less than 9 nor more than 9¼ inches in circumference."

(a) What should be the terminal velocity of a *smooth* baseball?
(b) Baseball aficionados know that the two cover pieces of a regulation baseball are hand-stitched together with exactly 216 raised cotton stitches. The seam and stitches add roughness to the surface and this lowers the drag coefficient by about 44 %, according to Adair (1990). What would this do to your calculated u_t?

How do your calculations compare with wind tunnel tests which give $u_t = 95$ miles/h?

8.12. *Baseball trivia*. Goose Gossage grumbled:

"Shucks, t' would have been over the 406 ft fence it it'd been warmer," as Willie Mays speared Goose's 400 ft drive on a wintry 0 °C day.

Common lore in baseball has it that a ball flies farther on a warmer day. To check this, estimate whether the batter would have gotten a home run had the day been warmer, say at 20 °C, instead of 0 °C.

Data: From the table at the back of this book, for air,

$$\text{At } 20\,°C \quad \rho = 1.205\,\text{kg/m}^3,\ \mu = 1.81 \times 10^{-5}\text{kg/m} \cdot \text{s}$$
$$\text{At } 0\,°C \quad \rho = 1.293\,\text{kg/m}^3,\ \mu = 1.72 \times 10^{-5}\text{kg/m} \cdot \text{s}$$

A hit ball travels at roughly 100 mph.

References

E. Achenbach, Experiments on the flow past spheres at very high Reynolds numbers. J. Fluid Mech. **54**(565) (1972)

R.K. Adair, *The Physics of Baseball*, Chap. 2, (Harper and Row, New York, 1990)

A. Haider, O. Levenspiel, Drag coefficient and terminal velocity of spherical and non-spherical particles. Powder Technol. **58**(63) (1989)

E.S. Pettijohn, E.B. Christiansen, Effect of particle shape on free-settling rates of isometric particles. Chem. Eng. Prog. **44**(157) (1948)

L. Schiller, *Hydro- und Aerodynamik Handbuch der Experimentalphysik*, Bd. IV, Teil 2, p. 335 (1932)

H. Schlichting, *Boundary Layer Theory*, 6th edn. (McGraw-Hill, New York, 1979), p. 17

J. Schmiedel, Experimentelle Untersuchungen über die Fallbewegung von Kugeln und Scheiben in reibenden Flussigkeiten. Phys. Z. **29**(593) (1928)

H. Wadell, The coefficient of resistance as a function of Reynolds number for solids of various shapes. J. Franklin Inst. **217**(459) (1934)

Part II
Heat Exchange

The second part of this volume deals with the exchange of heat from one flowing stream, whether it be solid, liquid, or gas, to another, and the many different kinds of devices, called heat exchangers, which can be used to do this. But before we get to this, we first introduce the three mechanisms of heat transfer and consider their interaction. Chapter 9 thus presents some of the findings on these three mechanisms of heat transfer. Chapter 10 then shows how to treat situations which involve more than one mechanism of heat transfer, and Chap. 11 considers unsteady state heating and cooling of objects. Chapter 12 onward then uses this information for the design of the three major types of heat exchangers: the recuperator, the direct contact exchanger, and the regenerator. Finally, Chap. 16 presents a collection of problems which uses ideas and findings from various chapters in this book.

Chapter 9
The Three Mechanisms of Heat Transfer: Conduction, Convection, and Radiation

In general, heat flows from here to there by three distinct mechanisms:

- By conduction, or the transfer of energy from matter to adjacent matter by direct contact, without intermixing or flow of any material.
- By convection, or the transfer of energy by the bulk mixing of clumps of material. In natural convection it is the difference in density of hot and cold fluid which causes the mixing. In forced convection a mechanical agitator or an externally imposed pressure difference (by fan or compressor) causes the mixing.
- By radiation such as light, infrared, ultraviolet, and radio waves which emanate from a hot body and are absorbed by a cooler body.

In turn, let us briefly summarize the findings on these three mechanisms of heat transfer.

9.1 Heat Transfer by Conduction

Conduction refers to the transfer of heat from the hotter to the colder part of a body by direct molecular contact, not by gross movement of clumps of hot material to the cold region. At steady state the rate of heat transfer depends on the nature of the material and the temperature differences and is expressed by Fourier's law as

$$\dot{q}_x = -kA\frac{dT}{dx} \quad \left[\frac{J}{s} = W\right] \tag{9.1}$$

where $\dot{q}_x$ is the rate of heat transfer in the x direction [W]; A is the area normal to the direction of heat flow [m²]; dT/dx is the temperature gradient in the x direction [K/m]; and k is the thermal conductivity, defined as the heat going through a cube of

© Springer Science+Business Media New York 2014
O. Levenspiel, *Engineering Flow and Heat Exchange*,
DOI 10.1007/978-1-4899-7454-9_9

Table 9.1 Short table of thermal conductivities for materials at room temperature[a]

Material	k, W/m K	Material	k, W/m K
Gases		Solids	
SO_2	0.009	Styrofoam	0.036
CO_2, H_2	0.018	Corrugated cardboard	0.064
H_2O	0.025	Paper	0.13
Air	0.026	Sand, dry	0.33
Liquids		Glass	0.35–1.3
Gasoline	0.13	Ice	2.2
Ethanol	0.18	Lead	34
Water	0.61	Steel	45
Mercury	8.4	Aluminum	204
Sodium	85	Copper	380

[a]For additional values, see Appendices A.15 and A.21

the material in question 1 m on a side resulting from a temperature difference on opposite faces of 1 °C. Table 9.1 gives k values for various materials [W/m K].

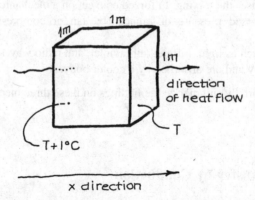

The minus sign in this equation tells that heat flows from regions of higher to lower temperature, not the other way round, and shows that the second law of thermodynamics is at work.

The complete equation for steady-state heat conduction in any arbitrary direction through an isotropic material, without heat generation, is

$$\dot{q} = -kA(\Delta T), \quad k = \text{constant} \tag{9.2}$$

Fourier's equation has been integrated for various simple geometries. Here are some steady-state solutions:

9.1.1   *Flat Plate, Constant* **k**

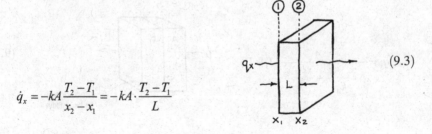

$$\dot{q}_x = -kA\frac{T_2 - T_1}{x_2 - x_1} = -kA \cdot \frac{T_2 - T_1}{L} \qquad (9.3)$$

9.1.2   *Flat Plate,* **k = k₀ (1 + βT)**

$$\dot{q}_x = -k_0 A\frac{(T_2 - T_1) + (\beta/2)(T_2^2 - T_1^2)}{x_2 - x_1} = -k_{\text{mean}}A\frac{T_2 - T_1}{L} \qquad (9.4)$$

where

$$k_{\text{mean}} = \frac{k_1 + k_2}{2}$$

9.1.3   *Hollow Cylinders, Constant* **k**

$$\dot{q}_r = -2\pi kL\frac{T_2 - T_1}{\ln(r_2/r_1)} \qquad (9.5)$$

9.1.4   *Hollow Sphere, Constant* **k**

$$\dot{q}_r = -4\pi kr_1 r_2\frac{T_2 - T_1}{r_2 - r_1} \qquad (9.6)$$

9.1.5   Series of Plane Walls

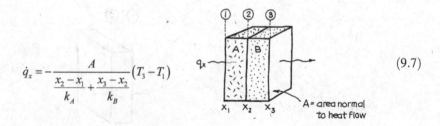

$$\dot{q}_x = -\frac{A}{\dfrac{x_2 - x_1}{k_A} + \dfrac{x_3 - x_2}{k_B}}(T_3 - T_1)$$

A = area normal to heat flow

(9.7)

9.1.6   Concentric Cylinders

$$\dot{q}_r = -\frac{2\pi L}{\dfrac{\ln(r_2/r_1)}{k_A} + \dfrac{\ln(r_3/r_2)}{k_B}}(T_3 - T_1)$$

(9.8)

9.1.7   Concentric Spheres

$$\dot{q}_r = -\frac{4\pi}{\dfrac{r_2 - r_1}{k_A r_1 r_2} + \dfrac{r_3 - r_2}{k_B r_2 r_3}}(T_3 - T_1)$$

(9.9)

9.1.8   Other Shapes

For nonsimple geometries or for nonuniform temperatures at the boundaries, the heat flow can only be obtained by solving Fourier's equation by numerical or graphical methods [see Welty (1978) or McAdams (1954)].

9.1.9   Contact Resistance

When heat flows across two touching plane walls, an extra resistance normally is found at the interface because the contacting surfaces are not quite smooth. This results in a sharp temperature drop at the surface. The heat flow across the interface can then be related to the temperature drop across the interface by

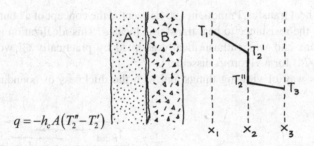

$$q = -h_c A\left(T_2'' - T_2'\right)$$

where h_c is defined as the contact heat transfer coefficient.

Overall, the heat flow across the two walls will then involve three resistances in series:

Across wall A : $\dot{q}_x = -k_A A \dfrac{T_2' - T_1}{x_2 - x_1}$

Across the interface : $\dot{q}_x = -h_c A\left(T_2'' - T_2'\right)$

Across wall B : $\dot{q}_x = -k_B A \dfrac{T_3 - T_2'}{x_3 - x_2}$

Noting that the $\dot{q}$'s are all equal, we can combine the above equations to eliminate the intermediate temperatures T_2' and T_2'' ending up with

$$\dot{q}_x = -\frac{1}{\frac{x_2 - x_1}{k_A} + \frac{1}{h_c} + \frac{x_3 - x_2}{k_B}} A (T_3 - T_1) \tag{9.10}$$

Equations analogous to the above can be developed for concentric spheres, concentric cylinders, and other shapes.

9.2   Heat Transfer by Convection

When hot fluid moves past a cool surface, heat goes to the wall at a rate which depends on the properties of the fluid and whether it is moving by natural convection, by laminar flow, or by turbulent flow. To account for

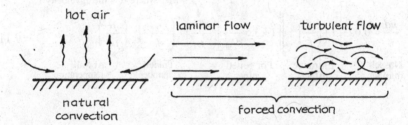

this form of heat transfer, Prandtl, in 1904, invented the concept of a boundary layer in which all the resistance to heat transfer is located. This idealization led to great simplifications and was enthusiastically adopted by practically all workers [see Adiutori (1974) for a vigorous dissenter].

With this way of viewing things and with the thickness of boundary layer δ, we have

$$\dot{q} = -kA\frac{T_{\text{fluid}} - T_{\text{wall}}}{\delta} = -kA\frac{\Delta T}{\delta}$$

Because δ cannot be estimated independently, we combine it with k to give

$$\dot{q} = -(k/\delta)A\Delta T = -hA\Delta T \qquad [\text{W}]$$

where, by definition,

$$h = \text{heat transfer coefficient, } [\text{W}/\text{m}^2\,\text{K}]$$

Note that h incorporates the thickness of an idealized boundary layer which will give the actual heat transfer rate. This quantity h is extremely useful since it is the rate coefficient which allows us to estimate the heat transfer rate in any particular situation.

Values of h have been measured in all sorts of situations, correlated with the properties of the fluid C_p, ρ, μ, k, the flow conditions u, and the system geometry d and compactly summarized in dimensionless form. The sampling of correlations which follows comes from McAdams (1954) or Perry and Chilton (1973) unless otherwise noted.

9.2.1   Turbulent Flow in Pipes

For both heating and cooling of most normal fluids (Pr = 0.7–700) in fully turbulent flow (Re > 10,000), moderate ΔT, and with physical properties measured at bulk conditions,

$$\frac{hd}{k} = 0.023\left[1+\left(\frac{d}{L}\right)^{0.7}\right]\left[1+3.5\frac{d}{d_{\text{coil}}}\right]\underbrace{\left(\frac{dG}{\mu}\right)^{0.8}}_{= u\rho, \text{ where } u = \text{mean velocity}}\left(\frac{C_p\mu}{k}\right)^{1/3}\left(\frac{\mu}{\mu_w}\right)^{0.14} \qquad (9.11)$$

$\underset{\text{Nusselt}}{\underbrace{\quad}} \quad \underset{\text{Entrance}}{\underbrace{\quad}} \quad \underset{\text{For coiled}}{\underbrace{\quad}} \qquad \underset{\text{Prandtl}}{\quad} \qquad \underset{\text{At wall}}{\quad}$

Nusselt number   Entrance effect   For coiled pipes   Prandtl number   At wall temperature

A simplified approximation for common gases (error $\pm 25\%$):

$$h = 0.0018 \frac{C_p G^{0.8}}{d^{0.2}} \quad [\text{W/m}^2\,\text{K}] \tag{9.12}$$

A simplified approximation for cooling or heating of water:

$$h = 91(T + 68)\frac{u^{0.8}}{d^{0.2}} \quad [\text{W/m}^2\,\text{K}] \quad \text{with } T \text{ in } ^\circ\text{C} \tag{9.13}$$

9.2.2 Turbulent Flow in Noncircular Conduits

1. *Rectangular cross section.* Use the equation for circular pipes, equation (9.11), with the following two modifications:

$$h_{\text{rect}} \cong 0.76\, h_{\text{pipes}} \tag{9.14}$$

and replace the pipe diameter with an equivalent diameter defined as

$$d_e = 4\left(\frac{\text{hydraulic}}{\text{radius}}\right) = 4\left(\frac{\text{cross-sectional area}}{\text{perimeter}}\right) = \frac{2 d_1 d_2}{d_1 + d_2} \tag{9.15}$$

2. *Annular passage.* For heat flow to the inner tube wall,

$$\frac{h_i d_e}{k} = 0.02 \left(\frac{d_e G}{\mu}\right)^{0.8} \left(\frac{C_p \mu}{k}\right)^{1/3} \left(\frac{d_o}{d_i}\right)^{0.53} \tag{9.16}$$

where

$$d_e = 4\left(\frac{\text{hydraulic}}{\text{radius}}\right) = d_o - d_i \tag{9.17}$$

To the outer tube wall, use equation (9.11) for circular pipes, but with the pipe diameter replaced by de of equation (9.17).

9.2.3 Transition Regime in Flow in Pipes

In the transition regime, $2{,}100 < \mathrm{Re} < 10{,}000$:

$$\frac{hd}{k} = 0.116\left[\left(\frac{dG}{\mu}\right)^{2/3} - 125\right]\left(\frac{C_p\mu}{k}\right)^{1/3}\left[1 + \left(\frac{d}{L}\right)^{2/3}\right]\left(\frac{\mu}{\mu_w}\right)^{0.14} \qquad (9.18)$$

9.2.4 Laminar Flow in Pipes (Perry and Chilton, pg. 168 (1984))

In the laminar flow regime, or $\mathrm{Re} < 2{,}100$, we have, for $\mathrm{Gz} < 100$,

$$\frac{hd}{k} = \left[3.66 + \frac{0.085\ \mathrm{Gz}}{1 + 0.047\ \mathrm{Gz}^{2/3}}\right]\left(\frac{\mu}{\mu_w}\right)^{0.14} \qquad (9.19)$$

where the Graetz number is defined as

$$\mathrm{Gz} = \mathrm{Re}\cdot\mathrm{Sc}\cdot\frac{d}{L} = \left(\frac{dG}{\mu}\right)\left(\frac{C_p\mu}{k}\right)\left(\frac{d}{L}\right) \qquad [\text{-}] \qquad (9.20)$$

For higher flow rates where $\mathrm{Gz} > 100$,

$$\frac{hd}{k} = 1.86\,\mathrm{Gz}^{1/3}\left(\frac{\mu}{\mu_w}\right)^{0.14} \qquad (9.21)$$

Perry and Chilton (1973) give numerous other expressions.

9.2.5 Laminar Flow in Pipes, Constant Heat Input Rate at the Wall (Kays and Crawford 1980)

When the velocity and temperature profiles are fully developed (away from the entrance region), axial dispersion theory predicts that

$$hd/k = 4.36 \qquad (9.22)$$

In evaluating h the term ΔT is defined as the difference in temperature between the wall at position x and the mixing cup temperature of the flowing fluid at the same position. This situation is found when using electrical resistance heating or radiant heating.

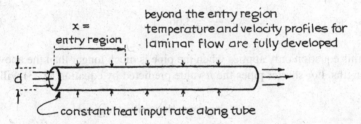

Theory shows that the laminar velocity profile is fully developed at about

$$x/d = 0.05\,\text{Re}$$

and that the thermal profile is fully developed at about

$$x/d = 0.05\,\text{Re}\cdot\text{Pr}$$

Thus, equation (9.22) only applies in tubes much longer than the larger of the above two entry lengths. Let us look at a few typical entry lengths at Re = 100:

Entry length, x/d				
Fluid	Pr	From velocity profile	From temperature profile	Slower developing profile
Liquid metal	0.01	5	0.05	Velocity
Water	1	5	5	Same for both
Oil	100	5	500	Temperatures

These values show that for liquid metals or ordinary aqueous fluids, the entry length is rather short. However, if oil or some other high Prandtl number fluid is flowing through the pipe, then the entry length may become substantial, and the value of h predicted by equation (9.22) will be too low. See Kays and Crawford, pg. 114 (1980), for h values for short pipes, and see Perry and Chilton (1973) for h values in other shaped ducts.

9.2.6 Laminar Flow in Pipes, Constant Wall Temperature (Kays and Crawford 1980)

This situation is approached when a process with high h occurs on the outside of the tubes (boiling, condensation, transfer to finned tubes). Here theory says that in the region of fully developed laminar velocity and temperature profiles,

$$\frac{hd}{k} = 3.66 \quad [\text{-}] \tag{9.23}$$

Again, this equation only applies when the pipe is much longer than the above two entry lengths. For shorter pipes the h value predicted by equation (9.23) will be

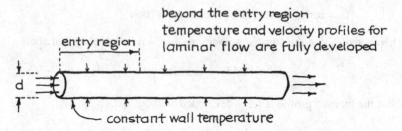

too low. Kays and Crawford, pg. 128 (1980), give h values for short pipes, and Perry and Chilton (1973) give h values for other shaped ducts.

9.2.7 Flow of Gases Normal to a Single Cylinder

Over a very wide range of Reynolds numbers, experimental results can be correlated by

$$\frac{hd}{k_f} = A \left(\frac{dG}{\mu_f} \right)^n \left(\frac{C_p \mu}{k_f} \right)^{0.3} \tag{9.24}$$

where subscript f refers to properties of the gas at the film temperature estimated as

$$T_f = \frac{T_{\text{bulk fluid}} + T_{\text{wall}}}{2}$$

and where the constants A and n are given in Table 9.2. For air at 93 °C and $Re = 1{,}000\text{--}50{,}000$, we have the following simplified equation:

$$h = 0.0018 \frac{C_p G^{0.6}}{d^{0.4}} \quad [\text{W/m}^2\,\text{K}] \tag{9.25}$$

Table 9.2 Constants in equation (9.24) for flow normal to single cylinders

$\frac{du\rho_f}{\mu_f}$	A	n	$\frac{hd}{k_f}$ for air, from equation (9.24)
1–4	0.960	0.330	0.890–1.42
4–40	0.885	0.385	1.40–3.40
40–4,000	0.663	0.466	3.43–29.6
4,000–40,000	0.174	0.618	29.5–121.
40,000–250,000	0.257	0.805	121.–528.

9.2.8 Flow of Liquids Normal to a Single Cylinder

For Re $= 0.1$–300, the data are correlated by

$$\frac{hd}{k_f} = \left[0.35 + 0.56\left(\frac{dG}{\mu_f}\right)^{0.52}\right]\left(\frac{C_p\mu}{k_f}\right)^{0.3} \tag{9.26}$$

9.2.9 Flow of Gases Past a Sphere

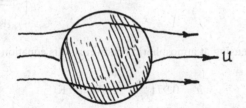

$$\frac{hd}{k_f} = 2 + 0.6\left(\frac{dG}{\mu_f}\right)^{0.5}\left(\frac{C_p\mu}{k}\right)_f^{1/3} \quad \text{for} \left(\frac{dG}{\mu_f}\right) < 325 \tag{9.27}$$

$$\frac{hd}{k_f} = 0.4\left(\frac{dG}{\mu_f}\right)^{0.6}\left(\frac{C_p\mu}{k}\right)_f^{1/3} \quad \text{for} \left(\frac{dG}{\mu_f}\right) = 325 - 70,000 \tag{9.28}$$

9.2.10 Flow of Liquids Past a Sphere

$$\frac{hd}{k_f} = \left[0.97 + 0.68\left(\frac{dG}{\mu_f}\right)^{0.52}\right]\left(\frac{C_p\mu}{k}\right)_f^{0.3} \tag{9.29}$$

9.2.11   Other Geometries

For tube banks, h values can be up to 50 % higher than for single tubes, the actual value depending on the number of rows and the geometry used. For tube banks, coiled tubes, tubes of noncircular cross section, finned tubes, and many other situations, see McAdams (1954), Chap. 10.

9.2.12   Condensation on Vertical Tubes

The theoretical equation derived by Nusselt in 1916 is still recommended today:

$$\frac{hL}{k_l} = 0.925\left(\frac{L^3\rho_l^2 g}{\mu_l \Gamma}\right)^{1/3} = 0.943\left(\frac{L^2\rho_l^2 g\lambda}{k_l\mu_l\Delta T}\right)^{1/4} \tag{9.30}$$

where

$$\Gamma = \left(\begin{array}{c}\text{flow rate of condensate} \\ \text{from the tube} \\ \hline \text{circumference}\end{array}\right) = \frac{F_l}{\pi d}\quad[\text{kg/s\,m}] \tag{9.31}$$

For steam condensing at atmospheric conditions, this equation reduces to

$$h = 0.97\left(\tfrac{d}{F_l}\right)^{1/3}\quad[\text{W/m}^2\,\text{K}] \tag{9.32}$$

9.2.13   Agitated Vessels to Jacketed Walls

For various types of agitators, we have the general expression

$$\frac{hd_{\text{jacket}}}{k} = a\left(\frac{L_p^2 N_r \rho}{\mu}\right)^{b}\left(\frac{C_p\mu}{k}\right)^{1/3}\left(\frac{\mu}{\mu_w}\right)^{m}$$

Reynolds number for agitated vessels

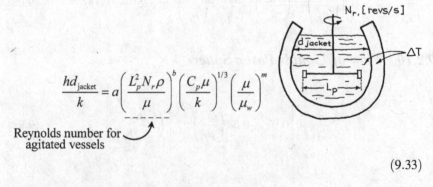

$$\tag{9.33}$$

where the constants a, b, and m are given in Table 9.3.

Table 9.3 Constants in equation (9.33) for heat transfer to the walls of agitated vessels

Type of agitator	a	b	m	Range of Re
Paddle	0.36	2/3	0.21	$300–3 \times 10^5$
Pitched blade turbine	0.53	2/3	0.24	80–200
Disk, flat blade turbine	0.54	2/3	0.14	$40–3 \times 10^5$
Propeller	0.54	2/3	0.14	2×10^3
Anchor	1.0	1/2	0.18	10–300
Anchor	0.36	2/3	0.18	300–40,000
Helical ribbon	0.633	1/2	0.18	$8–10^5$

9.2.14 Single Particles Falling Through Gases and Liquids (Ranz and Marshall 1952)

$$\frac{hd_p}{k} = 2 + 0.6\left(\frac{d_p u \rho}{\mu}\right)^{1/2}\left(\frac{C_p \mu}{k}\right)^{1/3} \tag{9.34}$$

9.2.15 Fluid to Particles in Fixed Beds (Kunii and Levenspiel, 1991)

(a) For beds of fine solids

with gases : $$\frac{hd_p}{k} = 0.012\,\mathrm{Re}_p^{1.6}\mathrm{Pr}^{1/3} \quad \text{for } \mathrm{Re}_p < 100 \tag{9.35}$$

with liquids : $$\frac{hd_p}{k} = 0.16\,\mathrm{Re}_p^{1.6}\mathrm{Pr}^{1/3} \quad \text{for } \mathrm{Re}_p < 10 \tag{9.36}$$

(b) For coarse solids with both gases and liquids

$$\frac{hd_p}{k} = 2 + 1.8\,\mathrm{Re}_p^{1/2}\mathrm{Pr}^{1/3} \quad \begin{cases} \text{for } \mathrm{Re}_p > 100,\ \text{gases} \\ \text{for } \mathrm{Re}_p > 10,\ \text{liquids} \end{cases} \tag{9.37}$$

where $\mathrm{Re}_p = (dpu_0\rho/\mu)$ and $u_0 =$ superficial velocity (upstream velocity or in vessel with no solids).

9.2.16   Gas to Fluidized Particles

The heat transfer coefficient is difficult to measure in this situation, so until reliable data becomes available, the following equation is suggested as a conservative estimate of h:

$$\frac{hd_p}{k} = 2 + 0.6\left(\frac{d_p u_0 \rho}{\mu}\right)^{1/2}\left(\frac{C_p \mu}{k}\right)^{1/3} \tag{9.38}$$

9.2.17   Fluidized Beds to Immersed Tubes

For beds of fine particles, or $\mathrm{Re}_{mf} < 12.5$, Botterill (1983) recommends the following simple dimensional expression (in SI units):

$$\frac{hd_p}{k_g} = 25\,\frac{d_p^{0.64}\rho_s^{0.2}}{\left[k_g(\text{at bed temperature})\right]^{0.4}} \tag{9.39}$$

For beds of large particles, or $\mathrm{Re}_{mf} > 12.5$, Botterill suggests using

$$\frac{hd_p}{k_g} = 0.7\left[d_p^{0.5}\left(d_p^*\right)^{1.17} + \left(d_p^*\right)^{0.45}\right] \tag{9.40}$$

where d_p^* is defined by equation (8.10).

9.2.18   Fixed and Fluidized Particles to Bed Surfaces

See Kunii and Levenspiel (1991), Chap. 13, equations (16)–(20).

9.2.19   Natural Convection

Slow-moving fluids passing by hot surfaces give larger than expected h values. This is because of natural convection. The particular variables which

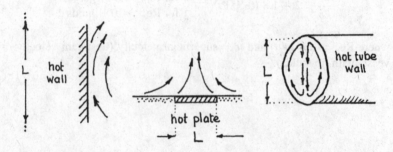

characterize natural convection are combined into a dimensionless group, the Grashof number, defined as

$$Gr = \frac{L^3 \rho_f^2 g \beta \Delta T}{\mu^2}$$

Characteristic length

At film conditions

Coefficient of volumetric expansion $\frac{1}{V}\left(\frac{\partial V}{\partial T}\right)_p \overset{ideal}{\underset{gas}{=}} \frac{1}{T}$

In main body

$T_{wall} - T_{bulk\ fluid}$

Correlations for natural convection are often of the form

$$Nu = A[Gr \cdot Pr]^B$$

or

$$\frac{hL}{k} = A\left[\underbrace{\left(\frac{L^3 \rho_f^2 g \beta \Delta T}{\mu_f^2}\right)\left(\frac{C_p \mu_f}{k_f}\right)}_{X}\right]^B$$

or

$$Y = AX^B$$

$\left.\begin{array}{c}\end{array}\right\}$ (9.41)

9.2.20 Natural Convection: Vertical Plates and Cylinders, $L > 1\ m$

$$\text{Laminar}: \quad Y = 1.36\,X^{1/5} \quad \text{for } X < 10^4 \tag{9.42}$$

$$\text{Laminar}: \quad Y = 0.55\,X^{1/4} \quad \text{for } X = 10^4 - 10^9 \tag{9.43}$$

$$\text{Turbulent}: \quad Y = 0.13\,X^{1/3} \quad \text{for } X > 10^9 \tag{9.44}$$

Simplified equations for air at room conditions:

$$h = 1.4\left(\frac{\Delta T}{L}\right)^{1/4} \quad [\text{W/m}^2\,\text{K}] \quad \text{for laminar regime} \tag{9.45}$$

$$h = 1.3(\Delta T)^{1/3} \quad [\text{W/m}^2\,\text{K}] \quad \text{for turbulent regime} \tag{9.46}$$

and for water at room conditions:

$$h = 120(\Delta T)^{1/3} \quad [\text{W/m}^2\,\text{K}] \quad \text{for } X > 10^9 \tag{9.47}$$

9.2.21   *Natural Convection: Spheres and Horizontal Cylinders,* $\mathbf{d} < 0.2$ *m*

$$\text{Laminar}: \qquad Y = 0.53 \, X^{1/4} \qquad \text{for } X = 10^3 - 10^9 \qquad (9.48)$$

$$\text{Turbulent}: \qquad Y = 0.13 \, X^{1/3} \qquad \text{for } X > 10^9 \qquad (9.49)$$

For $X < 10^4$, see Perry and Chilton (1973).

Simplified equations for air at room conditions:

$$h = 1.3(\Delta T/L)^{1/4} \quad [\text{W/m}^2 \, \text{K}] \quad \text{for laminar regime} \qquad (9.50)$$

$$h = 1.2(\Delta T)^{1/3} \quad [\text{W/m}^2 \, \text{K}] \quad \text{for turbulent regime, the usual case for pipes} \quad (9.51)$$

9.2.22   *Natural Convection for Fluids in Laminar Flow Inside Pipes*

In laminar flow, when $\text{Gr} > 1{,}000$, natural convection sets up an appreciable secondary flow of fluid in the pipe which in turn increases the

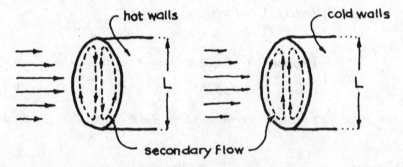

heat transfer coefficient. In this situation equations (9.19), (9.20), and (9.23) for laminar flow should include the additional multiplying factor:

$$0.87\left(1 + 0.015 \, \text{Gr}^{1/3}\right) \qquad (9.52)$$

For turbulent flow no such correction is needed because the tendency to set up a secondary flow pattern is effectively overwhelmed by the vigorous turbulent eddies.

9.2.23 Natural Convection: Horizontal Plates

(a) For heated plates facing up or cooled plates facing down:

$$\text{Laminar}: \quad Y = 0.54\, X^{1/4} \quad \text{for } X = 10^5 - 2 \times 10^7 \tag{9.53}$$

$$\text{Turbulent}: \quad Y = 0.14\, X^{1/3} \quad \text{for } X = 2 \times 10^7 - 3 \times 10^{10} \tag{9.54}$$

(b) For heated plates facing down or cooled plates facing up:

$$\text{Laminar}: \quad Y = 0.27\, X^{1/4} \quad \text{for } X = 3 \times 10^5 - 3 \times 10^{10} \tag{9.55}$$

(c) The three corresponding simplified equations for air at room conditions:

$$h = 1.3\left(\tfrac{\Delta T}{L}\right)^{1/4} \quad [\text{W/m}^2\,\text{K}] \quad \text{for laminar regime} \tag{9.56}$$

$$h = 1.5\Delta T^{1/3} \quad [\text{W/m}^2\,\text{K}] \quad \text{for turbulent regime} \tag{9.57}$$

$$h = 0.64\left(\tfrac{\Delta T}{L}\right)^{1/4} \quad [\text{W/m}^2\,\text{K}] \quad \text{for laminar regime} \tag{9.58}$$

9.2.24 Other Situations

Heat transfer coefficients for boiling, condensation, high-velocity gas flow (compressibility effects and supersonic flow), high-vacuum flow, and many other situations have been studied and reported in the vast heat transfer literature and are well condensed in McAdams (1954), in Perry and Chilton (1973), and in Cavaseno (1979).

9.3 Heat Transfer by Radiation

All materials emit, absorb, and transmit radiation to an extent which is strongly dependent on their temperature. Let

$$\alpha_{1 \leftarrow 2} = \frac{\text{energy absorbed by a surface at } T_1}{\text{energy incident coming from a source } T_2}, \quad \text{the absorptivity} \tag{9.59}$$

The absorptivity varies from 0 to 1. The perfect absorber has $\alpha = 1$ and is called a blackbody. Next, let

$$\varepsilon_1 = \frac{\text{energy emitted by a surface at } T_1}{\text{energy emitted by an ideal emitter, a blackbody, at } T_1}, \quad \text{emissivity} \quad (9.60)$$

and

$$\tau_1 = \frac{\text{energy transmitted through the body at } T_1}{\text{energy incident}}, \quad \text{transmittance} \quad (9.61)$$

Then, the fraction of energy reflected is $1 - \alpha - \tau$.

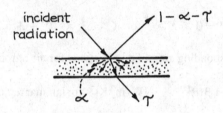

9.3.1   Radiation from a Body

The energy emitted from surface A_1 of a body is strongly dependent on the temperature and nature of the surface and is given by

$$\dot{q}_{1\rightarrow} = \sigma A_1 \varepsilon_1 T_1^4 \quad [\text{W}] \tag{9.62}$$

where the radiation constant

$$\sigma = 5.67 \times 10^{-8} \text{ W/m}^2 \text{ K}^4 \tag{9.63}$$

is called the Stefan–Boltzmann constant. Equation (9.62) is called the Stefan–Boltzmann law of radiation, and the fourth power of temperature is a consequence of the second law of thermodynamics.

9.3.2   Radiation onto a Body

The energy absorbed by a surface A_1 which is at T_1 from blackbody surroundings at T_2 is given by

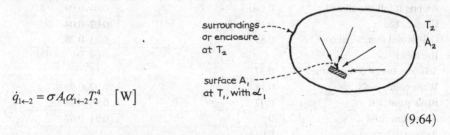

surroundings or enclosure at T_2

surface A_1 at T_1, with α_1

T_2
A_2

$$\dot{q}_{1\leftarrow2} = \sigma A_1 \alpha_{1\leftarrow2} T_2^4 \quad [\text{W}]$$

(9.64)

9.3.3   Energy Interchange Between a Body and Its Enveloping Surroundings

The energy interchange between a surface A_1 at T_1 and any kind of surroundings at T_2, from equations (9.62) and (9.64), is then

enveloping surroundings A_2, T_2

A_1, T_1, ε_1
$\alpha_{1\leftarrow2}$

$$\dot{q}_{12} = \sigma A_1 \left[\varepsilon_1 T_1^4 - \alpha_{1\leftarrow2} T_2^4 \right]$$

(9.65)

9.3.4   Absorptivity and Emissivity

If an object and its surroundings are both at T_1, then the object does not gain or lose heat. Thus, equation (9.65) becomes

$$\dot{q}_{12} = 0 = \sigma A_1 \left[\varepsilon_1 T_1^4 - \alpha_{1\leftarrow1} T_1^4 \right]$$

Now the value of α and of ε can vary greatly with the type of surface and with temperature, as shown in Table 9.4. However, at any particular temperature T_1, the above expression shows that

$$\varepsilon_1 = \alpha_{1\leftarrow1}$$

(9.66)

Table 9.4 Short table of absorptivities and emissivities of various materials[a]

Material	For solar radiation ($\sim 5,000$ K) onto a surface at room temperature, $\alpha_{room \leftarrow solar}$	For room temperature radiation, $\varepsilon_{room} = \alpha_{room \leftarrow room}$
Ag, polished	0.07	0.01
Al, bright foil or polished	0.1–0.3	0.04–0.09
Cu, polished	0.18	0.02–0.04
Galvanized iron, weathered	0.89	0.23–0.28
Hg, clean	—	0.09
Stainless steel # 301, polished	0.37	0.16
White paint, gloss	0.18	0.92–0.96
Black paint, flat	0.97	0.96–0.98
Aluminum paint	0.55	0.51–0.67
Asphalt pavement, clean	0.93	—
Concrete, rough	—	0.94
Earth, plowed field	0.75	—
Grass	0.75–0.80	—
Gravel	0.29	—
Red brick, rough	0.7–0.75	0.93
Roofing paper, black	—	0.91
White paper	0.28	0.95
Wood	—	0.90–0.04
Snow, clean	0.2–0.35	0.82
Ice	—	0.97
Water, deep	—	0.96

[a]Taken from references in this chapter

This means that the absorptivity of a surface for T_1 radiation equals the emissivity of that surface when it is at T_1.

9.3.5 Greybodies

An object whose absorptivity is the same for all temperature radiation is called a greybody. So for a greybody

$$\alpha = \varepsilon = \text{const, at all temperatures}$$

The greybody approximation is often used since it greatly simplifies difficult analyses.

9.3.6   Radiation Between Two Adjacent Surfaces

If the facing surfaces are close enough so that all radiation leaving one surface hits the other, then the heat interchange is

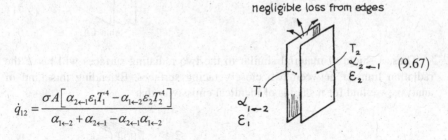

$$\dot{q}_{12} = \frac{\sigma A \left[\alpha_{2\leftarrow 1}\varepsilon_1 T_1^4 - \alpha_{1\leftarrow 2}\varepsilon_2 T_2^4 \right]}{\alpha_{1\leftarrow 2} + \alpha_{2\leftarrow 1} - \alpha_{2\leftarrow 1}\alpha_{1\leftarrow 2}} \qquad (9.67)$$

1. For *two facing grey surfaces*, $\alpha_{2\leftarrow 1} = \varepsilon_2$, $\alpha_{1\leftarrow 2} = \varepsilon_1$, so the above expression reduces to

$$\dot{q}_{12} = \frac{\sigma A \left[T_1^4 - T_2^4 \right]}{\frac{1}{\varepsilon_1} + \frac{1}{\varepsilon_2} - 1} \qquad (9.68)$$

2. For *concentric grey cylinders*, we obtain, similarly,

Always the smaller area

$$\dot{q}_{12} = \frac{\sigma A_1 \left[T_1^4 - T_2^4 \right]}{\frac{1}{\varepsilon_1} + \frac{A_1}{A_2}\left(\frac{1}{\varepsilon_2} - 1 \right)} \qquad (9.69)$$

9.3.7   Radiation Between Nearby Surfaces with Intercepting Shields

If two facing surfaces are separated by a very thin opaque shield, then if $\varepsilon_1 = \varepsilon_2$, while ε_s can be any value, we find

$$T_s^4 = \frac{T_1^4 + T_2^4}{2} \tag{9.70}$$

from which

$$\dot{q}_{12} = \frac{1}{2} \frac{\sigma A \left[T_1^4 - T_2^4 \right]}{\frac{1}{\varepsilon_1} + \frac{1}{\varepsilon_s} - 1} \tag{9.71}$$

Thus, a shield of material similar to the two radiating surfaces will halve the radiation transfer between two closely facing surfaces. Extending this kind of analysis, we find for n shields of identical emissivity that

$$\dot{q}_{12} = \frac{1}{n+1} \frac{\sigma A \left[T_1^4 - T_2^4 \right]}{\frac{1}{\varepsilon_1} + \frac{1}{\varepsilon_s} - 1} \tag{9.72}$$

Similarly, for a shield between pipes or around a sphere,

Always the inside pipe

$$\dot{q}_{12} = \frac{\sigma A_1 \left[T_1^4 - T_2^4 \right]}{\frac{1}{\varepsilon_1} + \frac{A_1}{A_2}\left(\frac{1}{\varepsilon_2} - 1\right) + \frac{A_1}{A_s}\left(\frac{2}{\varepsilon_s} - 1\right)} \tag{9.73}$$

In all cases radiation shields reduce the radiative heat interchange between bodies.

9.3.8   View Factors for Blackbodies

If both surfaces are black and not close together, then only a portion of the radiation leaving surface 1 is intercepted by surface 2. We call this the

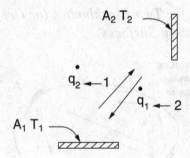

view factor F_{12}, and so the radiation leaving 1 which is intercepted by 2 is

$$\dot{q}_{2 \leftarrow 1} = \sigma A_1 F_{12} T_1^4 \tag{9.74}$$

Similarly, the radiation leaving 2 which is intercepted by 1 is

$$\dot{q}_{1 \leftarrow 2} = \sigma A_2 F_{21} T_2^4 \tag{9.75}$$

If both temperatures are equal, there can be no net transfer of heat between 1 and 2. Thus, we find

$$A_1 F_{12} = A_2 F_{21} \tag{9.76}$$

The net interchange of heat between these two surfaces is then

$$\dot{q}_{12} = \sigma A_1 F_{12} \left[T_1^4 - T_2^4 \right] \quad \text{with } A_1 F_{12} = A_2 F_{21} \tag{9.77}$$

If the surfaces are grey, not black, the heat exchange is approximated by

$$\dot{q}_{12} = \sigma A F'_{12} \left[T_1^4 - T_2^4 \right] \tag{9.78}$$

where

$$F'_{12} = \frac{1}{\frac{1}{F_{12}} + \left(\frac{1}{\varepsilon_1} - 1 \right) + \frac{A_1}{A_2} \left(\frac{1}{\varepsilon_2} - 1 \right)} \tag{9.79}$$

9.3.9 View Factor for Two Blackbodies (or GreyBodies) Plus Reradiating Surfaces

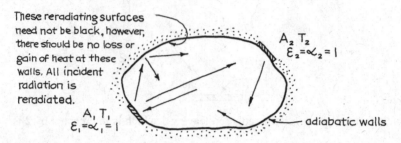

These reradiating surfaces need not be black, however, there should be no loss or gain of heat at these walls. All incident radiation is reradiated.

A, T_1
$\varepsilon_1 = \alpha_1 = 1$

$A_2\ T_2$
$\varepsilon_2 = \alpha_2 = 1$

— adiabatic walls

The net heat interchange between black surfaces 1 and 2 in the presence of adiabatic reradiating surfaces is given by

$$\dot{q}_{12} = \sigma A_1 \overline{F}_{12}\left(T_1^4 - T_2^4\right) \tag{9.80}$$

where $\overline{F}_{12}$ depends on F_{12} and the geometry of the reradiating surfaces. After making simplifying assumptions that surfaces 1 and 2 cannot see themselves and that the reradiating surfaces are all at one temperature, we find that

$$\overline{F}_{12} = \frac{A_2 - A_1 F_{12}^2}{A_1 + A_2 - 2A_1 F_{12}} \tag{9.81}$$

Because these reradiating surfaces return some of the radiation which would otherwise be lost suggests that $\overline{F}_{12}$ is always larger than F_{12}, and this is so.

If surfaces 1 and 2 are grey, then

$$\dot{q}_{12} = \sigma A_1 \mathscr{F}_{12}\left(T_1^4 - T_2^4\right) \tag{9.82}$$

where 1_{12} is the greybody view factor for systems with reradiating surfaces and is approximated by

$$\mathscr{F}_{12} = \frac{1}{\frac{1}{\overline{F}_{12}} + \left(\frac{1}{\varepsilon_1} - 1\right) + \frac{A_1}{A_2}\left(\frac{1}{\varepsilon_2} - 1\right)} \tag{9.83}$$

This is the most general of view factors.

The appendix in Siegel and Howell (1981) refers to view factors for over 200 different kinds of geometries and gives equations for 38 of these geometries. Figures 9.1, 9.2, 9.3, 9.4, and 9.5, from Jakob (1957), show the view factors for five simple geometries.

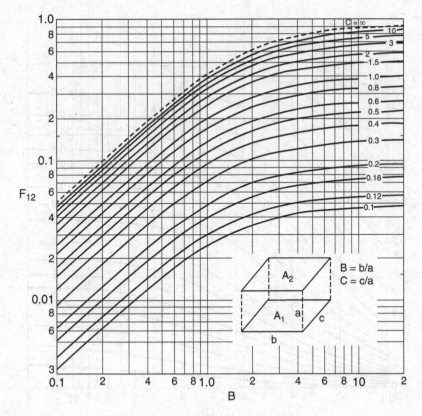

Fig. 9.1 View factor for two facing equal-sized rectangular surfaces

$$F_{12} = \frac{1}{\pi}\left[\frac{1}{BC}\ln\frac{1+B^2+C^2+B^2C^2}{1+B^2+C^2} + \frac{2}{C}\left(1+C^2\right)^{1/2}\tan^{-1}\frac{B}{\left(1+C^2\right)^{1/2}}\right.$$

$$\left. -\frac{2}{B}\tan^{-1}C - \frac{2}{C}\tan^{-1}B + \frac{2}{B}\left(1+B^2\right)^{1/2}\tan^{-1}\frac{C}{\left(1+B^2\right)^{1/2}}\right]$$

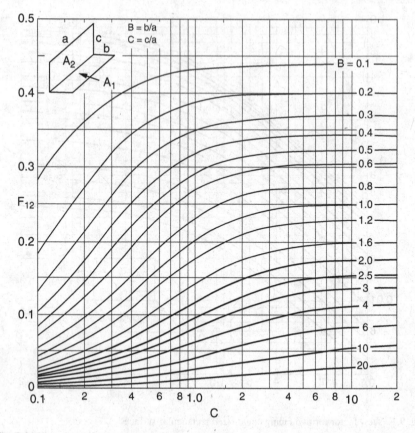

Fig. 9.2 View factor for two perpendicular rectangular surfaces having a common edge

$$F_{12} = \frac{1}{\pi B} \left\{ \frac{1}{4} \ln \left[\left(1 + B^2 + C^2\right)^{B^2 + C^2 - 1} \left(1 + B^2\right)^{1 - B^2} \right. \right.$$
$$\left. \times \left(1 + C^2\right)^{1 - C^2} \left(B^2\right)^{B^2} \left(C^2\right)^{C^2} \right]$$
$$- \frac{1}{4} \ln \left(B^2 + C^2\right)^{B^2 + C^2} + B \tan^{-1} \frac{1}{B} + C \tan^{-1} \frac{1}{C}$$
$$\left. - \left(B^2 + C^2\right)^{1/2} \tan^{-1} \frac{1}{\left(B^2 + C^2\right)^{1/2}} \right\}$$

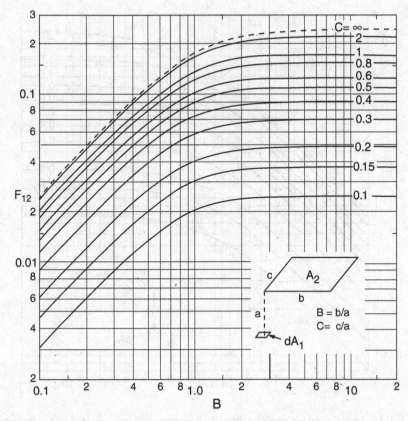

Fig. 9.3 Fraction of radiation leaving a differential surface which is intercepted by a facing rectangular surface, located as shown

$$F_{12} = \frac{1}{2\pi} \left(\frac{B}{(1+B^2)^{1/2}} \sin^{-1} \frac{C}{(1+B^2+C^2)^{1/2}} \right.$$
$$\left. + \frac{C}{(1+C^2)^{1/2}} \sin^{-1} \frac{B}{(1+B^2+C^2)^{1/2}} \right)$$

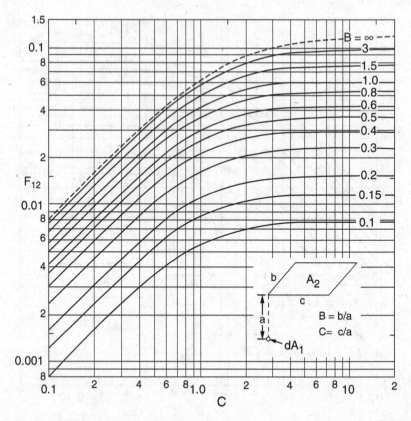

Fig. 9.4 Fraction of radiation leaving a differential sphere which is intercepted by a rectangular surface, located as shown

$$F_{12} = \frac{1}{4\pi} \sin^{-1} \frac{BC}{\left(1 + B^2 + C^2 + B^2 C^2\right)^{1/2}}$$

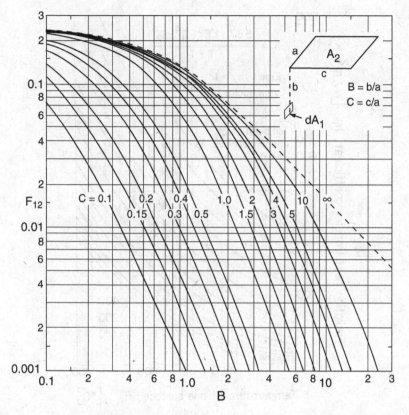

Fig. 9.5 Fraction of radiation leaving a differential surface which is intercepted by a rectangular surface perpendicular to it and located as shown

$$F_{12} = \frac{1}{2\pi}\left[\sin^{-1}\frac{1}{\left(1+B^2\right)^{1/2}} - \frac{B}{\left(B^2+C^2\right)^{1/2}}\sin^{-1}\frac{1}{\left(1+B^2+C^2\right)^{1/2}} \right]$$

9.3.10 Extensions

Gases consisting of molecules which are not symmetrical about all three principal axes (NH_3, CO, CO_2, H_2O, HCl, etc.) absorb and emit significant amounts of radiation at high temperature. Symmetrical molecules (O_2, N_2, H_2, etc.) do not absorb or emit significantly in the temperature range of practical interest.

Heat interactions between absorbing gases and surfaces are accounted for by a characteristic emissivity and a characteristic view factor, somewhat like two-surface systems. Clouds of fine particles, soot, luminous flames, etc., are treated in the same manner.

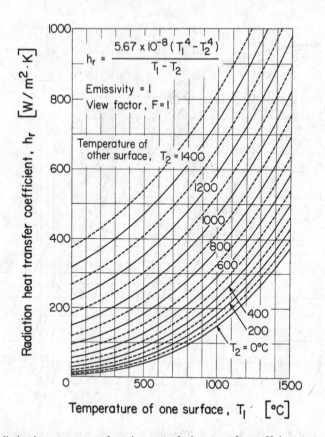

Fig. 9.6 Radiation between two surfaces in terms of a heat transfer coefficient

We do not take up these subjects here. The reader is referred to the references for further readings on these subjects.

9.3.11   Estimating the Magnitude of h_r

For design we want to know whether radiative transfer is appreciable compared to the other competing mechanisms of heat transfer and whether it need be considered at all in any analysis. Figure 9.6 is helpful for this purpose. It gives the radiation heat transfer coefficient between two closely facing $(F = 1)$ black surfaces $(\varepsilon = \alpha = 1)$. To find $h_{r,\text{actual}}$ for a particular situation, lower the h_r value given in this figure to account for $1 < 1$ and $\varepsilon < 1$. Thus,

$$h_{r,\text{actual}} = \varepsilon \,\mathscr{F}\, h_{r,\text{figure}}$$

If $h_r \ll h_{\text{convection}}$, ignore the radiation contribution to the overall heat transfer. A look at this figure shows that h_r becomes very large at high temperature.

Problems on Conduction, Convection, and Radiation

9.1. Which paint (black, white, or aluminum) can be roughly considered to be a blackbody, a greybody, or neither?

9.2. A pottery oven with 10-cm-thick walls has a number of 6 cm × 5 cm peep holes for monitoring the progress of the firing. How much heat will be lost from the 1,500 °C oven when a peep hole plug is removed?

9.3. A hot black painted pipe passes through a 20 °C room to heat it. At what pipe temperature is its convective heat loss and its radiative heat loss equal?

9.4. The Sun, a blackbody 1,392,000 km in diameter and 148×10^4 km away from the Earth, emits 6,150 K radiation. With the Sun directly overhead, what is the Sun's energy flux at Earth's surface (W/m_2)? Ignore absorption of radiation by clouds and atmosphere. Earth's diameter is 12,732 km.

References and Related Readings

Conduction

E.F. Adiutori, *The New Heat Transfer* (Venturo, Cincinnati, 1974)
J.R. Welty, *Engineering Heat Transfer*, 2nd edn. (Wiley, New York, 1978)
W.H. McAdams, Heat Transmission, 3rd edn. (McGraw-Hill, New York, 1954)

Convection

J.S.M. Botterill, Fluidized bed behavior, in *Fluidized Beds, Combustion and Applications*, ed. by J.R. Howard (Applied Science, New York, 1983)
V. Cavaseno (ed.), *Process Heat Exchange* (McGraw-Hill, New York, 1979), pp. 20, 101, 130, 140.
W.M. Kays, M.E. Crawford, *Convective Heat and Mass Transfer*, 2nd edn., Chapter 8 (McGraw-Hill, New York, 1980)
D. Kunii, O. Levenspiel, *Fluidization Engineering* (Krieger, Melbourne, 1979)
D. Kunii, O. Levenspiel, *Fluidization Engineering*, 2nd edn. (Butterworth, Boston, 1991)
R.H. Perry, C.H. Chilton, *Chemical Engineers' Handbook*, 5th edn., Sec. 10 (McGraw-Hill, New York, 1973); 6th edn., Sec. 10 (1984)
W.H. McAdams, Heat Transmission, 3rd edn. (McGraw-Hill, New York, 1954)
W.E. Ranz, W.R. Marshall Jr., Evaporation from drops. Chem. Eng. Prog. **48**, 141 (1952)

Radiation

H.C. Hottel, Radiant heat transmission. Mech. Eng. **52**, 699 (1930)

H.C. Hottel, A.F. Sarofim, *Radiative Transfer* (McGraw-Hill, New York, 1967)

M. Jakob, *Heat Transfer*, vol. 2 (Wiley, New York, 1957)

W.H. McAdams, *Heat Transmission*, 3rd edn., Chapter 4 (McGraw-Hill, New York, 1954)

M. Siegel, J.R. Howell, *Thermal Radiation Heat Transfer*, 2nd edn. (McGraw-Hill, New York, 1981)

Chapter 10
Combination of Heat Transfer Resistances

Heat loss from a warm room through a wall to the cold outside involves three heat transfer steps in a series: (i) convection at the inside surface of the wall, (ii) conduction through the wall, and (iii) convection on the outside of the wall. Next, consider a fireplace fire. Here heat reaches the room by radiation from the flames and also by convection of moving air. These processes occur in parallel. There are many processes like these which involve a number of heat transfer steps, sometimes in a series, sometimes in parallel, sometimes in a more involved way.

To find the overall effect of a number of heat transfer steps in a series and in parallel, we draw on the analogy to electrical theory. For *processes in series* the resistances are additive; thus,

$$R_{\text{overall}} = R_1 + R_2 + R_3 \tag{10.1a}$$

or, in terms of conductances,

$$\frac{1}{C_{\text{overall}}} = \frac{1}{C_1} + \frac{1}{C_2} + \frac{1}{C_3} \tag{10.1b}$$

For *processes in parallel*, it is the conductances which are additive. Thus,

$$C_{\text{overall}} = C_1 + C_2 + C_3 \tag{10.2a}$$

or, in terms of resistances,

$$\frac{1}{R_{\text{overall}}} = \frac{1}{R_1} + \frac{1}{R_2} + \frac{1}{R_3} \tag{10.2b}$$

© Springer Science+Business Media New York 2014
O. Levenspiel, *Engineering Flow and Heat Exchange*,
DOI 10.1007/978-1-4899-7454-9_10

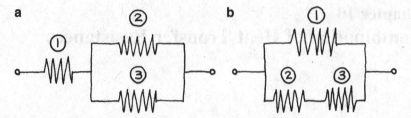

Fig. 10.1 Two series–parallel arrangements of resistances to heat transfer

In the *series–parallel situations* sketched in Fig. 10.1a, we have

$$\frac{1}{C_{\text{overall}}} = \frac{1}{C_1} + \frac{1}{C_2 + C_3} \tag{10.3}$$

and for the sketch of Fig. 10.1b,

$$C_{\text{overall}} = C_1 + \frac{1}{\frac{1}{C_2} + \frac{1}{C_3}} \tag{10.4}$$

For *processes in series* a glance at equation (10.1a and b) shows that the step with the largest resistance dominates and determines, in most part, the overall resistance. Resistances much smaller than this can be ignored.

For *processes in parallel* matters are quite different, for equation (10.2a and b) shows that the term with largest conductance (hence smallest resistance) dominates and for the most part determines the overall conductance of the process. Conductances much smaller than this can be ignored.

Whenever a particular step in the overall process dominates to the exclusion of all other steps, it is called the *rate controlling step* of the process.

In heat transfer, the individual conductances are represented by the convection coefficient h, the conductivity per unit length $k/\Delta x$, and the radiation coefficient h_r, while the overall conductance for the process is represented by the overall heat transfer coefficient, U. It is this overall transfer coefficient which is of prime importance in heat exchanger design. We now show the form of the overall transfer coefficient U for a few representative situations. The problems at the end of the chapter present other situations.

10.1   Fluid–Fluid Heat Transfer Through a Wall

As shown in Fig. 10.2 the wall material and both liquid films contribute resistance to heat transfer. Thus, we have

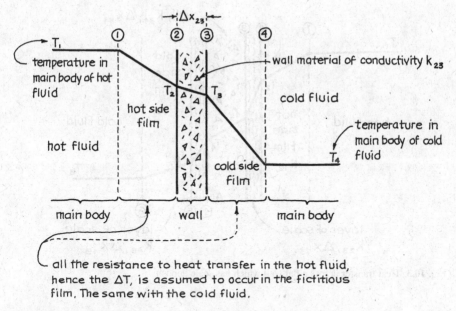

all the resistance to heat transfer in the hot fluid, hence the ΔT, is assumed to occur in the fictitious film. The same with the cold fluid.

Fig. 10.2 Heat transfer from fluid to fluid that are separated by a wall

$$\dot{q} = h_{12}A(T_1 - T_2)$$
$$\dot{q} = \frac{k_{23}A}{\Delta x_{23}}(T_2 - T_3)$$
$$\dot{q} = h_{34}A(T_3 - T_4)$$

Combining and eliminating the intermediate temperatures T_2 and T_3 gives

$$\dot{q} = -UA\Delta T \quad \text{where} \quad \frac{1}{U} = \frac{1}{h_{12}} + \frac{\Delta x_{23}}{k_{23}} + \frac{1}{h_{34}} \qquad (10.5)$$

More generally, if there are scale deposits on the surfaces of the separating wall, these deposits represent two more resistances in the series, as shown in Fig. 10.3. In this case we have

$$\frac{1}{U} = \frac{1}{h_2} + \frac{\Delta x_{23}}{k_{23}} + \frac{\Delta x_{34}}{k_{34}} + \frac{\Delta x_{45}}{k_{45}} + \frac{1}{h_{56}} \quad \left[\frac{m^2 \cdot K}{W}\right] \qquad (10.6)$$

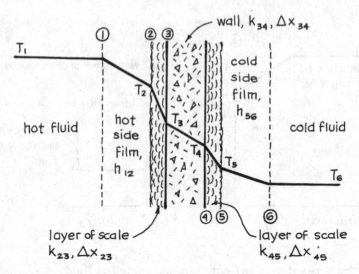

Fig. 10.3 Heat transfer across a flat wall which has scale deposits

Note that equations (10.5) and (10.6) are of the form of equation (10.1a and b) which represents resistances in a series. Thus, if any particular resistance step is very much larger than all the others (low h or low $k/\Delta x$), it will dominate and determine the overall resistance of the process.

10.2   Fluid–Fluid Transfer Through a Cylindrical Pipe Wall

Consider heat transfer from hot fluid at T_1 to cold fluid at T_6 across a pipe with thin scale coatings (thus, $A_2 \cong A_3$ and $A_4 \cong A_5$) as shown in Fig. 10.4.

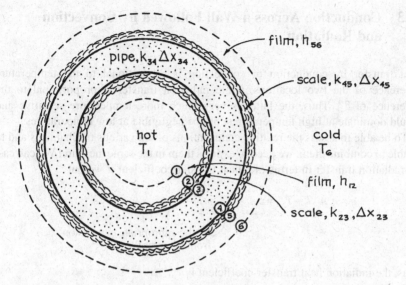

Fig. 10.4 Heat transfer across a scale-coated pipe wall

Noting that all the resistances occur in a series, we can show that

$$q = -UA\Delta T_{\text{overall}} = U_3A_3(T_1 - T_6) = U_4A_4(T_1 - T_6) \quad [\text{W}]$$

with

$$\frac{1}{U_3A_3} = \frac{1}{U_4A_4} = \frac{1}{h_{12}A_3} + \underbrace{\frac{\Delta x_{23}}{k_{23}A_3}}_{\text{Scale}} + \underbrace{\frac{\Delta x_{34}}{k_{34}A_{34,lm}}}_{\text{Wall}} + \underbrace{\frac{\Delta x_{45}}{k_{45}A_4}}_{\text{Scale}} + \frac{1}{h_{56}A_4} \quad \left[\frac{\text{K}}{\text{W}}\right] \tag{10.7}$$

where

$$A_{34,lm} = \frac{A_4 - A_3}{\ln\frac{A_4}{A_3}} \cong \frac{A_4 + A_3}{2} \quad \text{if} \frac{A_4}{A_3} < 2$$

Note that the value of U obtained depends on the area basis chosen, whether it is the inside area or the outside area of the pipe.

10.3   Conduction Across a Wall Followed by Convection and Radiation

Heat transfer by conduction or convection is proportional to the temperature difference of the two locations, while radiation transfer is proportional to the difference of T^4. Thus, the latter is very much more temperature sensitive and should dominate at high temperatures, but be negligible at low temperatures.

To be able to assess the relative contributions of the various mechanisms and to be able to combine them, we need to express them in the same measure. We can cast the radiation transfer in terms of a heat transfer coefficient if we put

$$q = h_r A_1 \; T_1 - T_2 \; = \sigma A_1 \mathscr{F}_{12} \; T_1^4 - T_2^4$$

Or the appropriate expression from Chapter 9

Thus, the radiation heat transfer coefficient is

$$h_r = \frac{\sigma \mathscr{F}_{12}\left(T_1^4 - T_2^4\right)}{T_1 - T_2} \tag{10.8}$$

Now consider the conduction of heat through a slab followed by convection to a fluid at T_3 and radiation to a facing surface at T_3, as sketched in Fig. 10.5. This is a series–parallel situation as shown in equation (10.3), or

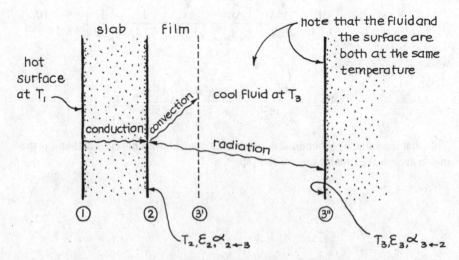

Fig. 10.5 Conduction through a wall followed by convection and radiation in parallel

$$\frac{1}{C_{total}} = \frac{1}{C_{12}} + \frac{1}{C_{23,conv} + C_{23,rad}}$$

In heat transfer language this combination of conductances gives

$$\dot{q}_{13} = -UA\Delta T = UA(T_1 - T_3) \quad [\text{W}]$$

where

$$\frac{1}{U} = \frac{\Delta x_{12}}{k_{12}} + \frac{1}{h_{23,conv} + h_{23,rad}} \quad \left[\frac{\text{m}^2 \cdot \text{K}}{\text{W}}\right]$$

and where, from equation (9.67) for parallel facing source and sink,

$$h_{23,rad} = \frac{\sigma[\alpha_{3\leftarrow2}\varepsilon_2 T_2^4 - \alpha_{2\leftarrow3}\varepsilon_3 T_3^4]}{(\alpha_{2\leftarrow3} + \alpha_{3\leftarrow2} - \alpha_{3\leftarrow2}\alpha_{2\leftarrow3})(T_2 - T_3)}$$

$$\left.\rule{0pt}{80pt}\right\}\quad(10.9)$$

10.4 Convection and Radiation to Two Different Temperature Sinks

Now consider a more complex case where heat is lost from surface 1 by convection to fluid at T_2, but also by radiation through the transparent fluid to a parallel surface at T_3, as shown in Fig. 10.6.

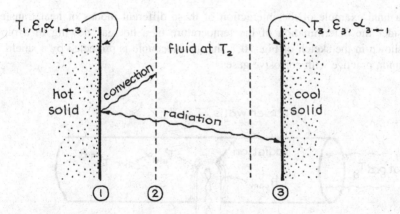

Fig. 10.6 Heat transfer to two different temperature sinks

The heat lost by surface 1 is given by

$$\dot{q}_1 = \dot{q}_{12,\text{conv}} + \dot{q}_{13,\text{rad}}$$
$$= h_{12,\text{conv}}A(T_1 - T_2) + h_{13,\text{rad}}A(T_1 - T_3)$$

where, from equation (9.67),

$$h_{13,\text{rad}} = \frac{\sigma A\left[\alpha_{3\leftarrow1}\varepsilon_1 T_1^4 - \alpha_{1\leftarrow3}\varepsilon_3 T_3^4\right]}{(\alpha_{1\leftarrow3} + \alpha_{3\leftarrow1} - \alpha_{1\leftarrow3}\alpha_{3\leftarrow1})(T_1 - T_3)}$$

In terms of $T_1 - T_2$, we get

$$\dot{q}_{1\rightarrow} = U_{12}A(T_1 - T_2) \quad \text{where } U_{12} = h_c + h_r\left(\frac{T_1 - T_3}{T_1 - T_2}\right) \qquad (10.10)$$

or, in terms of $T_1 - T_3$,

$$\dot{q}_{1\leftarrow} = U_{13}A(T_1 - T_3) \quad \text{where } U_{13} = h_c\left(\frac{T_1 - T_2}{T_1 - T_3}\right) + h_r \qquad (10.11)$$

Note that the temperature ratio appears in an h term whenever that h term is based on one ΔT, while the U and other h term are based on another ΔT.

10.5   Determination of Gas Temperature

As a final example of the interaction of these different modes of heat transfer, consider the determination of the temperature of a hot gas flowing in a pipe. As shown in the sketch of Fig. 10.7, the thermocouple is protected by a shield, a common practice with corrosive gases.

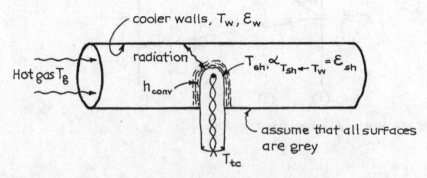

Fig. 10.7 Finding the temperature of a gas with a shielded thermocouple

First of all, no heat is lost by the thermocouple or the gas within the shield.

$$T_{tc} = T_{sh}$$

Now make a heat balance for the shield. Heat enters by convection from the hot gas; however, heat leaves by net radiation to the cooler walls. Thus, at steady state,

$$\dot{q} \text{ convection} + \dot{q} \text{ radiation interchange} = 0$$
$$\text{to shield} \quad \text{between shield and walls}$$

or

$$h_{conv} (T_g - T_{sh}) = \sigma \varepsilon_{sh} (T_{sh}^4 - T_w^4)$$

For gas flow
outside and normal
to a cylinder

For a completely enclosed greybody;
use equation (9.65)

and noting that $T_{sh} = T_{tc}$

$$T_g = T_{tc} + \frac{\sigma \varepsilon_{sh}}{h_{conv}} (T_{tc}^4 - T_w^4) \qquad (10.12)$$

$$\frac{W}{m^2 K^4}$$

$$\frac{W}{m^2 K}$$

The temperature of the shield (hence of thermocouple) will be somewhere between the gas and wall temperature. High h_{conv} and low ε_{sh} will result in the thermocouple measuring the gas temperature, while low h_{conv} and high ε_{sh} will result in the thermocouple measuring the wall temperature.

This type of radiation correction is very important at high temperatures.

10.6 Extensions

For other more complicated situations, either:

- Write out all the individual heat interchange terms and eliminate all intermediate temperatures. This procedure is illustrated in section A, above.
- Develop the electrical analog, find the overall conductance or overall resistance, and then replace with heat flow terms. This procedure is illustrated in section C, above.

Usually the latter procedure is simpler, but with more than one source and one sink, one must be careful in using this approach, as shown in section D, above.

Problems on Combining Resistances

10.1 *From battleships to skating rinks.* My neighbor can't resist auctions, and just last week he bought the World War II battleship *USS Iowa* for $277.00. He has great plans for this war relic and wants to implement them as soon as possible. One of his schemes is to build a neighborhood skating rink using ½-inch-thick steel plates instead of pipes to carry the refrigerant. Thus, the floor of the rink would consist of a double layer of steel plates a short distance apart with refrigerant flowing in between and ice above. However, it is important that no water form on the surface of the ice. With this restriction in mind, how thick should the ice layer be? The refrigerant is at − 18 °C. The air in the rink is at 25 °C.

   (a) Do your first calculation ignoring heat transfer by radiation and the resistance of the steel pipe.
   (b) Then include the resistance of the steel plate.

   Do you feel that the resistance of the steel plate can be reasonably ignored?

10.2 *Pottery kilns.* Business is so good at *Pottery West* that the master potter plans to construct a new and larger kiln about 2 m high in which to bake his artistic creations. Estimate:

   (a) The outside temperature of a vertical wall of this kiln
   (b) The heat loss through this wall

   Data: The inside temperature of the kiln wall will be 1,150 °C; room temperature is 20 °C. The kiln wall will be 20 cm thick, made of high-temperature firebrick ($k = 0.1$ W/m · K, $\varepsilon = 0.8$).
   Note: The findings of the previous problem suggest that you should not ignore the radiation from wall to room.

10.3 *Temperature of a space voyager.* Estimate the temperature of a spherical space probe as it passes Mars on its way to the outer planets.
   Data: Effective color temperature of the Sun = 6,150 K

   Radius of the Sun = 695,000 km
   Distance from the Sun to Earth = 148,000,000 km
   Distance from the Sun to Mars = 228,000,000 km
   The skin of the voyager is # 301 stainless steel

   For related information, see *Science* **127** 811 (1958) and **128** 208 (1958).

10.4 *Earth's temperature.* What would be the mean temperature of Earth if it were a gray body? See previous problem for data.

10.5 *The insulation of hot-air ducts.* Energy Savers, Inc., was upset to discover that the hot-air ducts under our building are completely uninsulated—just naked shiny 300-mm tin-plated pipes ($\varepsilon = 0.05$). What a waste of energy.

They urge that we immediately insulate because each minute of delay costs us money. We could opt for their preformed pop-on foam insulation. However, for a real first-class job, they strongly recommend their patented double protection formula—a 1.6-mm layer of especially thick insulating cardboard ($k = 0.15$ W/m K), glued snugly to the pipe, and then a coating of long-lasting, nonbiodegradable, insect-repellent, low-emissivity aluminum paint ($\varepsilon = 0.55$). Though more labor intensive and more costly, we are assured that this is the best that modern technology can offer.

I suppose they are right. However, before I sign a contract, I'd like to know whether the energy saving would really be substantial. So would you please determine what fraction of the original energy loss is saved with this insulation. For these calculations, take the temperature of the pipe walls to be 75 °C and the surrounding crawl space to be at 5 °C.

A thermocouple, protected by a stainless steel shield, is inserted in an air preheater duct. At the air velocity flowing in the duct, it is estimated that $h_{conv} = 100$ W/m^2 K. Find the temperature of the hot air:

10.6 If the thermocouple reads $T_{tc} = 400$ K and if the temperature of the steel walls is $T_w = 300$ K.

10.7 If $T_{tc} = 1{,}000$ K and if $T_w = 900$ K. Note how sharply the error in T_{tc} reading increases (because of the intrusion of radiation) as the temperature level rises.

10.8 *Heat transfer to the walls of gas/solid fluidized beds.* Observation shows that part of the time (fraction δ) the hot wall of a bubbling fluidized bed is bathed by the rising gas bubbles, the rest of the time by the gas/solid emulsion. In addition, the layer of emulsion right at the wall surface has somewhat different properties (larger voidage) than the rest of the emulsion.

Let us develop a bed/wall heat transfer model based on these observations, as follows. When bubbles bathe the surface, heat flows by convection h_1 and by radiation h_2 from the hot wall directly into the cold bed. However, when the emulsion bathes the surface, things are a bit more complicated because then heat flows by convection h_3 and radiation h_4 to the first layer of the emulsion. This first layer then passes the heat into the main body of the emulsion by unsteady-state heating with mean heat transfer coefficient h_5.

With this model, develop an expression to represent the overall heat transfer coefficient h in terms of δ and the five individual coefficients.

Chapter 11
Unsteady-State Heating and Cooling of Solid Objects

If a hot object is plunged into cold water, it cools, but not instantaneously. Two factors govern the cooling rate of the object:

- The film resistance at the surface of the object, characterized by the h value for that situation.
- The rate of heat flow out of the interior of the object. The governing differential equation for this conduction process is

$$\frac{\partial T_s}{\partial t} = \alpha \left(\frac{\partial^2 T_s}{\partial x^2} + \frac{\partial^2 T_s}{\partial y^2} + \frac{\partial^2 T_s}{\partial z^2} \right) \tag{11.1}$$

where

$$\alpha = \frac{k_s}{\rho_s C_s}, \text{thermal diffusivity} \left[m^2 / s \right]$$

$$\swarrow \frac{W}{mK}, \text{thermal conductivity}$$

and

$$T_s = \text{temperature at any point in the object } [K]$$

A dimensionless measure for conduction, which accounts for both the cooling time and the size of object, is given by the Fourier number

$$Fo = \frac{\alpha t}{L^2} = \frac{\alpha t}{(V/A)^2} = \frac{k_s}{\rho_s C_s} \cdot \frac{t}{(V/A)^2} \quad [\text{-}] \tag{11.2}$$

© Springer Science+Business Media New York 2014
O. Levenspiel, *Engineering Flow and Heat Exchange*,
DOI 10.1007/978-1-4899-7454-9_11

where the characteristic length of the object

$$L\begin{cases} = \dfrac{\text{volume}}{\text{surface}} = \dfrac{V}{A} & \text{in general} \\[2mm] = \dfrac{\text{thickness}}{2} & \text{for a slab} \\[2mm] = \dfrac{R}{2} & \text{for a cylinder} \\[2mm] = \dfrac{R}{3} & \text{for a sphere} \end{cases} \qquad (11.3)$$

The relative importance of the surface and the interior resistance terms is measured by the Biot number, a dimensionless group defined as

$$\text{Bi} = \left(\frac{\begin{array}{c}\text{interior}\\ \text{resistance}\end{array}}{\begin{array}{c}\text{surface}\\ \text{resistance}\end{array}} \right) = \frac{h(V/A)}{k_s} = \frac{hL}{k_s} \quad [\text{-}] \qquad (11.4)$$

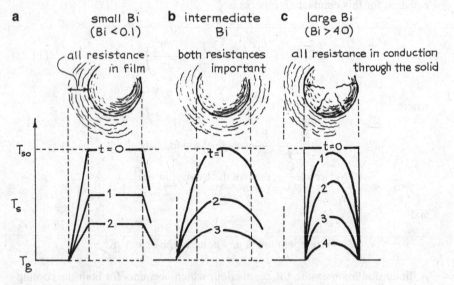

Fig. 11.1 Temperature–time history of a cooling particle for different ranges of Biot numbers

For a small Biot number, the main resistance is in the film; for a large Biot number, the main resistance is conduction of heat out of the body. Figure 11.1 shows the temperature–time history in various regimes for a spherical particle.

We first consider the two extreme cases and then the general case where both resistances are important. Design charts present compactly the whole range of situations for various shapes of particles.

11.1   The Cooling of an Object When All the Resistance Is at Its Surface ($\mathbf{Bi} = hL/k_s \to 0$)

This extreme views the object to be isothermal at any time, the whole object cooling (or heating) with time, as illustrated in Figs. 11.1a and 11.2. This type of analysis where the system in question has uniform properties throughout is called a *lumped parameter analysis*. A heat balance about the hot object being cooled then gives

$$-\dot{q} = \begin{pmatrix} \text{heat transfer rate} \\ \text{through the film} \end{pmatrix} = \begin{pmatrix} \text{rate of heat loss} \\ \text{from the object} \end{pmatrix} \quad [\text{W}]$$

$$= hA(T_s - T_g) = -\underset{\underset{V\rho_s}{\uparrow}}{W}C_s \frac{dT_s}{dT} \tag{11.5}$$

Separating and integrating at constant $hA/V\rho_s C_s$ gives

$$\frac{\Delta T}{\Delta T_{\max}} = \frac{T_s - T_g}{T_{s0} - T_g} = e^{-\text{Fo·Bi}} = e^{-(ht/L\rho_s C_s)} \tag{11.6}$$
$$\underset{L = V / A, \text{ characteristic length}}{} \qquad [\text{-}]$$

The instantaneous rate of heat loss from the object is found by combining equations (11.5) and (11.6), or

$$-\dot{q} = -\rho_s C_s V \frac{dT_s}{dt} = hA(T_{s0} - T_g)e^{-\text{Fo·Bi}} \quad [\text{W}] \tag{11.7}$$

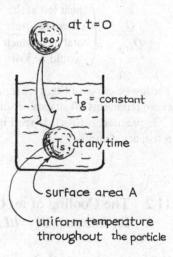

Fig. 11.2 Cooling of particle with all resistance in film

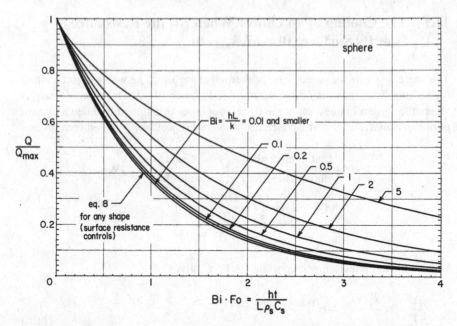

Fig. 11.3 The cooling of any shape of object when surface resistance controls (lowest curve) and the cooling of spheres in general (all the curves). The y-axis represents the fraction of heat remaining in the object

Also, the fractional cooling of the object is found either by integrating equation (11.7) or, more simply, by inspection of equation (11.6). Thus,

$$\frac{Q}{Q_{max}} = \begin{pmatrix} \text{heat left at the} \\ \text{object at time } t \\ \hline \text{total heat which} \\ \text{could be lost} \end{pmatrix} = \frac{T_s - T_g}{T_{s0} - T_g} = e^{-\text{Fo}\cdot\text{Bi}} = e^{-(ht/L\rho_sC_s)} \qquad (11.8)$$

The lowest curve of Fig. 11.3 represents the equation for this extreme. Note that one curve represents all shapes of solids.

In practice, if Bi <0.1 (see Fig. 11.3), then one can reasonably assume that film resistance controls.

11.2   The Cooling of an Object Having Negligible Surface Resistance (Bi = $hL/k_s \rightarrow \infty$)

In this extreme, when the hot object is plunged into cold fluid, its surface immediately drops to the temperature of the fluid, and conduction within the object is all important. This is illustrated in Fig. 11.1c and in Fig. 11.4. Solving equation (11.1)

Fig. 11.4 The cooling of an object of negligible surface resistance

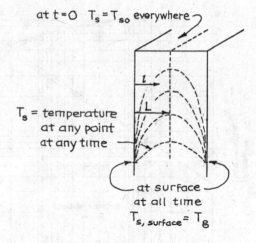

at t =0 T_s = T_{so} everywhere

T_s = temperature at any point at any time

at surface at all time
$T_{s, surface}$ = T_g

for a slab gives a rapidly converging infinite series for the temperature T_s at any location and any time:

$$\frac{\Delta T}{\Delta T_{max}} = \frac{T_s - T_g}{T_{s0} - T_g} = \frac{4}{\pi}\left(e^{-a}\sin b + \frac{1}{3}e^{-9a}\sin 3b + \frac{1}{5}e^{-25a}\sin 5b + \cdots\right) \quad (11.9)$$

where

$$a = \frac{\pi \alpha t}{4L^2}$$

and

$$b = \frac{\pi l}{2L}$$

The fraction of total heat remaining in the slab is then given by

$$\frac{Q}{Q_{max}} = \left(\frac{\begin{array}{c}\text{heat remaining}\\ \text{in slab}\end{array}}{\begin{array}{c}\text{total heat initially}\\ \text{in slab}\end{array}}\right) = \frac{8}{\pi^2}\left(e^{-a} + \frac{1}{9}e^{-9a} + \frac{1}{25}e^{-25a} + \cdots\right) \quad (11.10)$$

Similar equations have been derived for infinite cylinders and for spheres. Figure 11.5 shows both the fraction of heat remaining and the fraction of heat lost from these regular solids. By interpolation between these curves, one can estimate the extent of heating or cooling of any irregular solid.

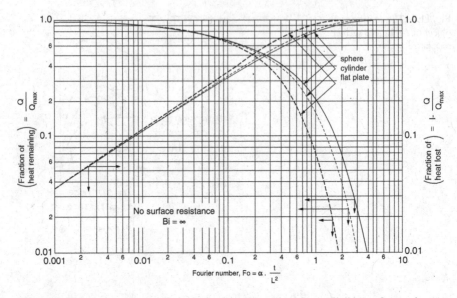

Fig. 11.5 Heat lost and heat remaining in a cooling object for negligible surface resistance [Prepared by Mator (1982)]

11.3   The Cooling of an Object Where Both Surface and Internal Resistances to Heat Flow Are Important (0.1 < Bi < 40)

Here (see Fig. 11.1b) we have the conduction equations of the just treated case B with the boundary condition at any time:

$$\begin{pmatrix} \text{Rate of heat flow out} \\ \text{from surface} \end{pmatrix} = -\dot{q} = hA\left(T_{s,\,l=L} - T_g\right) = k_s A\left(\frac{\partial T_s}{\partial l}\right)_{l=L} \quad (11.11)$$

The solutions to these equations have been derived for a number of shapes and are available in many heat transfer texts [e.g., see Gröber et al. (1961) and Boelter et al. (1956)]. In all cases these solutions involve slowly converging infinite series, which are tedious to evaluate. However, convenient graphical representations of these solutions have been prepared and are reproduced in Figs. 11.6, 11.7, 11.8, 11.9, 11.10, 11.11, 11.12, 11.13, and 11.14 in terms of the following dimensionless groups:

- An unaccomplished temperature change: $\frac{\Delta T}{\Delta T_{max}} = \frac{T_s - T_g}{T_{s0} - T_g}$
- The fraction of heat remaining in the solids: $\frac{Q}{Q_{max}}$
- A relative time: $\text{Fo} = \frac{\alpha t}{L^2}$
- A resistance ratio: $\text{Bi} = \frac{hL}{k_s}$
- A radius or distance ratio: $\frac{r}{R}$ or $\frac{l}{L}$

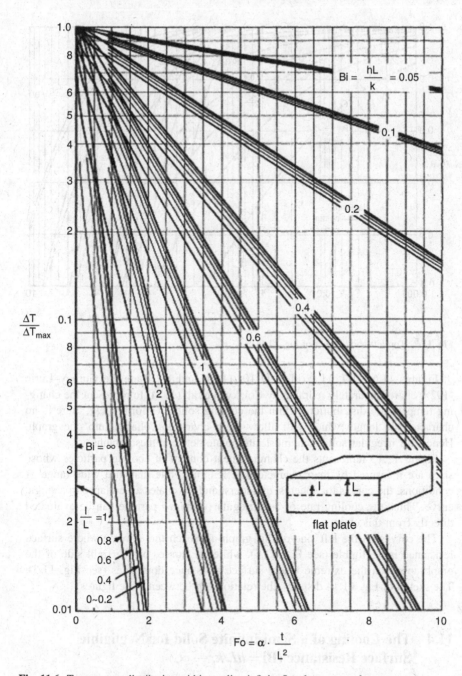

Fig. 11.6 Temperature distribution within cooling infinite flat plates, general case

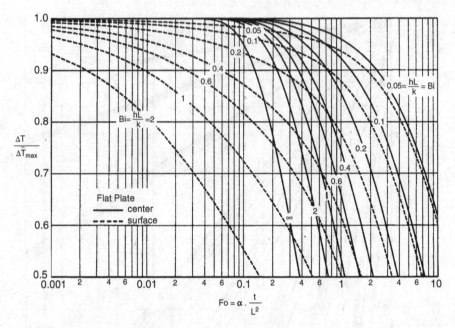

Fig. 11.7 Top left-hand corner of Fig. 11.6

Figures 11.6, 11.7, 11.8, 11.9, 11.10, 11.11, and 11.12 are the Gurney–Lurie (1923) charts accurately redrawn by Colakyan et al. (1984) to represent the changing temperature distribution within the cooling solids. Unfortunately, there is no characteristic length which will allow these charts to collapse into one graph. Hence, for irregular solids one must interpolate between these graphs.

Figure 11.13 represents the changing heat content of cooling particles whose sizes are measured by their characteristic lengths. Note that in a wide range of conditions, the curves for spheres, cylinders, and flat plates all collapse to a single curve. Thus, the cooling rate of any irregularly shaped particle can be evaluated directly from this chart.

The curves on the left side of this graph approach the extreme where surface resistance is negligible (see Fig. 11.5), while the curves on the right side of the graph approach the extreme where surface resistance dominates (see Fig. 11.3). The sketch of Fig. 11.14 displays the relationship between these figures.

11.4   The Cooling of a Semi-infinite Solid for Negligible Surface Resistance ($Bi = hL/k_s \rightarrow \infty$)

When a hot body at temperature T_{s0} is placed in contact with cold fluid at temperature T_g, the surface immediately drops to T_g, heat flows from the body, and it progressively cools as shown in Fig. 11.15. The governing differential

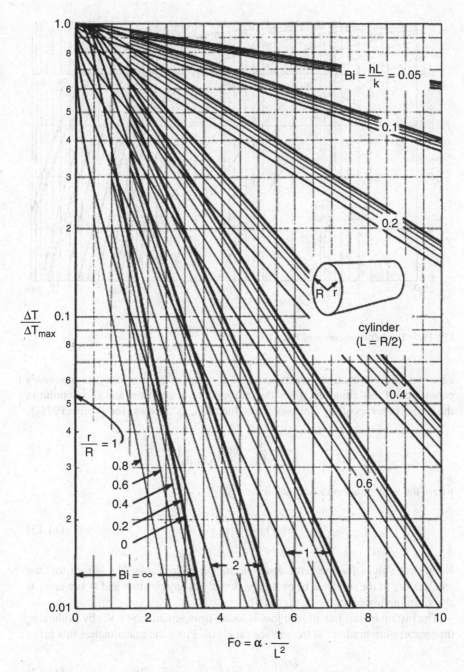

Fig. 11.8 Temperature distribution within cooling infinite cylinders, general case

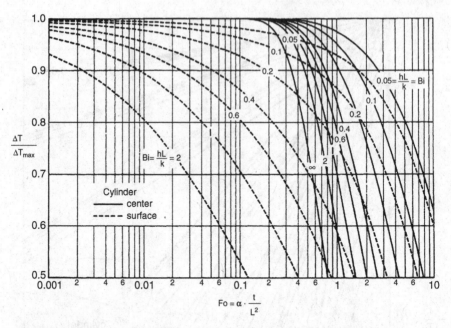

Fig. 11.9 Top left-hand corner of Fig. 11.8

equation for conduction, equation (11.1), when integrated for the boundary conditions of this situation, gives the temperature at any time and at any point in the object in terms of the Gaussian error function, as follows [see Welty (1974)]:

$$\frac{\Delta T}{\Delta T_{\max}} = \frac{T_s - T_g}{T_{s0} - T_g} = \mathrm{erf}\left(\frac{1}{\sqrt{4\alpha t}}\right) = \mathrm{erf}(y) \qquad (11.12)$$

where the error function is defined as

$$\mathrm{erf}(y) = \frac{2}{\sqrt{\pi}}\int_0^y e^{-x^2}dx \qquad (11.13)$$

Numerical values for the error function are given in Table 11.1. From this the temperature of the solid can be evaluated directly at any position and at any time, as shown in the lowest curve of Fig. 11.16.

The instantaneous rate of heat loss is found from equation (11.12) by evaluating the temperature gradient at the surface (at $l = 0$). From the mathematics this gives

$$-\dot{q} = k_s A \frac{\partial T_s}{\partial l}\bigg|_{l=0} = k_s A\left(\frac{T_{s0} - T_g}{\sqrt{\pi \alpha t}}\right) \quad [\mathrm{W}] \qquad (11.14)$$

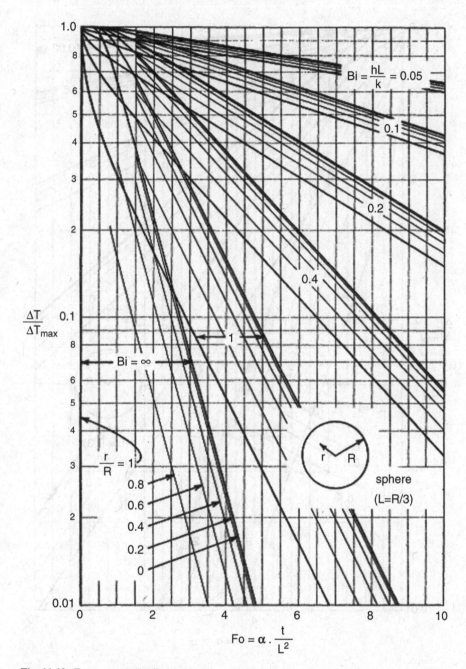

Fig. 11.10 Temperature distribution within cooling spheres, general case (also see Fig. 11.11)

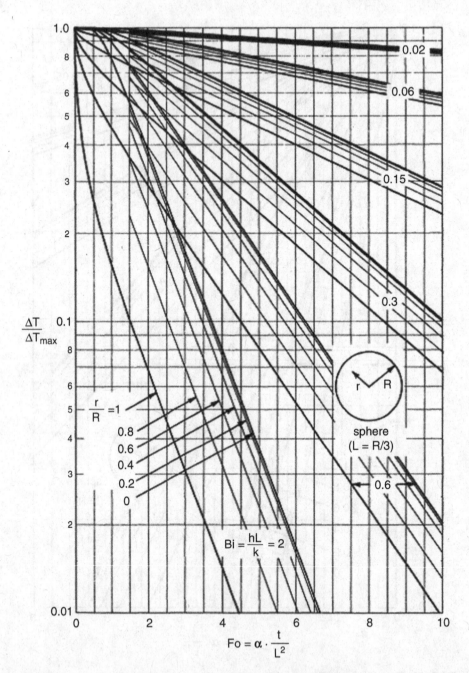

Fig. 11.11 Temperature distribution within cooling spheres, general case (also see Fig. 11.10)

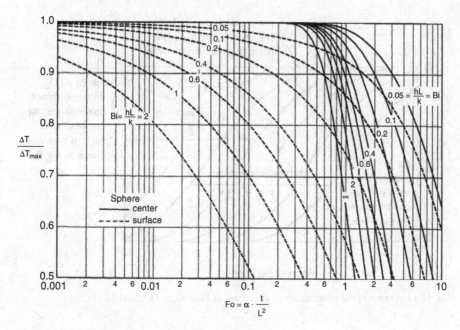

Fig. 11.12 Top left-hand corner of Figs. 11.10 and 11.11

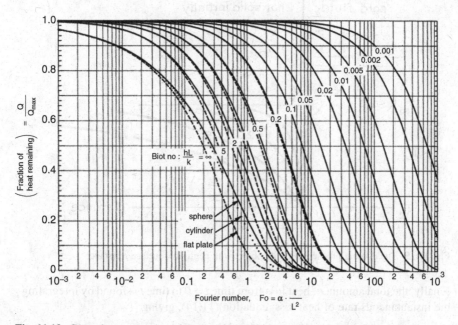

Fig. 11.13 General representation of the heat lost within a cooling sphere, infinite cylinder, and infinite flat plate [Prepared by Colakyan and Turton (1983)]

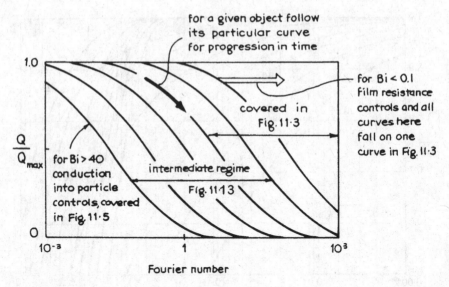

Fig. 11.14 Sketch of the relationship of the curves of Figs. 11.3, 11.5, and 11.13

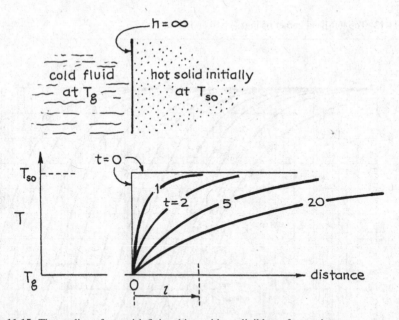

Fig. 11.15 The cooling of a semi-infinite object with negligible surface resistance

Finally, the total amount of heat loss from time $t = 0$ to time t is found by integrating the instantaneous rate of heat loss, equation (11.14), giving

$$Q_{\text{lost}} = \int_0^t (-\dot{q})dt = 2k_s A(T_{s0} - T_g)\sqrt{\frac{t}{\pi\alpha}} \quad [\text{J}] \qquad (11.15)$$

Table 11.1 Values of the error function[a] [This gives the solution to equation (11.12)]

$y = \frac{1}{\sqrt{4\alpha t}}$	$\operatorname{erf}(y) = \frac{\Delta T}{\Delta T_{max}}$	$y = \frac{1}{\sqrt{4\alpha t}}$	$\operatorname{erf}(y) = \frac{\Delta T}{\Delta T_{max}}$
0.0	0.0	0.70	0.678
0.05	0.056	0.75	0.711
0.10	0.112	0.80	0.742
0.15	0.168	0.85	0.771
0.20	0.223	0.90	0.797
0.25	0.276	0.95	0.821
0.30	0.329	1.0	0.843
0.35	0.379	1.2	0.910
0.40	0.428	1.4	0.952
0.45	0.475	1.6	0.976
0.50	0.520	1.8	0.989
0.55	0.563	2.0	0.995
0.60	0.604	2.5	0.9996
0.65	0.642	∞	1

[a]From *Tables of the Error Function and Its Derivative*. National Bureau of Standards. Applied Mathematics Series 41. Washington. D.C. (1954)

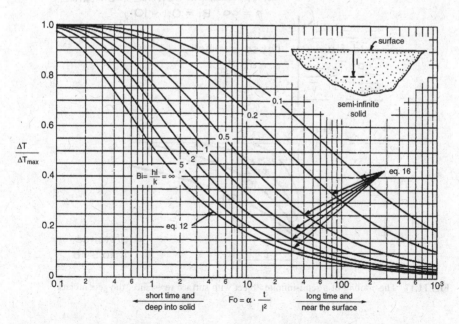

Fig. 11.16 Temperature distribution at any time and any position in a cooling semi-infinite object, general case [Equation (11.16)], and for no surface resistance [Equation (11.12)]

11.5   The Cooling of a Semi-infinite Body Including a Surface Resistance

This situation is similar to the previous case treated, but with an added surface resistance. Thus, we have a cooling behavior somewhat as sketched in Fig. 11.17. Integration of the conduction equation for this situation [see Sucec (1975)] gives the temperature of the solid at any position l from the surface at any time t as

$$\frac{\Delta T}{\Delta T_{max}} = \text{erf}\left(\frac{1}{2\sqrt{\text{Fo}}}\right) + \left[1 - \text{erf}\left(\frac{1}{2\sqrt{\text{Fo}}} + \text{Bi}\sqrt{\text{Fo}}\right)\right] \exp(\text{Bi} + \text{Bi}^2 \cdot \text{Fo})$$

(11.16)

By putting $\text{Bi} = \infty$, we see that equation (11.16) reduces directly to the special case expression of equation (11.12).

The cooling curves for both finite and infinite values of the Biot number are shown in Fig. 11.16. Note that at any point in the object, the approach to equilibrium (or the final temperature) takes a longer time for larger surface resistance (smaller values of the Biot number).

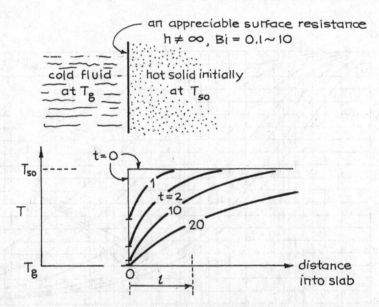

Fig. 11.17 The cooling of a semi-infinite object with surface resistance, the general case

11.6   Heat Loss in Objects of Size L for Short Cooling Times

Short cooling times mean that heat is only lost in the outer layer of the solid and that cooling has not yet penetrated deep into the object. In these situations the object can be treated as a semi-infinite solid, and its cooling behavior, whether it be slab, sphere, or what, only depends on the amount of surface in contact with the fluid, not its shape.

Thus, equations (11.12) and (11.16) and Fig. 11.16 represent the cooling in this time period. Boelter et al. (1956) and Schack (1965) both find that these simple "short time" solutions apply in the time period

$$\text{Fo} = \frac{\alpha t}{L^2} < 0.077$$

where L is the characteristic length of the particle.

11.7   The Cooling of Finite Objects Such as Cubes, Short Cylinders, Rectangular Parallelepipeds, and So On

The temperature at any point in these finite objects is related to the corresponding temperature in the three mutually perpendicular infinite bodies whose intersections produce the object in question. This relationship is a simple one and is given by Sucec (1975) as

$$\left(\frac{\Delta T}{\Delta T_{max}}\right)_{object} = \left(\frac{\Delta T}{\Delta T_{max}}\right)_x \left(\frac{\Delta T}{\Delta T_{max}}\right)_y \left(\frac{\Delta T}{\Delta T_{max}}\right)_z \tag{11.17}$$

where the terms on the right are evaluated from Figs. 11.6, 11.7, and/or 11.8, 11.9 for the point in question.

Similarly, the total heat loss from a finite object is related to the heat loss from the bounding infinite slabs and cylinders by

$$\left(\frac{Q}{Q_{max}}\right)_{object} = \left(\frac{Q}{Q_{max}}\right)_x \left(\frac{Q}{Q_{max}}\right)_y \left(\frac{Q}{Q_{max}}\right)_z \tag{11.18}$$

where the terms on the right are evaluated from Fig. 11.13.

Example 11.2 shows how to use these equations.

11.8   Intrusion of Radiation Effects

When heat enters or leaves a body by both convection and radiation, the coefficient h used in the Biot numbers throughout this chapter should be the overall coefficient accounting for both these mechanisms of heat transfer, or

$$h_{overall} = h_{convection} + h_{radiation}$$

This radiation coefficient may change considerably as the surface temperature of the particle changes. To determine whether the radiation contribution is appreciable compared to convection and thus needs to be considered, use Fig. 9.6 with a correction for view factor and emissivity.

11.9   Note on the Use of the Biot and Fourier Numbers

In most texts the Fourier and the Biot numbers for spheres and cylinders are defined in terms of the radius of the object, rather than the characteristic size of the object, V/A. Be careful not to confuse these measures in the charts from other books:

$$\text{for flat plates}: Fo_{here} = Fo_{other}; \quad Bi_{here} = Bi_{other}$$

$$\text{for cylinders}: Fo_{here} = 4Fo_{other}; \quad Bi_{here} = \frac{1}{2} Bi_{other}$$

$$\text{for spheres}: \quad Fo_{here} = 9Fo_{other}; \quad Bi_{here} = \frac{1}{3} Bi_{other}$$

The advantage of using V/A over R is that the curves for various shapes of objects often lie close to each other or collapse to a single curve. In addition, with the definition used here, the Biot and Fourier numbers are the actual ratios of resistances as defined in equations (11.3) and (11.4).

Example 11.1 Verifying a Key Assumption in the Analysis of Fluidized Bed Heat Exchangers

When a stream of hot solids is contacted by cold gas in a fluidized bed heat exchanger, the simple treatment of Chap. 14 assumes that:

(a)   Cold entering gas heats up instantaneously to the bed temperature.
(b)   Each particle of entering solid cools down instantaneously to the bed temperature.
(c)   Both gas and solid leave the bed at the same temperature.

(continued)

(continued)

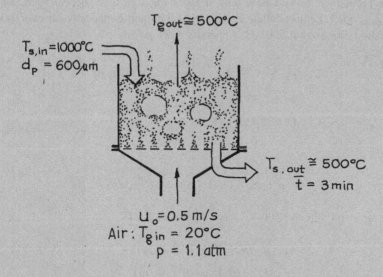

The term "instantaneous" as used above is reasonably approximated when the time needed for the two feed streams to get close to the bed temperature is much shorter than the time of stay of those streams in the exchanger. Preliminary considerations in Chap. 14 showed that assumption (a) is reasonably well met in practice. Let us look at assumption (b) here.

Suppose that hot sand ($d_p = 600\,\mu m$) at 1,000 °C flows continuously into a fluidized bed exchanger where it is cooled by air at room temperature and that gas and solid leave at about 500 °C. For a fluidizing velocity of $u_0 = 0.5$ m/s at 1.1 atm, find how long it takes for an incoming particle to cool to 550 °C (90 % approach).

(a) Assume that conduction within the particles controls.
(b) Assume that film resistance at the surface of the particles controls.
(c) Account for both resistances.

Compare these times to the mean residence time of solids in the exchanger, about 3 min.

Solution

The problem is to find how long it takes for a particle to cool such that $Q/Q_{max} = 0.1$. Let us tabulate all the physical properties needed to answer this question. For sand at 500 ° C, Appendix A.21 gives

$$\rho_s = \frac{\rho_{s,\text{bulk}}}{1 - \varepsilon} = \frac{1,500}{1 - 0.42} = 2,600 \, \text{kg/m}^3$$
$$k_s = 0.33 \, \text{W/m K}$$
$$C_s = 800 \, \text{J/kg K}$$

Thus,

$$\alpha = k_s/\rho_s C_s = 1.59 \times 10^{-7} \text{m}^2/\text{s}$$

For air at 1.1 atm and 500 °C, from Appendix A.21

$$\mu_g = 36.19 \times 10^{-6} \, \text{kg/m s}$$
$$k_g = 57.45 \times 10^{-3} \, \text{W/m K}$$
$$C_g = 1,093 \, \text{J/kg k}$$

and from Appendix A.12

$$\rho_g = \frac{p_A(mv)}{RT} = \frac{(101,325)(1.1)(0.0289)}{(8.314)(773)} = 0.50 \, \text{kg/m}^3$$

The heat transfer coefficient between a fluidized particle and its surroundings is given by equation (9.38). Replacing values gives

$$h = \frac{k_g}{d_p}\left[2 + 0.6\left(\frac{d_p u_0 \rho_g}{\mu_g}\right)^{1/2}\left(\frac{C_g \mu_g}{k_g}\right)^{1/3}\right]$$

$$= \frac{57.45 \times 10^{-3}}{6 \times 10^{-4}}\left\{2 + 0.6\left[\frac{(6 \times 10^{-4})(0.5)(0.5)}{36.19 \times 10^{-6}}\right]^{1/2}\left[\frac{(1,093)(36.19 \times 10^{-6})}{57.45 \times 10^{-3}}\right]^{1/3}\right\}$$

$$= 295 \, \text{W/m}^2\text{K}$$

We are now ready to solve for the cooling time based on the three different assumptions.

11.9.1   Assumption A. Particle Conduction Controls: $Bi \to \infty$

For spherical particles ($L = d_p/6 = 10^{-4}$ m) and $Q/Q_{max} = 0.1$, Fig. 11.5 gives

$$Fo = \alpha \frac{t}{L^2} = 1.7$$

from which

$$t = 1.7 \frac{L^2}{\alpha}$$

Replacing values gives

$$t = \frac{(1.7)(10^{-4})^2}{1.59 \times 10^{-7}} = 0.11\,s$$

11.9.2   Assumption B. Film Resistance Controls: $Bi \to 0$

Method A. Use equation (11.8). From equation (11.8) rearranged, we have

$$
\begin{aligned}
t &= \frac{L\rho_s C_s}{h} \ln\frac{Q_{max}}{Q} \\
&= \frac{(10^{-4})(2{,}600)(800)}{295} \ln 10 = 1.62\ s
\end{aligned}
$$

Method B. Use Fig. 11.3. From the lowest curve of this figure, we have

$$Bi \cdot Fo = \frac{ht}{L\rho_s C_s} = 2.32$$

Thus,

$$
\begin{aligned}
t &= 2.32 \left(\frac{L\rho_s C_s}{h}\right) \\
&= 2.32 \left[\frac{(10^{-4})(2{,}600)(800)}{295}\right] = 1.64\,s
\end{aligned}
$$

11.9.3   Assumption C. Accounting for Both Resistances

For this we must first evaluate the Biot number for the cooling particles. Thus,

$$Bi = \frac{hL}{k_s} = \frac{(295)(10^{-4})}{0.33} = 0.0893$$

Then, Fig. 11.13 shows that

$$\text{Fo} = \alpha \frac{t}{L^2} = 28$$

from which

$$t = 28\frac{L^2}{\alpha} = \frac{28(10^{-4})^2}{1.59 \times 10^{-7}} = 1.76\,\text{s}$$

Note that Fig. 11.3 can be used for this solution in place of Fig. 11.13. However, if you try using the design charts of Figs. 11.10, 11.11, and 11.12, you will have problems.

Comments
As expected, the correct solution which accounts for both resistances gives a longer time, 1.76 s, than either of the solutions which only considers one or the other of the two resistances.

Comparing the two resistances, we see that the film dominates (1.64 s vs. 0.11 s) and just about controls (1.64 s vs. 1.76 s). Now both Figs. 11.1 and 11.14 point out that when $\text{Bi} < 0.1$, then one can assume that film resistance controls. In this problem, $\text{Bi} = 0.089$, which just meets this condition, and so the results are as expected.

The cooling time found here (1.76 s) is very much shorter than the mean residence time in the exchanger (3 min); hence, the assumption that the hot incoming particles cool instantaneously to the bed temperature is a reasonable idealization of the state of affairs.

Table 14.1 gives the relaxation times for a whole range of materials and particle sizes, calculated by the method of this example. Note where the particles of this problem fit into Table 14.1.

Example 11.2 Deep-Fried Fish Sticks
A cod fillet, about $6 \times 1 \times 2$ cm, is taken from a cooler at 0 °C and slipped in hot oil at 180 °C.

(a)   What is the center point temperature of the fillet after 5 min?
(b)   How much heat has been taken up by the fillet during this time?

Data: For cod,

$$k = 0.5\,\text{W/m K}$$

$$\alpha = 0.17 \times 10^{-6}\,\text{m}^2/\text{s}$$

(continued)

(continued)

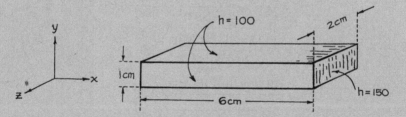

For the fillet in the deep-fat fryer,

$$h = 150 \text{ W/m}^2 \text{ K for the two small end faces}$$
$$h = 100 \text{ W/m}^2 \text{ K for the four long faces}$$

Solution
The fillet can be represented by the intersection of three mutually perpendicular planes 6 cm, 1 cm, and 2 cm thick. Thus,

$$L_x = 0.06/2 = 0.03 \text{ m}$$
$$L_y = 0.01/2 = 0.005 \text{ m}$$
$$L_z = 0.02/2 = 0.01 \text{ m}$$

The individual Biot numbers are

$$\text{Bi}_x = \frac{h_x L_x}{k} = \frac{150(0.03)}{0.5} = 9$$
$$\text{Bi}_y = \frac{100(0.005)}{0.5} = 1$$
$$\text{Bi}_z = \frac{100(0.01)}{0.5} = 2$$

The individual Fourier numbers are

(continued)

(continued)

$$Fo_x = \frac{\alpha t}{L_x^2} = \frac{(0.17 \times 10^{-6})(300)}{(0.03)^2} = 0.055$$

$$Fo_y = \frac{(0.17 \times 10^{-6})(300)}{(0.005)^2} = 2.0$$

$$Fo_z = \frac{(0.17 \times 10^{-6})(300)}{(0.01)^2} = 0.50$$

A. *Center Point Temperature.* From Figs. 11.6 and 11.7, for the three midplanes, we have

$$\left(\frac{\Delta T}{\Delta T_{max}}\right)_x \cong 1$$

$$\left(\frac{\Delta T}{\Delta T_{max}}\right)_y = 0.25$$

$$\left(\frac{\Delta T}{\Delta T_{max}}\right)_z = 0.65$$

We are now ready to replace in equation (11.17). Thus, at the intersection of the three midplanes, in effect the center point, we have

$$\left(\frac{\Delta T}{\Delta T_{max}}\right)_{centerpoint} = \frac{T_{oil} - T_{center}}{T_{oil} - T_{cooler}} = \left(\frac{\Delta T}{\Delta T_{max}}\right)_x \left(\frac{\Delta T}{\Delta T_{max}}\right)_y \left(\frac{\Delta T}{\Delta T_{max}}\right)_z \quad (i)$$

$$= 1(0.25)(0.65) = 0.1625$$

but

$$\left(\frac{\Delta T}{\Delta T_{max}}\right)_{centerpoint} = \frac{T_{oil} - T_{center}}{T_{oil} - T_{cooler}} = \frac{180 - T_{conter}}{180 - 0} \quad (ii)$$

Combining equations (i) and (ii) gives

$$T_{center} = 151 \,^\circ C$$

B. *Heat Absorbed.* First of all, for cod fish fillets

$$\rho = 1,050 \,\text{kg/m}^3 (\text{approximately})$$

$$C_p = \frac{k}{\rho \alpha} = \frac{0.5}{(1,050)(0.17 \times 10^{-6})} = 2,801 \,\text{J/kg K}$$

(continued)

(continued)

The maximum amount of heat which can be absorbed

$$Q_{max} = WC_p(T_{oil} - T_{cooler})$$
$$= [(1,050)(0.06 \times 0.01 \times 0.02)](2,801)(180 - 0) = 6,353\,J$$

(iii)

For the three intersecting infinite slabs, each with its own Biot and Fourier number, Fig. 11.13 shows that

$$\left(\frac{Q}{Q_{max}}\right)_x = 0.76 \quad \text{at Bi} = 9 \text{ and Fo} = 0.055$$

$$\left(\frac{Q}{Q_{max}}\right)_y = 0.23 \quad \text{at Bi} = 1 \text{ and Fo} = 2$$

$$\left(\frac{Q}{Q_{max}}\right)_z = 0.54 \quad \text{at Bi} = 9 \text{ and Fo} = 0.5$$

Thus, for the fish fillet, equation (11.18) becomes

$$\left(\frac{Q}{Q_{max}}\right) = (0.76)(0.23)(0.54) = 0.094$$

(iv)

and with equation (iii) the heat remaining to be absorbed is

$$Q = (0.094)(6,353) = 997$$

So the heat which has been absorbed is

$$Q = 6,353 - 997 = 5,756\,J$$

Note: One can find the average temperature of the fillet directly from these Q values.

Problems on Unsteady-State Heating and Cooling of Solid Objects

11.1. Table 14.1 states that a 1-cm PVC plastic sphere moving through air at 1 m/s would have a thermal relaxation time of 170 s. Verify this figure.

11.2. A cold (0 °C), long, instrumented copper cylinder 5 cm o.d. is quickly plunged into a fluidized bed maintained at 100 °C, and the cylinder's center point temperature reads 40 °C, 60 °C, and 80 °C after 60, 110, and 200 s. Find the heat transfer coefficient between cylinder and bed.

11.3. *Roasting peanuts.* One method of preparing fat-free roasted peanuts involves lowering a wire basket of raw shelled peanuts into a vat of molten mannitol and sorbitol (nonsweet sugars) instead of into hot oil. When the peanuts are well roasted, they are removed, drained, lightly salted, and ready for packing. If peanuts, originally at 15 °C, are lowered into the 165 °C roasting medium:

(a) Find the time needed for their centers to reach 105 °C.
(b) How hot does the surface temperature of the peanuts become?

Data and assumptions: Assume that the peanuts are close to spherical with diameters of 7.5 mm and have the following properties:

$$k_s = 0.5 \text{ W/m K}$$
$$\rho_s = 1,150 \text{ kg/m}^3$$
$$C_s = 1,700 \text{ J/kg K}$$

Between peanuts and molten sugar, take $h = 80 \text{ W/m}^2 \text{ K}$.

11.4. *More on roasting peanuts.* Another way to dry-roast peanuts is to dump a batch of peanuts into a fluidized bed of mannitol particles kept at 140 °C (melting point of mannitol = 160 °C) and then remove them when their centers reach 105 °C. This process does not leave a coating of hexose on the peanuts (see previous problem).

(a) Find the time needed to roast a batch of peanuts this way.
(b) Estimate the surface temperature of the just roasted peanuts.

Data: See previous problem for the thermal properties of peanuts. Between peanuts and fluidized bed, take

$$h = 200 \text{ W/m}^2 \text{ K}$$

11.5. The thermal properties of a sand pile are to be evaluated by quickly pouring a bucketful of hot sand into a long 12-cm-i.d. aluminum pipe which is immersed in water at 20 °C. An axial thermocouple reads 140 °C just after the sand is poured in, 32 °C after 100 min, and 25 °C after 135 min.

From this information, evaluate the thermal conductivity and thermal diffusivity of the sand pile.

11.6. *Restarting a fluidized bed incinerator.* If the airflow to a fluidized bed incinerator stops because of a power outage or some such reason, the solids will collapse to form a "slumped" bed, which then will cool slowly. If the bed is still hot enough when the flow of cold fluidizing air is resumed, it will reignite spontaneously. However, if the bed temperature drops below the ignition temperature, then a long involved procedure must be followed to reignite it.

Let us estimate how long the bed can remain slumped and still restart spontaneously when airflow is resumed. In restarting:

(a) Assume that only a little air is fed to the bed at the beginning so that the bed remains slumped while the hot center region reignites and then spreads. Then, airflow is turned up and the bed refluidizes.

(b) Assume that full airflow is used immediately so that the slumped bed refluidizes right away.

Can you think of any of the advantages and drawbacks of these alternatives? *Data:* The slumped bed is 1 m deep and 4 m in diameter. The hot operating fluidized bed is at 850 °C. The ignition temperature is 600 °C. The thermal diffusivity of the slumped bed is 10^{-6} m²/s. Assume that the top and bottom surfaces of the slumped bed drop to room temperature in about an hour and that the bed walls are well insulated.

11.7. *Heating of carbon particles.* Hot spherical carbon particles ($d_p = 3$ mm) are needed for an experiment. To prepare this feed material, carbon spheres at 0 °C are allowed to fall one at a time through a large diameter heated pipe containing nitrogen at 1 atm. Wall and gas are kept at 500 °C.

(a) Determine the length of pipe needed for the leaving particles to be at an average temperature of 300 °C.

(b) What are the surface temperature and the midpoint temperature of these leaving particles?

Data: For carbon particles,

$$\rho_s = 550 \text{ kg/m}^3$$
$$C_s = 1{,}415 \text{ J/kg K}$$
$$k_s = 0.18 \text{ W/m K}$$

Assume that the particles are falling at their terminal velocity when they enter the heating chamber.

11.8. A very viscous chocolate syrup is to be heated by forcing it through a scraped wall heat exchanger. If the blades wipe the walls clean and bring fresh syrup to the surface twice a second, estimate the heat transfer coefficient at the walls of the exchanger.

Data: Thermal properties of chocolate syrup are estimated to be

$$k_s = 0.5 \text{ W/mK}$$
$$\rho_s = 1{,}200 \text{ kg/m}^3$$
$$C_s = 3{,}600 \text{ J/kg K}$$

11.9. *From hot dogs to knackwurst.* Why make hot dogs when it costs about the same to make knackwurst, which sells for about twice the price? The process is nearly the same—a few different spices and ingredients and minor adjustments to the slicers and dicers and mashers and smashers. The only problem that I see is with the sterilizer.

Sanitary standards require that every part of the product be heated to 105 °C for proper sterilization. For hot dogs this is done by forcing the paste-like mass in plug flow through a long 18-mm tube whose wall is kept at 120 °C by outside condensing steam. Knackwursts are fatter than hot dogs, so we have to replace this 18-mm tube with a 28-mm tube. For the same processing rate of product (tons/day), how long should this tube be?

11.10. *Liver paste for sandwiches.* Chubby liver sausages (assume cylindrical, 5 cm across, 10 cm long), originally at 20 °C, are to be processed in an autoclave kept at 115 °C, and we want every part of the sausage to reach 105 °C. Estimate the lowest temperature of the sausage after 3 h in the autoclave.

Data: For food products containing a water fraction x, we have the following estimates:

$$C_p = 4{,}184x + 800(1 - x), \text{ J/kg} \cdot \text{K}$$
$$k = 0.56x + 0.25(1 - x), \text{ W/m} \cdot \text{K}$$

Liver sausage has about the same water content as canned dog food or 73 %; its density is about 1,050 kg/m^3, and in the autoclave, $h = 7.6$ W/m^2 K.

11.11. *Saving money at the university.* Our university's steam heating plant keeps our campus buildings at a comfortable 22 °C, day and night, 7 days a week. To save on the heating bill, it is proposed that steam to the buildings be shut off every afternoon at 6:00 p.m. and be turned on again the following morning at 6:00 a.m. However, irrespective of whether the heat is off or on, the forced air circulation fans will be kept running in all the buildings to keep the temperature in the buildings uniform throughout.

Test runs at various seasons of the year show that when the heat is turned off, the buildings all cool about halfway to the temperature of the surroundings by 6:00 a.m. How much heating steam would the university save if it followed this on/off procedure?

11.12. *The age of the Earth.* It is well known that the ground becomes hotter and hotter as one digs 1, 2, 5, or more kilometers down into the Earth. The thermal gradient dT/dl varies from place to place, under continent and ocean, in tunnel and bore hole, but on the average it has been found that the temperature increases by about 1 °C for each 24 m of depth. From this information Fourier (in 1820) and later Kelvin (in 1864) made rough estimates on when the Earth began to cool from its molten state. Let us try to do this too.

(a) Taking the temperature of the Earth's surface to be 20 °C and the freezing temperature of rock as 1,200 °C, estimate when the Earth began to solidify. Ignore surface resistance.

(b) Check whether it is reasonable to ignore surface resistance to heat transfer.

(c) This calculation ignores any heat that may be generated in the Earth's interior by radioactivity. If we accounted for this added factor, would it increase or decrease our time estimate?

For references and discussions, see H. S. Carslaw and J. C. Jaeger, *The Conduction of Heat in Solids*, 2nd ed., p. 85, Oxford, 1959.

11.13. A hot granular material ($\alpha_{bulk} = 2 \times 10^{-7}$ m^2/s) is to be cooled by having it slide down a vertical tube 8 m long and 40 mm i.d. whose walls are kept at 100 °C. The hot solids will be fed to the cooler at 300 °C and are expected to slide in plug flow down the tube at 16 mm/s. At what mean temperature will the solids leave the cooler?

References and Notes

L.M.K. Boelter, V.H. Cherry, H.A. Johnson, R.C. Martinelli, *Heat Transfer Notes* (McGraw-Hill, New York, 1956). Gives detailed derivations of the many equations for unsteady state conduction used in this chapter

M. Colakyan, R. Turton, O. Levenspiel, Unsteady-state transfer to various shaped objects. Heat Transf. Eng. **5**, 82 (1984)

H. Gröber, S. Erk, U. Gringull, *Fundamentals of Heat Transfer* (translated from the German by J. R. Moszynski), (McGraw-Hill, New York, 1961). Also a good source book for many of the underlying equations of this chapter

H.P. Gurney, J. Lurie, Charts for estimating temperature distributions in heating and cooling solid shapes. Ind. Eng. Chem. **15**, 1170 (1923)

J. Mator, M.S. Project, *Chemical Engineering Department* (Oregon State University, Corvallis, 1982)

A. Schack, *Industrial Heat Transfer* (translated from the 6th German ed., by I. Gutman), (Wiley, New York, 1965)

J. Sucec, *Heat Transfer* (Simon and Schuster, New York, 1975)

J.R. Welty, *Engineering Heat Transfer* (Wiley, New York, 1974), p. 135

Chapter 12
Introduction to Heat Exchangers

Heat exchangers are devices for transferring heat from a hot flowing stream to a cold flowing stream. There are three broad types of exchangers:

- The recuperator or the through-the-wall nonstoring exchanger
- The direct-contact nonstoring exchanger
- The regenerator, accumulator, or heat-storing exchanger

The type chosen in any situation depends in large part on the nature of the transferring phases, whether gas–gas, gas–liquid, gas–solid, liquid–liquid, liquid–solid, or solid–solid, and on the mutual solubility of the phases involved.

Let us present some examples of these three kinds of exchangers.

12.1 Recuperators (Through-the-Wall Nonstoring Exchangers)

In recuperators the two flowing streams are separated by a wall and heat has to pass through this wall. Many different contacting patterns are used for the two fluids and we will study a number of them in later chapters. The sketches of Fig. 12.1 indicate some of these many different contacting patterns.

Recuperators are certainly less effective than the direct-contact exchanger because the presence of the wall hinders the flow of heat. But this type of exchanger is used where the fluids are not allowed to contact each other, as with gas–gas systems, miscible liquids, dissolving solids, or reactive chemicals.

© Springer Science+Business Media New York 2014
O. Levenspiel, *Engineering Flow and Heat Exchange*,
DOI 10.1007/978-1-4899-7454-9_12

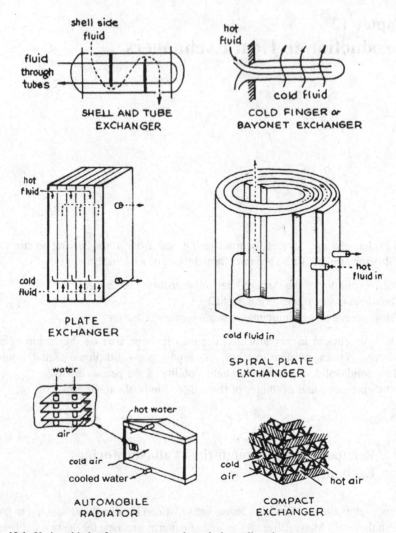

Fig. 12.1 Various kinds of recuperators or through-the-wall exchangers

12.2   Direct-Contact Nonstoring Exchangers

In direct-contact nonstoring exchangers, the streams contact each other intimately, the hotter stream giving up its heat directly to the colder stream.

Naturally this type of exchanger finds use when the two contacting phases are mutually insoluble and do not react with each other. Thus, it cannot be used with gas–gas systems.

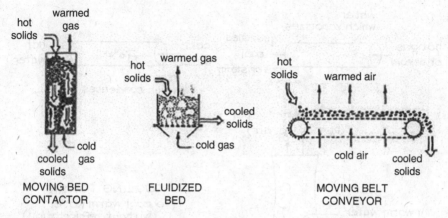

Fig. 12.2 Gas–solid direct-contact nonstoring exchangers

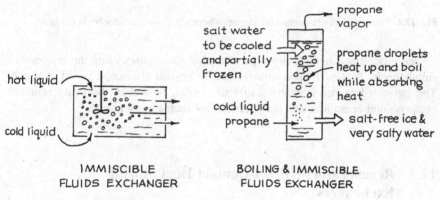

Fig. 12.3 Fluid–fluid direct-contact nonstoring exchangers

Direct-contact exchangers are of three broad types. First, there are the gas–solid exchangers, various forms of which are shown in Fig. 12.2.

Then, there are the fluid–fluid exchangers where the two contacting fluids are mutually immiscible. These are shown in Fig. 12.3.

Finally, it is not always necessary that the two contacting fluids be mutually insoluble, and Fig. 12.4 shows exchangers where one of the flowing fluids does dissolve in the other. In particular, in air–water systems the direct-contact exchanger is of prime importance just because one of the phases (water) does dissolve, or evaporate, in the other phase (air). The water cooling tower, shown in Fig. 12.4, is such an example and, in fact, represents the most widely used type of heat exchanger in industry.

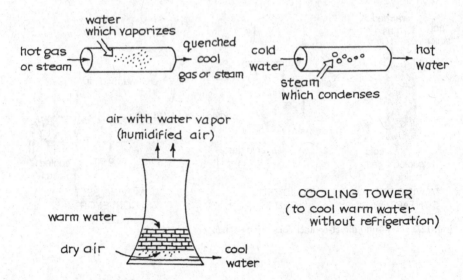

Fig. 12.4 Fluid–fluid direct-contact exchangers where one phase can dissolve in the other

The proper treatment of this type of exchanger requires using the methods of simultaneous heat and mass transfer and is beyond the scope of this volume. The interested reader is referred to Fair (1972a, b) and to the many standard books on unit operations and mass transfer for such a presentation.

12.3   Regenerators (Direct-Contact Heat Storing Exchangers)

With regenerators a hot stream of gas transfers its heat to an intermediary, usually a solid, which later gives up this stored heat to a second stream of cold gas. There are a number of different ways of doing this, as shown in Fig. 12.5.

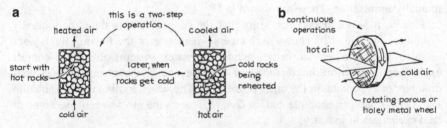

Fig. 12.5 Heat regenerators or heat storing exchangers: (**a**) Heat storing solids are stationary; (**b**) heat storing solids continuously circulate between hot and cold streams

12.4 Exchangers Using a Go-Between Stream

In a number of difficult situations or when the two heat exchanging locations are far apart, a third stream may be used as go-between, to take up heat from the hot stream and then deliver it to the cold stream. This go-between stream may consist of solid particles or it may be a fluid.

Consider a few processes which use this idea.

12.4.1 The Heat Pipe for Heat Exchange at a Distance

The heat pipe (see discussion after Problem 3.24) very effectively transports heat from one location to another, and since the primary resistance to heat transfer is at the two ends of the pipe, where heat is taken up and released, finned tubes are normally used in these sections, as shown in Fig. 12.6. The fluid in the pipe which boils at one end and condenses at the other acts as the go-between in the transfer of heat.

Examples of the use of heat pipes are in the recovery of stored heat in solar home heating (see Problem 3.27), in space capsules as a means of transferring heat from the hot side (facing the sun) to the cold side of the capsule, and in microelectronic and hi-fi equipment to draw off heat from critical components and dissipate it into the air, thereby avoiding overheating. Finally, close to 100,000 heat pipes were built into the Alaska pipeline supports to keep heat away from the support footings which had to be imbedded in permafrost soil.

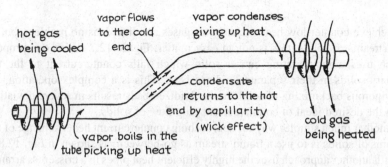

Fig. 12.6 The heat pipe transfers heat from one place to another, often far apart

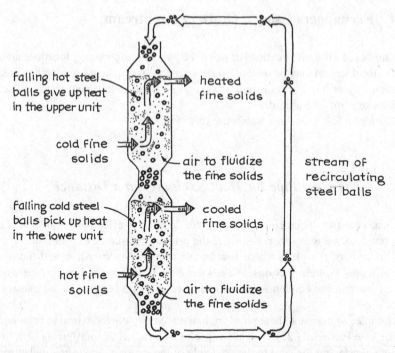

Fig. 12.7 The SPHER process [J. E. Gwyn et al., *Chem. Eng. News*, p. 42 (Sept. 15, 1980)] for the recovery of heat from spent shale and its transfer to cold fresh shale. This is a counterflow solid–solid heat exchanger which uses a third stream of solids as go-between

12.4.2   Solid–Solid Heat Transfer

To achieve counterflow heat exchange of gases and liquids is no problem, but for two streams of solids, this is not an easy matter. Figure 12.7 shows one proposal which uses a recirculating carrier solid which falls countercurrent to the two primary solids in their separate fluidized beds. This is a complex operation, and the vigorous backmixing of solids in the fluidized beds results in severe deviations from the desired ideal of countercurrent plug flow of solids.

Another much simpler way for approaching countercurrent heat exchange of two streams of solids is to use a liquid stream as go-between, as shown in Fig. 12.8.

Still another approach uses the highly efficient heat pipes in a crisscross arrangement to yield countercurrent heat transfer. This is shown in Fig. 12.9.

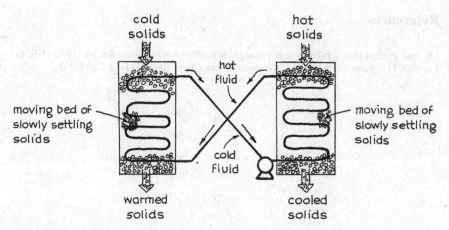

Fig. 12.8 Counterflow solid–solid heat exchanger using a liquid go-between

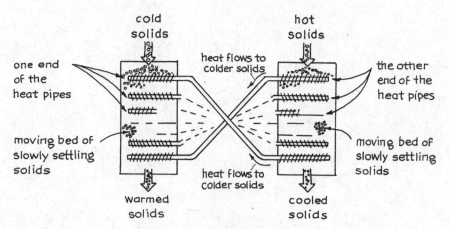

Fig. 12.9 Counterflow solid–solid heat exchanger using properly arranged heat pipes as go-between [O. Levenspiel and R. T. Chan, U.S. Pat. No. 4,408,656]

12.4.3   Comments

What these designs show is that there are many different ways of transferring heat from one flowing stream to another, and the important first decision is to choose the right type of exchanger. Often this is a clean-cut decision, but sometimes one has to compare the economics of quite different kinds of exchangers.

In the following chapters we consider in turn the various kinds of exchangers and their important design parameters and simple design methods.

References

J.R. Fair, Process heat transfer by direct fluid phase contact. AIChE Symp. Ser. **68**(118), 1 (1972a)

J.R. Fair, Designing direct-contact cooler/condensers, Chem. Eng. 91 (June 12, 1972b)

Chapter 13
Recuperators: Through-the-Wall Nonstoring Exchangers

Recuperators are heat exchangers wherein the two fluids transferring heat are kept away from each other by a wall. Their design requires two pieces of information.

1. *The overall heat transfer coefficient U* which accounts for the overall resistance to transfer caused by the wall. This includes the individual film resistances and the wall resistance. From Chap. 10 we have

$$\frac{1}{UA} = \frac{1}{h_1 A_1} + \frac{\Delta x}{k A_{\text{mean}}} + \frac{1}{h_2 A_2} .$$

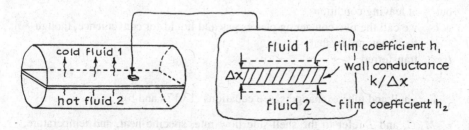

2. *The contacting pattern of the two phases*. This can be very complex and difficult to evaluate. To make the analysis of exchangers tractable, we idealize real contacting patterns by considering them to be combinations of: (i) plug flow, (ii) mixed flow, (iii) batch completely mixed fluid, and (iv) batch unmixed and stagnant fluid.

© Springer Science+Business Media New York 2014
O. Levenspiel, *Engineering Flow and Heat Exchange*,
DOI 10.1007/978-1-4899-7454-9_13

Let us use the following nomenclature for this chapter:

A	area of exchanger [m^2]
C	specific heat of a fluid [J/kg K]
$\mathscr{F}$	$\overline{\Delta T}/\Delta T_{lm}$, mean temperature driving force in the exchanger compared to that of countercurrent flow, which is the very best possible [−]
$\dot{m}$	flow rate of a fluid [kg/s]
NTU	number of transfer units, based on a particular phase [−] (see Fig. 13.14)
P	$\Delta T_i/\Delta T_{max}$, temperature change of phase i compared to the maximum possible temperature change [−] (see Fig. 13.4)
$\dot{q}$	rate of heat gain by a phase [W]
Q	heat lost or gained by a fluid up to a given point in the exchanger [J/kg of one of the phases]
R	ratio of temperature changes or heat flows of the fluids [−] (see Fig. 13.4)
ΔT	temperature difference between phases (or temperature driving force) at a given location in the exchanger [K]
ΔT_g	temperature change of gas phase [K]
ΔT_l	temperature change of liquid phase [K]
ΔT_{lm}	logarithmic mean temperature driving force [K] (see equation (13.5))
ΔT_{max}	maximum temperature difference between phases [K]
U	overall heat transfer [W/m^{2}K]

Subscripts:

0	at zero time
1, 2	at one end or the other end of the exchanger
in	at entering conditions
out	at leaving conditions
g, l	we call the two contacting phases gas and liquid for convenience, though they may not be that
tx	transferred

For shell and tube exchangers, see equations (13.17a and b):

- $\dot{M}$, C, and T refer to the shell-side flow rate, specific heat, and temperature, respectively.
- $\dot{m}$, c, and t refer to the tube-side flow rate, specific heat, and temperature, respectively.
- η is the efficiency of heat removal from a phase.

Let us now develop performance equations for various contacting patterns.

13.1 Countercurrent and Cocurrent Plug Flow

These two contacting patterns are idealization and approximations of the flows in double-pipe heat exchangers. To develop the performance equation for these contacting patterns, and in general for any other contacting pattern, we draw on the following relationship:

$$\begin{pmatrix} \text{Rate of} \\ \text{heat loss} \\ \text{by gas} \end{pmatrix} = \begin{pmatrix} \text{rate of} \\ \text{heat gain} \\ \text{by liquid} \end{pmatrix} = \begin{pmatrix} \text{rate of heat} \\ \text{transfer from} \\ \text{gas to liquid} \end{pmatrix} \tag{13.1a}$$

or

$$\dot{q}_{\text{leaving gas}} = \dot{q}_{\text{entering liquid}} = \dot{q}_{\text{transferred}} \tag{13.1b}$$

We treat two broad cases: first, fluids with no phase change and with constant specific heat and, second, exchangers wherein phase changes such as boiling or condensation do occur.

13.1.1 No Phase Change, C_p Independent of Temperature

Figure 13.1 sketches the distance versus temperature diagram and the more useful Q versus T diagram for these two idealized flow patterns. Note that Q represents the amount of heat exchanged as the fluids pass through the exchanger. This quantity is conveniently measured either by the enthalpy change of one of the fluids [J/kg] or as a fraction of the total heat transferred.

1. *Countercurrent exchangers.* Around the whole exchanger the heat balance of equations (13.1a and b) can be written as

$$\boxed{\dot{q} = -\dot{m}_g C_g \left(T_{g,\text{out}} - T_{g,\text{in}} \right) = \dot{m}_l C_l (T_{l,\text{out}} - T_{l,\text{in}}) = UA\overline{\Delta T}} \tag{13.2}$$

Similarly, for a differential slice of exchanger of interfacial area dA, we have

$$d\dot{q} = \underbrace{-\dot{m}_g C_g dT_g}_{\text{I}} = \underbrace{\dot{m}_l C_l dT_l}_{\text{II}} = \underbrace{U\Delta T dA}_{\text{III}} \tag{13.3}$$

where $\Delta T = T_g - T_l$ is called the temperature driving force. Combining terms I and III, II and III; adding the results together; and integrating while keeping U constant gives, after considerable manipulation,

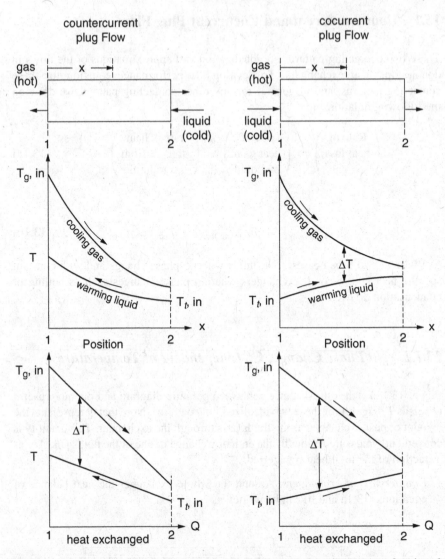

Fig. 13.1 Temperature profiles in ideal countercurrent and cocurrent exchangers, as a function of position and as a function of the heat transferred between streams

$$\dot{q} = \dot{m}_g C_g \left(T_{g,\text{in}} - T_{g,\text{out}} \right) = \dot{m}_l C_l (T_{l,\text{out}} - T_{l,\text{in}}) = UA\Delta T_{lm} \tag{13.4}$$

where the proper temperature driving force is the logarithmic mean temperature driving force, defined as

$$\Delta T_{lm} = \frac{\Delta T_2 - \Delta T_1}{\ln \frac{\Delta T_2}{\Delta T_1}} = \frac{\left(T_{g,\text{out}} - T_{l,\text{in}} \right) - \left(T_{g,\text{in}} - T_{l,\text{out}} \right)}{\ln \frac{T_{g,\text{out}} - T_{l,\text{in}}}{T_{g,\text{in}} - T_{l,\text{out}}}} \tag{13.5}$$

Rearranging equations (13.4) and (13.5) gives, after even more manipulation, the efficiency of heat exchange η, as

$$\eta_l = \frac{\Delta T_1}{\Delta T_{max}} = \frac{T_{l,out} - T_{l,in}}{T_{g,in} - T_{l,in}} = \frac{1 - K}{(\dot{m}_l C_l / \dot{m}_g C_g) - K} \tag{13.6}$$

or

$$\eta_g = \frac{\Delta T_g}{\Delta T_{max}} = \frac{T_{g,in} - T_{g,out}}{T_{g,in} - T_{l,in}} = \frac{1 - K}{1 - (\dot{m}_g C_g / \dot{m}_l C_l)K} \tag{13.7}$$

where

$$K = \exp\left[-UA\left(\frac{1}{\dot{m}_g C_g} - \frac{1}{\dot{m}_l C_l}\right)\right] \tag{13.8}$$

In the special case of equal heat flows $(\dot{m}_g C_g = \dot{m}_l C_l = \dot{m}C)$, the operating lines on the Q versus T chart become parallel and

$$\Delta T_{lm} \to \Delta T \text{ a constant throughout the exchanger}$$

In this situation equations (13.6) and (13.7) reduce to

$$\eta_g = \eta_l = \frac{\Delta T_g}{\Delta T_{max}} = \frac{\Delta T_l}{\Delta T_{max}} = \frac{UA/\dot{m}C}{1 + (UA/\dot{m}C)} \tag{13.9}$$

Equations (13.4) and (13.5) are useful for finding the size of an exchanger given the desired inlet and outlet temperatures. Equations (13.6), (13.7), (13.8), and (13.9) are useful for the inverse problem, to find the outlet temperatures in a given system.

2. *Cocurrent exchangers.* Similar to countercurrent contacting, the size of exchanger (or flow rate) to give desired terminal temperatures is found to be

$$\boxed{\dot{q} = \dot{m}_g C_g \left(T_{g,in} - T_{g,out}\right) = \dot{m}_l C_l (T_{l,out} - T_{l,in}) = UA\Delta T_{lm}} \tag{13.10}$$

where

$$\Delta T_{lm} = \frac{\Delta T_2 - \Delta T_1}{\ln \frac{\Delta T_2}{\Delta T_1}} = \frac{(T_{g,out} - T_{l,out}) - (T_{g,in} - T_{l,in})}{\ln \frac{T_{g,out} - T_{l,out}}{T_{g,in} - T_{l,in}}} \tag{13.11}$$

From these expressions the efficiency of heat exchange (and thus the terminal temperatures) in a given exchanger is found to be

$$\eta_l = \frac{T_{l,\text{out}} - T_{l,\text{in}}}{T_{g,\text{in}} - T_{l,\text{in}}} = \frac{\dot{m}_g C_g}{\dot{m}_g C_g + \dot{m}_l C_l} \left(1 - K'\right) \qquad (13.12)$$

or

$$\eta_g = \frac{T_{g,\text{in}} - T_{g,\text{out}}}{T_{g,\text{in}} - T_{l,\text{in}}} = \frac{\dot{m}_l C_l}{\dot{m}_g C_g + \dot{m}_l C_l} \left(1 - K'\right) \qquad (13.13)$$

where

$$K' = \exp\left[-UA\left(\frac{1}{\dot{m}_g C_g} + \frac{1}{\dot{m}_l C_l}\right)\right] \qquad (13.14)$$

For equal heat flows $\left(\dot{m}_g C_g = \dot{m}_l C_l = \dot{m}C\right)$, the above efficiency expressions reduce to

$$\eta_g = \eta_l = \frac{\Delta T_g}{\Delta T_{\text{max}}} = \frac{\Delta T_l}{\Delta T_{\text{max}}} = \frac{1}{2}\left[1 - e^{-(UA/2\dot{m}C)}\right] \qquad (13.15)$$

3. *Cocurrent or countercurrent contacting with changing U.* If U varies linearly with temperature, integration of equation (13.3) gives

$$(U\Delta T)_{lm} = \frac{U_2 \Delta T_1 - U_1 \Delta T_2}{\ln \frac{U_2 \Delta T_1}{U_1 \Delta T_2}} \quad \text{with } U = a + bT \qquad (13.16)$$

to be used in equation (13.4) or equation (13.10).

4. *Conclusion.* Countercurrent flow is always more efficient than cocurrent flow.

13.1.2 Exchangers with a Phase Change

Suppose cold water is used to condense hot atmospheric steam. For countercurrent flow the Q versus T diagram will then be as shown in Fig. 13.2 with three distinct regimes, each with its particular U value. For this situation the overall heat loss/heat gain expression, equation (13.2), remains unchanged; however, the rate expression, equation (13.3), must be integrated either graphically, numerically, or by considering the three sections of the exchanger separately, while using equations (13.4) and (13.5) for each section.

The Q versus T graph is a useful representation for phase change systems, for finding intermediate conditions, and for helping one see what is happening.

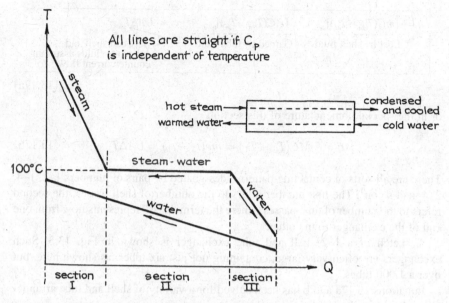

Fig. 13.2 Q vs. T diagram for countercurrent plug flow with a phase change

The values of Q for this graph can be found directly from the enthalpy tables for the fluids because in the absence of shaft work, the energy balance for a flowing fluid, equation (1.4), reduces to

$$Q = \Delta H$$

Of course, the lines representing the hot fluid and the cold fluid can never cross each other on the Q versus T graph.

13.2   Shell and Tube Exchangers

With shell and tube exchangers, we have various flow patterns besides simple cocurrent and countercurrent plug flow. All are less efficient than countercurrent plug flow. By this we mean that they need more exchanger surface area for given end conditions. Despite this drawback these exchangers are widely used in industry because they often are more convenient, more compact, and less expensive to construct for a given duty.

We treat these exchangers as countercurrent plug flow units with a fudge factor $\mathscr{F}$, somewhere between 0 and 1, to account for the lowered contacting efficiency. Thus, the performance equations for these exchangers, assuming constant U, become

$$\dot q = -[\dot mC(T_{\text{out}} - T_{\text{in}})]_{\text{shell}} \qquad = [\dot mC(T_{\text{out}} - T_{\text{in}})]_{\text{tube}} \qquad = UA\Delta T_{lm}\mathscr{F}$$

Lost by shell fluid   Gained by tube fluid   Between shell and
                                                        tube-side fluids assuming
                                                        countercurrent flow

(13.17a)

or in the special nomenclature of this section

$$\dot q = -\dot MC(T_2 - T_1) = \dot mc(t_2 - t_1) = UA\overline{\Delta T} \qquad (13.17\text{b})$$

There are all sorts of contacting patterns, designated by pairs of numbers 1–2, 1–4, 2–4, and so on[1]. The first number refers to the number of shell passes; the second refers to the number of tube passes, where the term "pass" represents flow from one end of the exchanger to the other.

A sketch of a 1–2 shell and tube exchanger is shown in Fig. 13.3. Such exchangers are often very large, containing not just six tubes, as shown here, but over a 1,000 tubes.

Equations 13.17a and b has been solved for a variety of shell and tube arrangements based on the following assumptions:

- Ideal plug flow of all streams
- No temperature gradients across the flow path of any fluid
- Same heat transfer area for each pass
- Constant U value throughout the exchanger

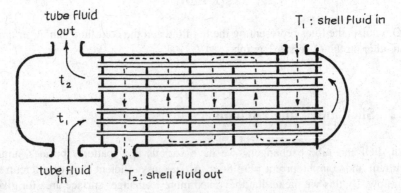

Fig. 13.3 The insides of a 1–2 shell and tube exchanger, considerably simplified

---

[1] Note that uppercase letters $\dot M$, C, and T represent the shell-side fluid; lowercase letters $\dot m$, c, and t represent the tube-side fluid; $\Delta T = T - t$ represents the temperature driving force; and subscripts 1 and 2 stand for the entering and leaving fluid stream, respectively.

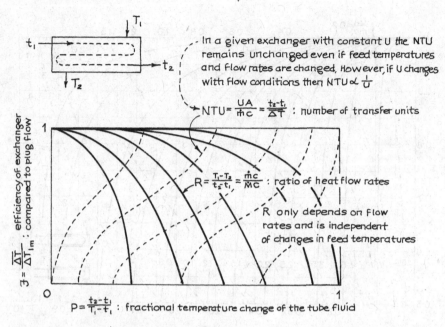

In a given exchanger with constant U the NTU remains unchanged even if feed temperatures and flow rates are changed. However, if U changes with flow conditions then $NTU \propto \frac{1}{U}$

$NTU = \frac{UA}{\dot{m}c} = \frac{t_2 - t_1}{\overline{\Delta T}}$: number of transfer units

$R = \frac{T_1 - T_2}{t_2 - t_1} = \frac{\dot{m}c}{\dot{M}C}$: ratio of heat flow rates

R only depends on flow rates and is independent of changes in feed temperatures

efficiency of exchanger compared to plug flow

$\frac{\overline{\Delta T}}{\Delta T_{lm}} = \mathcal{E}$

$P = \frac{t_2 - t_1}{T_1 - t_1}$: fractional temperature change of the tube fluid

Fig. 13.4 Interrelationship of variables of equations 13. 17a and b for shell and tube exchangers. $\dot{M}$, C, and T refer to the shell-side fluid; $\dot{m}$, c, and t refer to the tube-side fluid; $\Delta T = T - t$ is the temperature driving force

- Constant specific heat and no phase change for the fluids, hence no vaporization or condensation within the exchanger
- Negligible heat losses to the surroundings
- No conduction along the exchanger walls in the directions of flow of fluids

The resulting expressions are cumbersome, but with the aid of computer graphics, the relationship between all the variables can be displayed compactly in a chart of the type shown in Fig. 13.4. Figures 13.5, 13.6, 13.7, 13.8, 13.9, and 13.10, prepared by Turton et al. (1984), show this relationship for various combinations of shell and tube passes.

Comments

1. Shell and tube exchangers are used primarily for liquid–liquid systems.
2. Routing of fluids—here are some considerations:

   - The inside of tubes is much easier to clean than the shell side so the scum- or scale-forming fluid should flow through the tubes.
   - The corrosive fluid should flow through the tubes to avoid the expense of special metals for both shell and tubes.
   - The tube-side Δp usually is greater than the shell-side Δp, so the less viscous fluid should pass through the tubes.
   - An odd number of tube passes, 3, 5, or 7, is rarely used because the expansion and stress problems associated with temperature changes are hard to deal with.

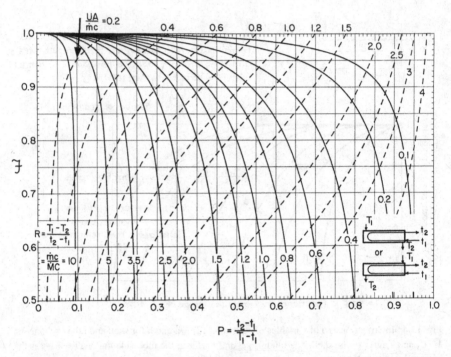

Fig. 13.5 Performance chart for a 1–2 shell and tube exchanger

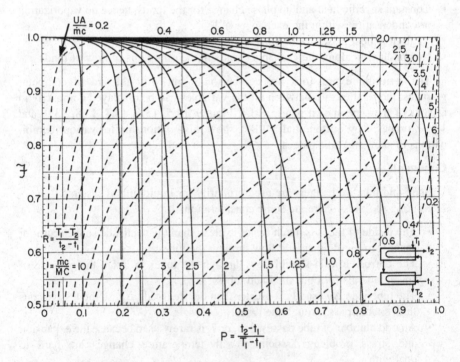

Fig. 13.6 Performance chart for a 2–4 shell and tube exchanger

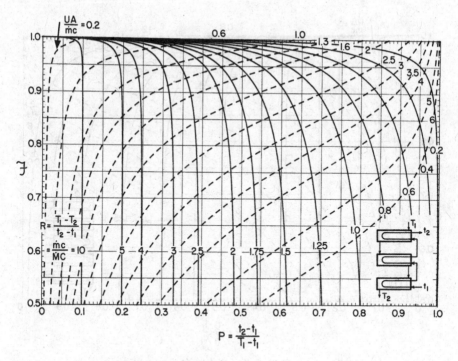

Fig. 13.7 Performance chart for a 3–6 shell and tube exchanger

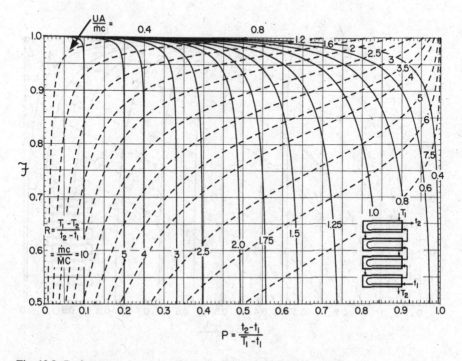

Fig. 13.8 Performance chart for a 4–8 shell and tube exchanger

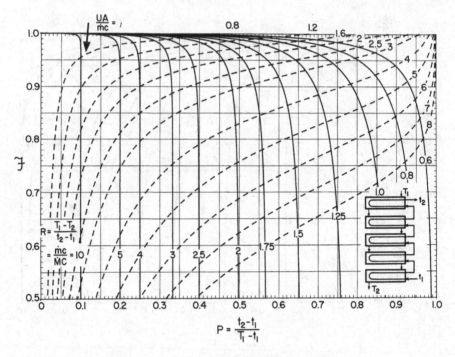

Fig. 13.9 Performance chart for a 5–10 shell and tube exchanger

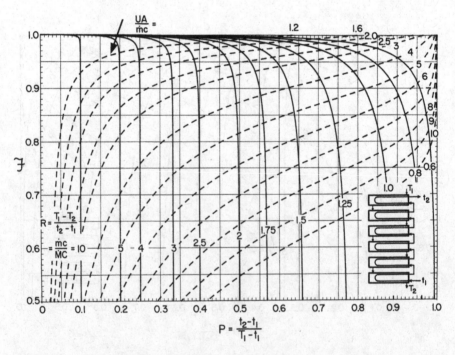

Fig. 13.10 Performance chart for a 6–12 shell and tube exchanger

- There are various ways of contacting the phases for a given number of shell passes and tube passes. Always try to approach countercurrent flow.
- These charts show that the larger the number of shell and tube passes, the closer does flow approach ordinary countercurrent flow with its largest temperature driving force.

3. The actual design of shell and tube exchangers is a complex business. The interested reader should look up the specialized books devoted to this subject.

13.3 Crossflow and Compact Exchangers

These exchangers come in all shapes and sizes as sketched in Fig. 13.11. Here we recognize two sorts of flow for a phase: either well mixed laterally or unmixed. For example, consider the temperature of the hot liquid at position A and at position B in the sketch of Fig. 13.12. With no lateral mixing, hot fluid will take separate and parallel paths through the exchanger in which case the horizontally flowing hot fluid at B will be much cooler than the hot fluid at A. With lateral mixing all the fluid along AB will be at one and the same temperature. We show these two schemes as in Fig. 13.13.

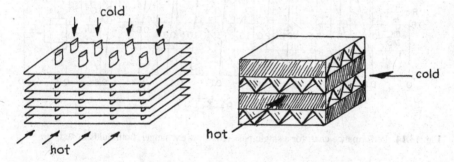

Fig. 13.11 Some of the many kinds of crossflow and compact heat exchangers

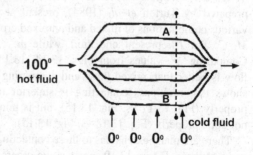

Fig. 13.12 Crossflow heat exchanger

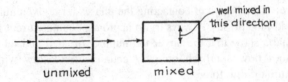

Fig. 13.13 For horizontally flowing fluid

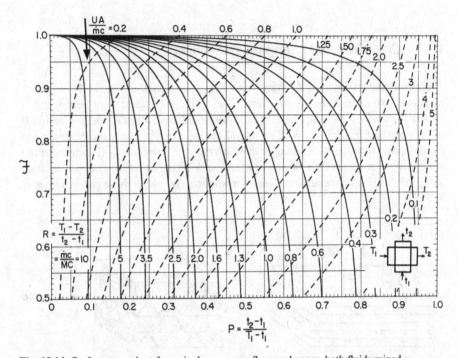

Fig. 13.14 Performance chart for a single-pass crossflow exchanger, both fluids mixed

The interrelationship of variables for various combinations of ideal mixed and ideal unmixed flow has been evaluated theoretically and has been displayed in design charts similar to Fig. 13.4. Figures 13.14, 13.15, 13.16, 13.17, and 13.18, prepared by Turton *et al.* (1984), present accurate computer-drawn charts for various combinations of mixed and unmixed crossflow exchangers. In these charts, M, C, and T represent one fluid, while m, c, and t represent the other fluid. Comparing $\mathscr{F}$ values from Figs. 13.14, 13.15, and 13.16 shows that unmixed flow is better than mixed flow; and comparing Fig. 13.15 with 13.17, and 13.18 shows that multipass contacting is superior to single-pass contacting, if done properly (Fig. 13.17 vs. Fig. 13.15), but is poorer than single-pass contacting, if not done properly (Fig. 13.18 vs. Fig. 13.15).

There are many extensions to these contacting patterns. Mueller (1973) presents some of these. Figure 13.19 shows a two others.

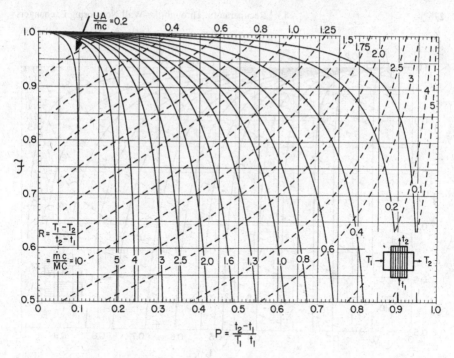

Fig. 13.15 Performance chart for a single-pass crossflow exchanger, one fluid mixed, one unmixed

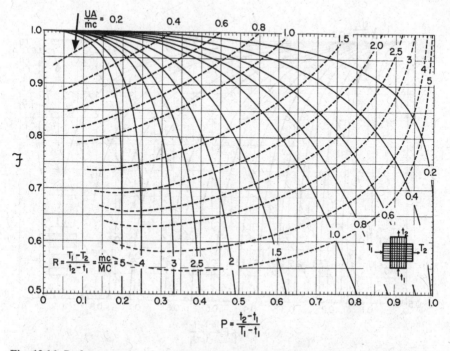

Fig. 13.16 Performance chart for a single-pass crossflow exchanger, both fluids unmixed [See Baclic (1978) for the mathematics of this contacting pattern]

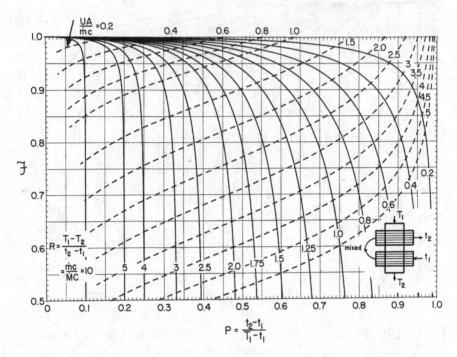

Fig. 13.17 Performance chart for a two-pass crossflow exchanger, countercurrent arrangement

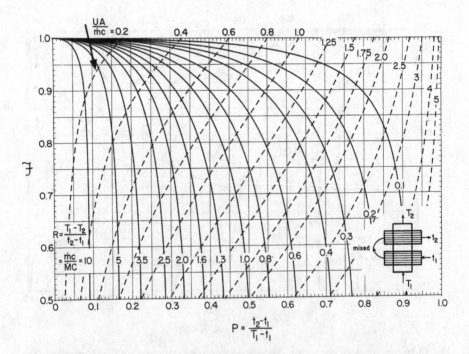

Fig. 13.18 Performance chart for a two-pass crossflow exchanger, cocurrent arrangement

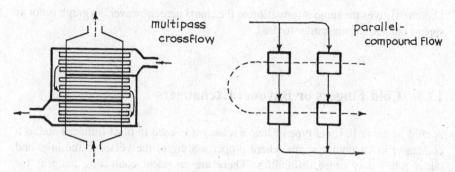

Fig. 13.19 Extensions to simple crossflow exchanger

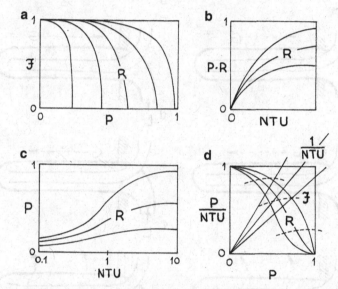

Fig. 13.20 Various ways of displaying the relationship between variables of a heat exchanger: (**a**) from Bowman et al. (1940), also presented in TEMA (1978); (**b**) from Kays and London (1963); (**c**) shown in TEMA (1978); (**d**) from Muller (1973)

Remarks The sketches of Fig. 13.20 show four other ways of displaying the interrelationship of variables in heat exchangers. Sketch (a) is identical to the design charts of this chapter except that the variable NTU is left out.

Consequently, although it allows one to find the size of exchanger directly, given the terminal temperatures of the exchanger, the reverse problem of finding the exit temperature of fluids in a given exchanger requires a tedious trial-and-error solution. Sketches (b) and (c) are essentially equivalent to charts of this chapter, in that they allow one to solve any kind of problem without trial and error. However, these figures do not display the $\mathcal{F}$ values, which must be found with an extra calculation.

Sketch (d) gives the same information as the charts here; however, its graph is not as spread out and thus is harder to read.

13.4   Cold Fingers or Bayonet Exchangers

A cold finger or bayonet type of heat exchanger is used in blast furnaces and as a condenser in vacuum systems where proper sealing of the vessel to the inlet and outlet tubes may cause difficulties. There are six ideal contacting patterns for bayonet exchangers as shown in Fig. 13.21.

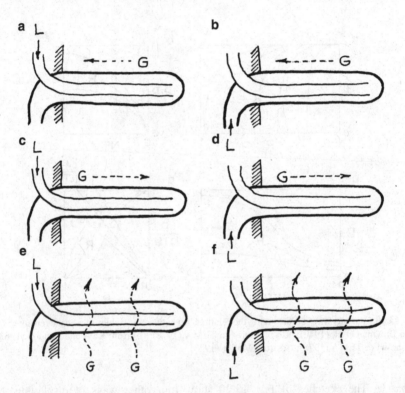

Fig. 13.21 Various possible contacting patterns for bayonet exchangers

Let us evaluate the proper mean driving force $\overline{\Delta T_o}$ knowing the temperatures of the incoming streams. This value is then introduced into the performance expression:

$$\dot{q} = U_o A_o \overline{\Delta T_o} \tag{13.18}$$

We adopt the following nomenclature:

g hot gas flowing on the outside
h, U film and overall heat transfer coefficient [W/m^2 K]
l cool liquid flowing in the bayonet
T_l' temperature in the inner tube [K]
T_l'' temperature in the annulus [K]
A_i, A_o surface area of inner and of outer tube [m^2]
P_i, P_o perimeter of inner and of outer tube [m]

Now consider arrangement (a), as shown in Fig. 13.22. Ignoring heat losses at the ends of the exchanger, we have:

Rate of heat transfer into annulus

$$\left(T_g - T_l''\right)U_oP_odx + \left(T_l' - T_l''\right)U_iP_idx = -\dot{m}_lC_ldT_l'' \tag{13.19}$$

Rate of heat transfer into inner tube

$$\left(T_l'' - T_l'\right)U_iP_idx = \dot{m}_lC_ldT_l' \tag{13.20}$$

Rate of total heat gain by liquid

$$\left(T_g - T_l''\right)U_oP_odx = \dot{m}_lC_l\left(dT_l' - dT_l''\right) \tag{13.21}$$

Overall heat balance

$$\dot{m}_lC_l(T_{l,\text{out}} - T_{l,\text{in}}) = -\dot{m}_gC_g\left(T_{g,\text{out}} - T_{g,\text{in}}\right) \tag{13.22}$$

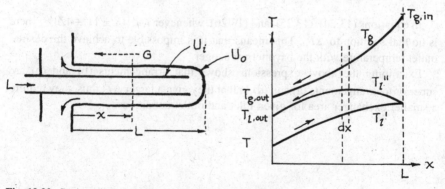

Fig. 13.22 Setting up the heat balance for the bayonet exchanger

Solving these equations is tedious [see Hurd (1946)], and in terms of

$$X = \frac{T_{g,\text{in}} - T_{g,\text{out}}}{T_{l,\text{out}} - T_{l,\text{in}}}, \quad Y = \frac{(T_{g,\text{in}} + T_{g,\text{out}}) - (T_{l,\text{out}} + T_{l,\text{in}})}{(T_{l,\text{out}} - T_{l,\text{in}})},$$

$$Z = \frac{U_i A_i}{U_o A_o}$$

(13.23)

gives

$$\dot{q} = U_o A_o \overline{\Delta T}_o, \text{ where } \overline{\Delta T}_o = \frac{(T_{l,\text{out}} - T_{l,\text{in}})\left[(X+1)^2 + 4Z\right]^{1/2}}{\ln \frac{Y + \left[(X+1)^2 + 4Z\right]^{1/2}}{Y - \left[(X+1)^2 + 4Z\right]^{1/2}}}$$

(13.24)

This equation holds for arrangement (a) and also for arrangement (d).
   For arrangements (b) and (c), a similar analysis gives

$$\dot{q} = U_o A_o \overline{\Delta T}_o, \text{ where } \overline{\Delta T}_o = \frac{(T_{l,\text{out}} - T_{l,\text{in}})\left[(X-1)^2 + 4Z\right]^{1/2}}{\ln \frac{Y + \left[(X-1)^2 + 4Z\right]^{1/2}}{Y - \left[(X-1)^2 + 4Z\right]^{1/2}}}$$

(13.25)

Arrangements (e) and (f) represent boiling or condensation or crossflow or very high flow rate of fluid outside the bayonet. For these arrangements simply put $X = 0$ in the above expressions to obtain

$$\dot{q} = U_o A_o \overline{\Delta T}_o, \text{ where } \overline{\Delta T} = \frac{(T_{l,\text{out}} - T_{l,\text{in}})(1 + 4Z)^{1/2}}{\ln \frac{Y + (1+4Z)^{1/2}}{Y - (1+4Z)^{1/2}}}$$

(13.26)

In equations (13.24), (13.25), and (13.26), whenever $Y < [(X \pm 1)^2 + 4Z]^{1/2}$, there is no real solution to $\overline{\Delta T}_o$. This means that it is impossible to achieve the desired outlet temperature with the bayonet exchanger.
   Examining the above expressions shows that arrangements (b) and (c) are superior to arrangements (a) and (d) in that they give a larger $\overline{\Delta T}$; thus, they require a smaller exchanger area for given inlet and outlet conditions.

13.5 Mixed Flow *L*/Plug Flow *G* Exchangers

The heating vat with immersed heat exchanger coils can be idealized by this contacting pattern. Its sketch and Q versus T diagram are shown in Fig. 13.23. The key here is to recognize that $T_{l,\text{out}}$ represents the liquid temperature contacted by all the gas. The performance equation, equation (13.1), then becomes

$$\boxed{\dot{q} = -[\dot{m}C(T_{\text{out}} - T_{\text{in}})]_g = [\dot{m}C(T_{\text{out}} - T_{\text{in}})]_l = UA\Delta T_{lm}} \qquad (13.27)$$

where

$$\Delta T_{lm} = \frac{\left(T_{g,\text{in}} - T_{l,\text{out}}\right) - \left(T_{g,\text{out}} - T_{l,\text{out}}\right)}{\ln\frac{T_{g,\text{in}} - T_{l,\text{out}}}{T_{g,\text{out}} - T_{l,\text{out}}}} \qquad (13.28)$$

from which

$$T_{g,\text{out}} = \frac{\left[(1-K)\dot{m}_g C_g + K\dot{m}_l C_l\right]T_{g,\text{in}} + (1-K)\dot{m}_l C_l T_{l,\text{in}}}{(1-K)\dot{m}_g C_g + \dot{m}_l C_l} \qquad (13.29)$$

$$T_{l,\text{out}} = \frac{(1-K)\dot{m}_g C_g T_{g,\text{in}} + \dot{m}_l C_l T_{l,\text{in}}}{(1-K)\dot{m}_g C_g + \dot{m}_l C_l} \qquad (13.30)$$

and where

$$K = \frac{T_{g,\text{out}} - T_{l,\text{out}}}{T_{g,\text{in}} - T_{l,\text{out}}} = \exp\left[-(UA/\dot{m}_g C_g)\right] \qquad (13.31)$$

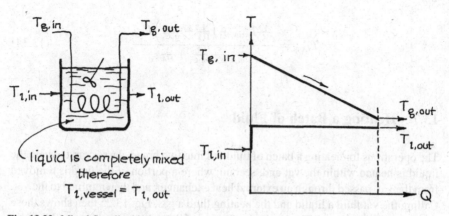

Fig. 13.23 Mixed flow liquid/plug flow gas

13.6 Mixed Flow *L*/Mixed Flow *G* Exchangers

A stirred vat where the heating fluid swirls about or condenses in a double shell is best approximated by mixed flow/mixed flow, shown in Fig. 13.24. Recognizing that the temperature of the leaving streams represents all the fluid in the vat equation (13.1) becomes

$$\dot{q} = -[\dot{m}C(T_{\text{out}} - T_{\text{in}})]_g = [\dot{m}C(T_{\text{out}} - T_{\text{in}})]_l = UA(T_{g,\text{out}} - T_{l,\text{out}}) \qquad (13.32)$$

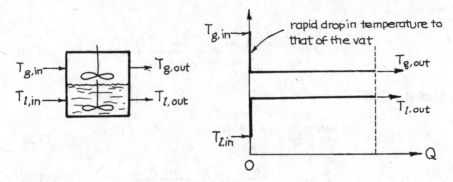

Fig. 13.24 Mixed flow liquid/mixed flow gas

from which

$$T_{g,\text{out}} = \frac{\left(\frac{1}{UA} + \frac{1}{\dot{m}_l C_l}\right)T_{g,\text{in}} + \frac{1}{\dot{m}_g C_g}T_{l,\text{in}}}{\frac{1}{\dot{m}_g C_g} + \frac{1}{\dot{m}_l C_l} + \frac{1}{UA}} \qquad (13.33)$$

and

$$T_{l,\text{out}} = \frac{\frac{1}{\dot{m}_l C_l}T_{g,\text{in}} + \left(\frac{1}{UA} + \frac{1}{\dot{m}_g C_g}\right)T_{l,\text{in}}}{\frac{1}{\dot{m}_g C_g} + \frac{1}{\dot{m}_l C_l} + \frac{1}{UA}} \qquad (13.34)$$

13.7 Heating a Batch of Fluid

The operations for heating a batch of fluid fall into two broad classes; first, where the liquid is heated within the vat and, second, where a portion of the fluid is removed from the vat, passed through an external heat exchanger, and then returned to the vat. Calling the vat fluid a liquid and the heating fluid a gas, Fig. 13.25 then shows these two types of operations. The particular feature of the external exchanger is that one

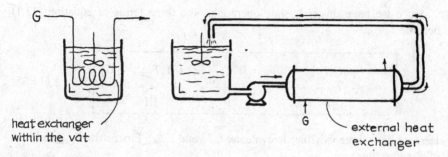

heat exchanger
within the vat

external heat
exchanger

Fig. 13.25 Two ways of heating a batch of liquid with a stream of gas

can have a high liquid velocity through the exchanger tubes, hence high U value. This is a big advantage for viscous, difficult to stir liquids.

With each class of exchanger, we can have a variety of contacting patterns. We consider a few simple cases here.

13.8   Uniformly Mixed Batch *L*/Mixed Flow *G* Exchangers

This ideal represents a batch of liquid heated by hot gas swirling through the shell of the vessel. We sketch this operation in Fig. 13.26.

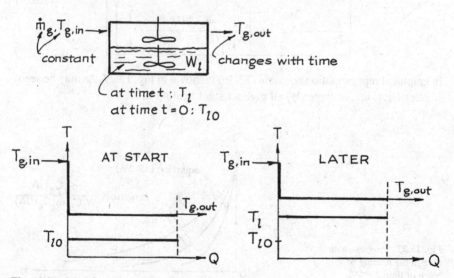

Fig. 13.26 Heating a batch of liquid with mixed flow of gas

Since we have unsteady-state operations, the three terms of equation (13.1) become

$$\underbrace{\dot{q} = -[\dot{m}C(T_{out} - T_{in})]_g}_{\text{I}} = \underbrace{W_l C_l \frac{dT_l}{dt}}_{\text{II}} = \underbrace{UA(T_{g,out} - T_l)}_{\text{III}}$$

(13.35)

Here we have three variables to evaluate: t, T_l, and $T_{g,out}$. First, write $T_{g,out}$ in terms of T_l and t; thus,

$$T_{g,out} = \frac{\dot{m}_g C_g T_{g,in} + UAT_l}{\dot{m}_g C_g + UA}$$

Eliminating $T_{g,out}$ in I and II of equation (13.35) and rearranging give one differential equation in two variables:

$$\frac{dT_l}{dt} = \frac{\dot{m}_g C_g UA(T_{g,in} - T_l)}{W_l C_l(\dot{m}_g C_g + UA)}$$

Separating and integrating from $t=0$ to t give

$$\left. \begin{array}{l} \ln\left[\dfrac{T_{g,in} - T_l}{T_{g,in} - T_{l0}}\right] = -\dfrac{\dot{m}_g C_g UA}{W_l C_l(\dot{m}_g C_g + UA)} \cdot t \\[3mm] \text{or} \\[2mm] \dfrac{T_{g,in} - T_l}{T_{g,in} - T_{l0}} = \exp\left[\dfrac{-\dot{m}_g C_g UA}{W_l C_l(\dot{m}_g C_g + UA)} \cdot t\right] \end{array} \right\}$$

(13.36)

In graphical representation equation (13.36) is shown in Fig. 13.27. As may be seen, smaller U, A, $\dot{m}_g$, or larger W_l all give a longer heating time.

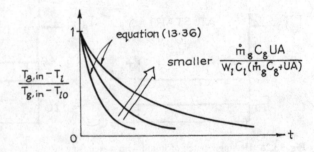

Fig. 13.27 Progress with time in the heating of a batch of liquid

13.9   Uniformly Mixed Batch *L*/Isothermal, Mixed Flow *G* (Condensation or Boiling) Exchangers

The two typical operations represented by this ideal are shown in the sketches of Fig. 13.28. Consider the first of these, in which a batch of liquid is heated by a stream of condensing steam. The sketches of Fig. 13.29 represent this situation. The analysis of the second of these operations of Fig. 13.28 follows in a similar fashion. For this first case the three terms of equation (13.1) become

$$\dot{q} = \dot{m}_g \lambda_g = W_l C_l \frac{dT_l}{dt} = UA(T_g - T_l) \qquad (13.37)$$

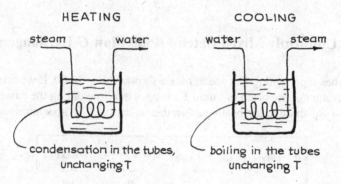

Fig. 13.28   Heating and cooling of a batch of liquid

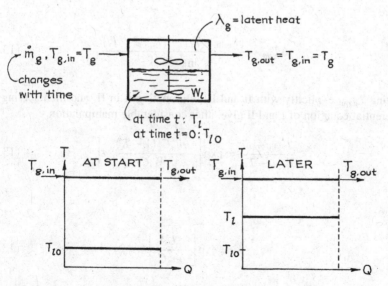

Fig. 13.29   Heating a batch of liquid with condensing vapor

Here there are just two variables in II and III. So combining and integrating give

$$\ln\left[\frac{T_g - T_l}{T_g - T_{l0}}\right] = -\frac{UA}{W_l C_l} \cdot t \quad \text{or} \quad \frac{T_g - T_l}{T_g - T_{l0}} = \exp\left[-\frac{UA}{W_l C_l} \cdot t\right] \qquad (13.38)$$

The quantity of liquid condensed W_g until any time t is given by a heat balance. Thus,

$$W_g \lambda_g = W_l C_l (T_l - T_{l0}); \text{ where} \begin{cases} \lambda_g > 0 \text{ for boiling} \\ \lambda_g < 0 \text{ for condensation} \end{cases} \qquad (13.39)$$

The temperature–time diagram for this case is very much like the figure for the previous case, Sect. 13.8 of this chapter.

13.10   Uniformly Mixed Batch *L*/Plug Flow *G* Exchangers

The sketches representing this operation are shown in Fig. 13.30. If we assume that the temperature of the batch of liquid T_l changes but little during the passage of an element of gas through the heating coils, then equation (13.1) becomes

$$\dot{q} = \dot{m}_g C_g \left(T_{g,\text{in}} - T_{g,\text{out}}\right) = W_l C_l \frac{dT_l}{dt} = UA\Delta T_{lm} \qquad (13.40)$$

$$\underbrace{\phantom{\dot{q} = \dot{m}_g C_g \left(T_{g,\text{in}} - T_{g,\text{out}}\right)}}_{\text{I}} \quad \underbrace{\phantom{W_l C_l \frac{dT_l}{dt}}}_{\text{II}} \quad \underbrace{\phantom{UA\Delta T_{lm}}}_{\text{III}}$$

where

$$\Delta T_{lm} = \frac{\left(T_{g,\text{in}} - T_l\right) - \left(T_{g,\text{out}} - T_l\right)}{\ln\frac{T_{g,\text{in}} - T_l}{T_{g,\text{out}} - T_l}} \qquad (13.41)$$

Writing $T_{g,\text{out}}$ explicitly with II and III, eliminating it in II, and then solving the differential equation of I and II give, after considerable manipulation,

$$\frac{T_{g,\text{in}} - T_l}{T_{g,\text{in}} - T_{l0}} = \exp\left[-\frac{\dot{m}_g C_g (1 - K)}{W_l C_l} \cdot t\right] \qquad (13.42)$$

where

$$K = \exp\left[-\frac{UA}{\dot{m}_g C_g}\right] \qquad (13.43)$$

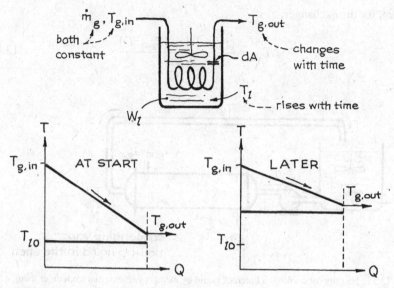

Fig. 13.30 Heating a batch of liquid with plug flow of hot gas

The temperature of exit gas when the liquid is at temperature T_l is then found to be

$$T_{g,\text{out}} = T_l + K\left(T_{g,\text{in}} - T_l\right) \tag{13.44}$$

Again, the graphical representation is similar to that of Sect. 13.8 of this chapter.

13.11 Uniformly Mixed Batch *L*/External Exchanger with Isothermal *G*

For this arrangement the sketch with nomenclature is shown in Fig. 13.31. A heat balance and rate expression for exchanger and vat give

$$\dot{q} = \underbrace{m_l C_l\left(T_l' - T_l\right)}_{\text{I}} = \underbrace{W_l C_l \frac{dT_l}{dt}}_{\text{II}} = \underbrace{UA\Delta T_{lm}}_{\text{III}} \tag{13.45}$$

I	II	III
Heat entering the vat by recirculating liquid	Heat accumulation in the vat	Transfer rate in the external exchanger

where, for the exchanger,

$$\Delta T_{lm} = \frac{(T_g - T_l) - (T_g - T_l')}{\ln \frac{T_g - T_l}{T_g - T_l'}}$$ (13.46)

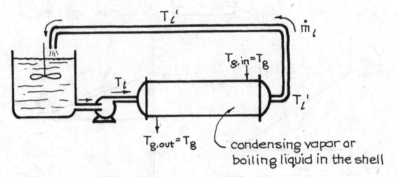

Fig. 13.31 Heating (or cooling) a batch of liquid by means of an external exchanger using a flow of condensing gas (or boiling liquid)

Combining II and III gives T_l'; replacing in I and then solving I and II give

$$\frac{T_g - T_l}{T_g - T_{l0}} = \exp\left[-\frac{\dot{m}_l(1 - K)}{W_l} \cdot t\right]$$ (13.47)

where

$$K = \exp\left[-\frac{UA}{\dot{m}_l C_l}\right]$$ (13.48)

13.12 Uniformly Mixed Batch *L*/External Shell and Tube Exchanger

Here the nomenclature is as shown in Fig. 13.32. The heat balances then become

$$\dot{q} = \dot{m}_l C_l (T_l' - T_l) = W_l C_l \frac{dT_l}{dt} = -\dot{m}_g C_g (T_{g,\text{out}} - T_{g,\text{in}}) = UA\Delta T_{lm}\mathscr{F}$$

I	II	III	IV
Heat	Accumulation	Heat loss	Transfer
added	in vat	by gas	rate
to vat			

(13.49)

where

$$\Delta T_{lm} = \frac{(T_{g,\text{in}} - T_l') - (T_{g,\text{out}} - T_l)}{\ln \frac{T_{g,\text{in}} - T_l'}{T_{g,\text{out}} - T_l}} \tag{13.50}$$

And $\mathscr{F}$ ...

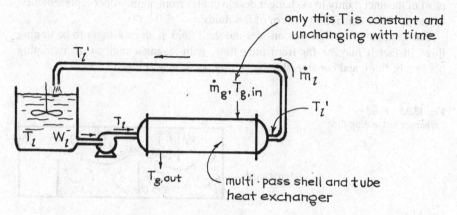

Fig. 13.32 Heating a batch of liquid by means of an external multipass shell and tube exchanger

and where $\mathscr{F}$ is the correction factor for multipass shell and tube exchangers.

These equations contain four variables: T_l, T_l', $T_{g,\text{out}}$, and t. First, eliminate two of them, T_l' and $T_{g,\text{out}}$, from ΔT_{lm}. Thus, from I and II, we write

$$T_l' = T_l + \frac{W_l C_l}{\dot{m}_l C_l} \frac{dT_l}{dt} \tag{13.51}$$

Then, from I and III

$$T_{g,\text{out}} = T_{g,\text{in}} - \frac{W_l C_l}{\dot{m}_g C_g} \frac{dT_l}{dt} \tag{13.52}$$

Replacing equations (13.51) and (13.52) into I and IV and integrating give, after considerable manipulation,

$$\frac{T_{g,\text{in}} - T_l}{T_{g,\text{in}} - T_{l0}} = \exp\left[-\frac{1 - K}{W_l}\left(\frac{\dot{m}_g \dot{m}_l C_g}{\dot{m}_l C_l - K\dot{m}_g C_g}\right)t\right] \tag{13.53}$$

where

$$K = \exp\left[-UA\mathscr{F}\left(\frac{1}{\dot{m}_g C_g} - \frac{1}{\dot{m}_l C_l}\right)\right] \tag{13.54}$$

Again, the temperature–time curve for this operation is similar in shape to the curve of Sect. 13.8 of this chapter.

13.13 Final Comments

By far, most of the research in heat transfer aims at finding accurate and reliable values for the heat transfer coefficient h for a variety of situations. Often, however, most of the uncertainty in exchanger design comes from an improper representation of the flow and contacting pattern of the fluids.

For example, in design we consider the shell fluid in an exchanger to be in plug flow. In fact it may be far from plug flow, with stagnant regions, recirculating pockets of fluid, and considerable bypassing, as shown in Fig. 13.33.

Fig. 13.33 Fluid in exchanger not in plug flow

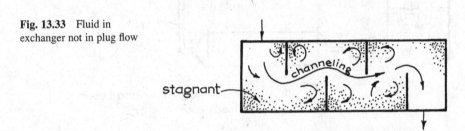

On the tube side, the fluid has a choice of flow path among the many parallel tubes in the bundle. Slightly different frictional resistance from tube to tube in clean exchangers will lead to different flow rates of fluid, different tube temperatures, and different buildup rates of scale in these tubes. The final result can be a very wide distribution of residence times of fluid in the tubes, hence large deviation from plug flow.

The $\mathscr{F}$ charts, all based on ideal contacting, can thus be too optimistic. This whole question needs to be carefully looked into. Also, in practical design we must try our best to minimize these deviations from plug flow.

Another all but ignored phenomenon, but one which could lead to a substantial error in prediction, concerns the transport of heat from hot to cold regions of the exchanger by the metal walls themselves. To illustrate this phenomenon, sketch (a) of Fig. 13.34 shows the temperature distribution in an ideal countercurrent exchanger. But heat transport along the metal walls will give a more uniform temperature in the metal and thus distort the temperature profile in the fluid, somewhat as shown in sketch (b). The overall effect of this phenomenon is to lower the exchanger efficiency $\mathscr{F}$, the extreme being to cause the efficiency of any shell and tube exchanger to approach that of the stirred vat of Sect. 13.6, above. Jakob et al. (1957) considers and presents an analysis of this phenomenon.

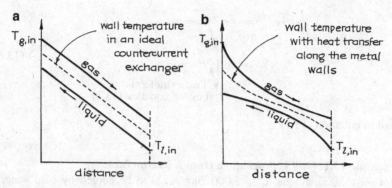

Fig. 13.34 Heat transfer along the metal walls of a recuperator will distort the temperature profile and lower the contact efficiency $\mathscr{F}$

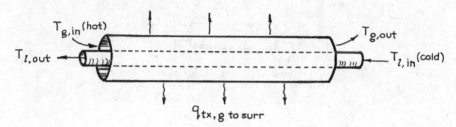

Fig. 13.35

Table 13,1 Range of U values in recuperators[a]		
Type of recuperator	U, W/m² K	
Gas to gas	10–35	
Water to water	850–1,700	
Oil to oil	100–300	
Condensing vapor to air (steam pipes in air)	35–90	
Condensing steam to boiling organic	280–2,300	
Condensing steam to boiling water	1,700–4,500	

[a]From Perry (1950). Pages 480–482 give numerous additional values

In this chapter we have only looked at some of the many possible contacting patterns. Equations and $\mathscr{F}$ charts for other contacting patterns may be found in specialized books on the subject.

When heat losses to the surroundings are to be considered, the governing expression, equation (13.1), has to be modified somewhat. For example, in the double-pipe exchanger with heat loss to the surroundings, as shown in Fig. 13.35, the basic equations become

Finally, the overall heat transfer coefficients U have orders of magnitude as shown in Table 13.1.

$$q_{\text{lost by } g} = q_{\text{gained by } l} + q_{\text{gained by surr}}$$

$$\parallel \qquad\qquad \parallel$$

$$q_{tx, g \, to \, l} \qquad q_{tx, g \, to \, surr}$$

Two extra terms
beyond equation (13.1)

(13.55)

instead of equation (13.1).

Example 13.1. Exit Temperature from a Recuperator

Hot oil (150°C, 1 kg/s, $C_p = 2{,}000$ J/kg K) is to be cooled by cold water (25°C, 1.026 kg/s, $C_p = 4{,}200$ J/kg K) in a countercurrent concentric tube heat exchanger ($A = 4.87$ m², $U = 500$ W/m² K). Find the temperature of the leaving oil.

Solution

Calling the hot fluid g and the cold fluid l, we have from equations (13.7) and (13.8)

$$\frac{T_{g,\text{in}} - T_{g,\text{out}}}{T_{g,\text{in}} - T_{l,\text{in}}} = \frac{1 - \exp\left[-UA\left(\frac{1}{\dot{m}_g C_g} - \frac{1}{\dot{m}_l C_l}\right)\right]}{1 - \frac{\dot{m}_g C_g}{\dot{m}_l C_l}\exp\left[-UA\left(\frac{1}{\dot{m}_g C_g} - \frac{1}{\dot{m}_l C_l}\right)\right]}$$

Replacing values gives

$$\frac{150 - T_{g,\text{out}}}{150 - 25}$$

$$= \frac{1 - \exp\left[-(500)(4.87)\left(\dfrac{1}{(1)(2{,}000)} - \dfrac{1}{(1.026)(4{,}200)}\right)\right]}{1 - \dfrac{2{,}000}{(1.026)(4{,}200)}\exp\left[-(500)(4.87)\left(\dfrac{1}{(1)(2{,}000)} - \dfrac{1}{(1.026)(4{,}200)}\right)\right]}$$

(continued)

(continued)

from which the temperature of the leaving cooled oil is

$$T_{g,\text{out}} = 71\,°C$$

Note: See example 8.10 of Todd and Ellis (1982) for a quite different approach to this problem.

Example 13.2. Heating a Batch of Liquid

Thirty cubic meters of liquid benzene in a storage tank is required to be under pressure and at 150 °C for a batch extraction. The storage temperature is 0 °C. A pump connected to the tank is able to circulate 4.8 kg/s of benzene to the tube side of a 1-2 shell and tube heat exchanger ($U = 240$ W/m^2 K, $A = 40$ m^2), while heating gas at 2.0 kg/s and 200 °C passes through the shell side of the exchanger.

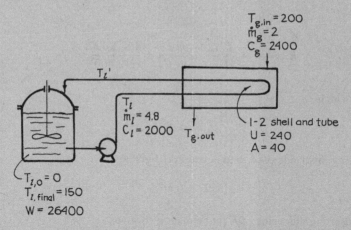

How long will it take to heat the stirred batch of benzene using this external heat exchanger?

Data:

$$\rho_{\text{benzene}} = 880 \text{ kg/m}^3, \ C_p(\text{benzene}) = 2{,}000 \text{ J/kg K}$$
$$C_p(\text{heating gas}) = 2{,}400 \text{ J/kg K}$$

Solution

This setup represents a mixed batch liquid/external shell and tube exchanger which is treated in Sect. 13.12. Thus, equations (13.53) and (13.54) apply. With values given these equations become

$$\frac{200 - 150}{200 - 0} = \exp\left[-\frac{1-K}{26,400} \left\{ \frac{2(4.8)(2,400)}{4.8(2,000) - K(2)(2,400)} \right\} t \right] \qquad \text{(i)}$$

where

$$K = \exp\left\{ -240(40)\mathscr{F} \left[\frac{1}{2(2,400)} - \frac{1}{4.8(2,000)} \right] \right\} \qquad \text{(ii)}$$

Now evaluate $\mathscr{F}$ for the 1–2 exchanger. Referring to Fig. 13.5

$$\text{NTU} = \frac{UA}{\dot{m}_l C_l} = \frac{240(40)}{4.8(2,000)} = 1$$

and

$$R = \frac{T_{g,\text{in}} - T_{g,\text{out}}}{T_l' - T_l} = \frac{\dot{m}_l C_l}{\dot{m}_g C_g} = \frac{4.8(2,000)}{2(2,400)} = 2$$

from which

$$\mathscr{F} = 0.76$$

Inserting this value of $\mathscr{F}$ into equation (ii) gives

$$K = 0.4677$$

and inserting this value of K into equation (i) gives

$$t = 21{,}948 \text{ s} \cong 6 \text{ h } 6 \text{ min}$$

This is the time needed to heat the batch of benzene from 0 °C to 150 °C.

Problems on the Design of Recuperators

13.1 Which of the pair of schemes shown on the opposite page do you think is better in the sense that it has a larger $\mathscr{F}$ value? Use qualitative arguments

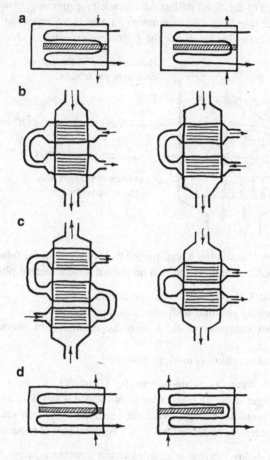

only, give your reasons, and remember that:

- Approach to countercurrent plug flow is desired.
- It is more important to improve the end of the exchanger where ΔT is smaller because it contributes more to the exchanger size.

13.2 The showers in the changing rooms for our gymnasium will need a continuous supply of 1 t/min of hot water. One way of doing this is by heating city water from 10 °C to 90 °C in a 1–3 pass heat exchanger where water flows through the tubes while waste saturated steam at 1 atm condenses in the shell, leaving as liquid at 100 °C. If $U = 1,500$ W/m^2 K, find:

(a) The area of exchanger needed
(b) The amount of steam needed (in kg/h)

Data:

$$\lambda_{condensation} = 2.29 \times 10^6 J/kg, \; C_{p,\,water} = 4,184 \; J/kg \; K$$

13.3 In the design of fluidized boilers for electricity-generating power stations, the heat transfer rate from bed to immersed tubes is an important design parameter. To estimate this let us run the following experiment. Joe Shlemiel

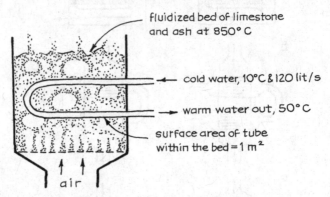

calculates U assuming a well-mixed flow of water in the tube. You feel that this is a poor assumption and that he should have assumed plug flow.

(a) What did J. S. find with his assumption?
(b) What would you find with your assumption?
(c) What percentage error did J. S. make assuming, of course, that you are right?
(d) Which assumption is more reasonable?

13.4 Clean cold water is heated from 0 °C to 60 °C in a countercurrent heat exchanger using an equal flow of hot waste water which enters at 90 °C. This 60 °C is not hot enough. What must be done to the flow rate of hot water if the cold water is to be heated to 70 °C? Assume that U remains unchanged.

13.5 The figure below shows a countercurrent concentric pipe heat exchanger ($A = 8 \, m^3$) for water with a rather unusual piping arrangement. Also shown are all the inlet flow rates and inlet temperatures. If $U = 523 \; W/m^2 \, K$, evaluate the temperatures T_3 and T_4. Take $C_{p,\,H_2O} = 4,184 \; J/kg \; K$ at all temperatures.

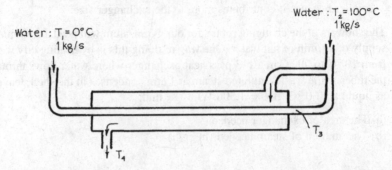

Hot oil A (300 °C) is cooled with fluid B (0 °C) in a double-pipe heat exchanger and both fluids leave at 200 °C. That is not good enough, so consider the following modifications. Find the outlet temperature of oil A:

13.6 If we add a second identical exchanger in series with the first, in effect doubling the exchanger area.

13.7 If we double the flow rate of coolant. Assume no change in heat transfer coefficient.

13.8 A certain process generates steam at atmospheric pressure and when super-heated to 600 °C. It is desired to recover the heat content of this steam by countercurrent heat exchange with cold dry air entering at 0 °C.

(a) Construct a plot of Q versus T for steam at atmospheric pressure from 0 °C to 600 °C, expressing Q in J/mol.
(b) With the aid of this plot, calculate the largest quantity of cold air which can be heated to 600 °C using perfect countercurrent heat exchange (mol air/mol steam).
(c) What percent of the total heat in the original steam is recovered by the process of part (b)?
(d) Calculate the smallest number of moles of air which by countercurrent flow can extract all the heat from 1 mol of steam entering the exchanger; in effect, cool this stream down to 0 °C.
(e) To what temperature will the air in (d) be heated?

Data: From thermodynamic tables,

$$C_{p,\text{air}} = 29.29 \text{ J/mol K}$$
$$H(\text{steam, 600 °C}) = 3,704.7 \times 10^3 \text{ J/kg}$$
$$H(\text{steam, 100 °C}) = 2,675.5 \times 10^3 \text{ J/kg}$$
$$H(\text{water, 100 °C}) = 4,17.5 \times 10^3 \text{ J/kg}$$
$$H(\text{water, 0 °C}) = 0$$

13.9 A water-cooled countercurrent flow exchanger in a refrigeration plant is used for condensing ammonia. Superheated ammonia enters the condenser at 50 °C and condenses at 25 °C, and liquid NH_3 is then cooled to 6 °C before leaving the exchanger. Cooling water enters at 5 °C and leaves at 15 °C, flowing at 20 kg/min. Calculate the condenser area needed if

$$U(\text{between } NH_3 \text{ vapor and } H_2O) = 60 \text{ W/m}^2 \text{ K}$$
$$U(\text{between condensing } NH_3 \text{ and } H_2O) = 360 \text{ W/m}^2 \text{ K}$$
$$U(\text{between liquid } NH_3 \text{ and } H_2O) = 120 \text{ W/m}^2 \text{ K}$$

Data: From thermodynamic tables,

$$H(NH_3 \text{ vapor at } 50\,°C) = 1,537.7 \times 10^3 \text{ J/kg}$$
$$H(\text{saturated } NH_3 \text{ vapor, } 25\,°C) = 1,465.0 \times 10^3 \text{ J/kg}$$
$$H(\text{boiling } NH_3 \text{ liquid, } 25\,°C) = 298.8 \times 10^3 \text{ J/kg}$$
$$H(NH_3 \text{ liquid at } 6\,°C) = 208.9 \times 10^3 \text{ J/kg}$$
$$H(H_2O \text{ at } 5\,°C) = 21,000 \text{ J/kg}$$
$$H(H_2O \text{ at } 15\,°C) = 63,000 \text{ J/kg}$$

Note: This question is similar to the one asked in the Professional Engineers examination, given by the state of Oregon in the 1950s, except that they did not give the enthalpies.

13.10 For the end conditions shown below, compare the heat exchanger efficiency of the following contacting schemes with respect to countercurrent plug flow. In effect determine $\mathscr{F}$ for each scheme:

(a) Cocurrent flow
(b) 1–2 shell and tube exchanger
(c) 2–4 shell and tube exchanger
(d) Crossflow, unmixed exchanger
(e) Crossflow, mixed gas, unmixed liquid exchanger
(f) Crossflow, unmixed gas, mixed liquid exchanger

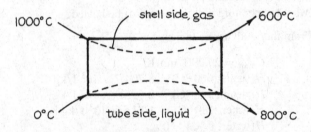

13.11 A steam of cold air at 0 °C is heated to 800 °C in a 1–2 shell and tube heat exchanger using 2½ times as much hot air entering at 1,000 °C. Which of the following two piping arrangements would result in a smaller exchanger area, to have the hot gas flow through the shell side or through the tube side? Show calculations to support your answer.

13.12 Light lubricating oil at 110 °C is cooled with brackish water at 10 °C in a 3–6 shell and tube heat exchanger with oil in the shell side. When the exchanger is clean, the oil leaves at 35 °C and the water leaves at 60 °C. With time, however, scale forms in the tubes and the efficiency of heat transfer becomes lower. Today the exchanger is quite dirty and the oil leaves at 50 °C. Find $U_{\text{dirty}}/U_{\text{clean}}$ and $\mathscr{F}_{\text{dirty}}/\mathscr{F}_{\text{clean}}$. (problem prepared by Carol Magnusson)

An oil (150 °C, 10 kg/s, $C_p = 3{,}750$ J/kg K) is to be cooled to as low a temperature as possible using cooling water (35 °C, 22.5 kg/s).

13.13 In a 1–2 shell and tube exchanger ($U = 750$ W/m² K, $A = 100$ m²), should oil be in the shell or the tube side? Determine the exit oil temperature for these two alternatives.

13.14 We have available both a 1–2 and a 4–8 exchanger. Both have $U = 750$ W/m² K and $A = 100$ m². With oil in the shell side, which exchanger gives a cooler leaving oil? Evaluate these temperatures.

13.15 Determine the length of bayonet exchangers needed in a 21-tube bundle which is to be used to heat 12.5 t/h of cold fluid ($C_p = 2$ kJ/kg K) from 0 °C to 40 °C by shell fluid ($C_p = 4$ kJ/kg K) cooling from 120 °C to 40 °C. For the inner tube $d = 3$ cm, $A_i = 7$ cm², and $U = 1{,}000$ W/m² K. For the outer tube $d = 5.2$ cm, $A_o = 14$ cm², and $U = 500$ W/m² K.

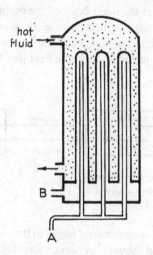

hot
Fluid

B

A

(a) Assume that cold fluid enters at A (see above).
(b) Assume that cold fluid enters at B (see above).
(c) Which routing is better?

13.16 Whales, dolphins, and porpoises are able to maintain surprisingly high body temperatures even though they are immersed continuously in very cold water. Since the extremities of these animals (tails, fins, flukes) have a large surface, much of the heat loss occurs there.

(a) Now an engineering student designing a dolphin from first principles would probably view the heat loss from a flipper somewhat as shown here:

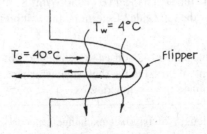

Let us suppose that blood at 40 °C enters a flipper at 0.3 kg/s, feeds the flipper, is cooled somewhat, and then returns to the main part of the body. The dolphin swims in 4 °C water; the flipper area is 3 m² and $U = 418.4$ W/m² K between the flipper and water. At what temperature does blood reenter the main body of the dolphin?

(b) Frankly, an ordinary engineering student would design a lousy dolphin. Let us try to do better; in fact, let us learn from nature and see whether we can reduce some of the undesirable heat loss by transferring heat

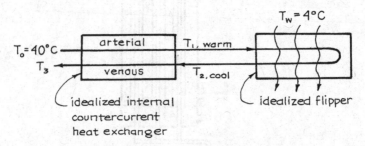

from the outgoing warm arterial blood to the cooled venous blood. Such a scheme is idealized above. Assume for this internal exchanger that $A = 2.4$ m² and $U = 523$ W/m² K. With this extra exchanger find the temperature of the blood returning to the main part of the body and, in addition, the fraction of original heat loss which is saved.

Data: Approximate the properties of blood by water.

A continuous stream of 6 t/h of cold water is to be heated from 0 °C to 95 °C using a stream of hot steam at 260 °C and 1 atm which cools, condenses, and leaves as liquid at very close to 100 °C. Find the exchanger area needed and the hourly steam rate:

13.17 For countercurrent contacting

13.18 If the water is heated in a big stirred vat containing a long immersed coil for the cooling and condensing steam

13.19 If the steam bubbles and condenses into its own condensate in a big vat, while the water to be heated passes through a long coiled tube immersed in the hot condensate

13.20 If steam bubbles and condenses in its condensate in a jacketed vat, while the water to be heated passes in mixed flow through the jacket of the vat
   Data: Use the following approximate values:

$$H(\text{water 0 °C}) = 0$$
$$H(\text{water, 95 °C}) = 400 \times 10^3 \text{ J/kg}$$
$$H(\text{water, 100 °C}) = 420 \times 10^3 \text{ J/kg}$$
$$H(\text{steam, 100 °C}) = 2{,}680 \times 10^3 \text{ J/kg}$$
$$H(\text{steam, 260 °C}) = 3{,}000 \times 10^3 \text{ J/kg}$$
$$U_{\text{condensing section}} = 2{,}000 \text{ W/m}^2 \text{ K}$$
$$U_{\text{steam-water}} = 500 \text{ W/m}^2 \text{ K}$$
$$U_{\text{water-water section}} = 1{,}275 \text{ W/m}^2 \text{ K}$$

13.21 Repeat example 13.2 with the following changes: the liquid flow rate through the exchanger is tripled; the gas flow rate is quadrupled, in which case the overall heat transfer coefficient doubles.

13.22 Repeat example 13.2 with just one change: the hot gas used to heat the benzene enters the exchanger at saturated conditions at 200 °C, condenses there, and leaves at 200 °C. In this situation U becomes 600 W/m^2 K. Now how long does it take to heat the benzene?

My mountain cabin sits close to a mineral spring—smelly but hot. I can't use this water directly for baths, but I can use it to heat up fresh cold water in my unique bath tub—an old discarded well-insulated jacketed kettle from a plastic factory. Hot water at 67.5 °C enters the jacket at 34.4 lit/min, swirls about, and leaves. The tub is filled with 860 lit of fresh cold water for a family bath.

13.23 Find the time needed to heat this tubful of cold water from 5 °C to a comfortable toasty 55 °C.
   Data: Between hot and cold water: $U = 400$ W/m^2 K, $A = 3$ m^2. The outside of the vat is well insulated so assume that $U = 0$ to surroundings.

13.24 Find the time needed to heat this tubful of cold water from 5 °C to 55 °C if the insulation is torn off to expose an attractive copper surface.
   Data: Without insulation: $U = 240$ W/m^2 K, $A = 5$ m^2 to the surroundings which are at 9 °C.

13.25 We plan to cool one-ton batches of oil ($T = 300$ °C, $C_p = 3{,}000$ J/kg K) down to 120 °C in 2 h in a stirred vat by pumping nearly boiling water through coiled tubes which are immersed in the hot oil and withdrawing steam at roughly 100 °C.

(a) How much exchanger surface do we need?

(b) How much steam is generated during this 2-h process?

Data: For the cooling coil,

$$U = 1,000 \, \text{W/m}^2 \, \text{K}$$

To boil water

$$\lambda = 2.29 \times 10^6 \, \text{J/kg}$$

13.26 A one-ton batch of hot oil ($C_p = 3,000$ J/kg) is to be cooled from 300 °C to 120 °C in a stirred vat by passing cold water (0 °C, 1.0 kg/s) in plug flow through heat exchanger coils immersed in the hot oil.

(a) Find how long it will take to cool the oil.

(b) What is the temperature of the water leaving the coils at the beginning and at the end of the run?

Data: For the exchanger coil,

$$U = 1000 \, \text{W/m}^2 \, \text{K}, \quad A = 1.5 \, \text{m}^2$$

For water

$$C_p = 4184 \, \text{J/kg} \, \text{K}$$

13.27 We wish to design an adiabatic heat exchanger to cool 100 t/h of hot water from 90 °C to 50.5 °C by contacting with an equal amount of cold water at 10 °C. The velocities are such that $U_{\text{overall}} = 600 \, \text{W/m}^2 \cdot \text{K}$. Calculate the area of exchanger surface needed for each of the following contacting patterns:

(a) Parallel flow (cocurrent plug flow)

(b) Countercurrent flow

(c) 1–2 shell and tube exchanger

(d) rossflow exchangers having 1 shell and 1 tube pass

References and Related Readings

B.S. Baclic, A simplified formula for cross flow for heat exchanger effectiveness. ASME J. Heat Transf. **100**, 746 (1978)

R.A. Bowman, A.C. Mueller, W.M. Nagle, Mean temperature difference in design. Trans. ASME **62**, 283 (1940)

V. Cavesano (ed.), *Process Heat Exchange*, Sec. 1 (McGraw-Hill, New York, 1979). Presents the collective wisdom and practical art of real live designers of recuperators

N.L. Hurd, Mean temperature difference in the field or bayonet tube. Ind. Eng. Chem. **38**, 1266 (1946)

M. Jakob, *Heat Transfer*, Chapter 34 (Wiley, New York, 1957)

W.A. Kays, A.L. London, *Compact Heat Exchangers*, 2nd edn. (McGraw-Hill, New York, 1964)

D.Q. Kern, *Process Heat Transfer* (McGraw-Hill, New York, 1950). An excellent primary reference to recuperators of all types

A.C. Mueller, in *Handbook of Heat Transfer*, ed. by W.M. Roshenow, J.P. Hartnett, Sec. 18, (McGraw-Hill, New York, 1973)

J.H. Perry, *Chemical Engineers' Handbook*, 3rd edn. (McGraw-Hill, New York, 1950)

TEMA, *Standards of Tubular Exchanger Manufacturers Association*, 6th edn. (Tubular Exchanger Manufacturers Association, Inc, New York, 1978)

J.P. Todd, H.B. Ellis, *Applied Heat Transfer* (Harper & Row, New York, 1982), p. 430

R. Turton, C.D. Ferguson, O. Levenspiel, Performance and design charts for heat exchangers. J. Heat Transf. **106**, 893 (1984). *Chem. Eng.* **93**, 81, Aug. 18, 1986

Chapter 14
Direct-Contact Gas–Solid Nonstoring Exchangers

In this chapter, we first take up the fluidized bed heat exchanger and then various other devices such as moving grates and moving bed exchangers.

14.1 Fluidized Bed Heat Exchangers: Preliminary Considerations

Consider the fluidized bed of Fig. 14.1 wherein a stream of cold particles and a stream of hot gas enter, exchange heat, and then leave. Let us examine some of the properties of this stream of solids. From these findings we will be in a position to choose a reasonable set of assumptions to represent the gas–solid fluidized bed exchanger.

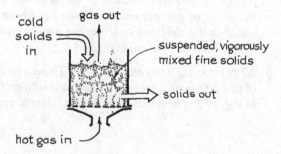

Fig. 14.1 Solids heating in a fluidized bed

1. ΔT *within single particles.* Suppose a particle, initially at T_{s0}, suddenly finds itself in surroundings at T_g. Let us ask how long it would take for the particle to approach the temperature of its surroundings. This we call the thermal relaxation time of the particle.

© Springer Science+Business Media New York 2014
O. Levenspiel, *Engineering Flow and Heat Exchange*,
DOI 10.1007/978-1-4899-7454-9_14

As an estimate of the relaxation time, let us determine the time needed for a 90 % approach to equilibrium conditions, of particles of various sizes moving through 100 °C air at 1 m/s. Following the calculation method of Example 11.1, we obtain the values of Table 14.1. Clearly, for particles about $d_p = 1$ mm or smaller, the relaxation time is very small, certainly an order of magnitude smaller than the mean residence time of solids in the vessel.

Table 14.1 Thermal relaxation times for spherical particles passing through air at 1 m/s

	Relaxation time, t			
	$d_p = 100\,\mu$m	$d_p = 1$ mm	$d_p = 10$ mm	$d_p = 100$ mm
Copper	130 ms[a]	7.5 s[a]	320 s[a]	190 min[a]
	(0.03 ms)[b]	(0.003 s)[b]	(0.3 s)[b]	(0.5 min)[b]
Iron	140 ms[a]	7.8 s[a]	330 s[a]	190 min[a]
Stainless steel	140 ms[a]	7.8 s[a]	330 s[a]	190 min[a]
Glass	64 ms[a]	3.7 s[a]	150 s[a]	110 min
Sand	80 ms[a]	4.5 s[a]	240 s	160 min
PVC plastic	56 ms[a]	3.8 s	170 s	150 min
Wood	40 ms[a]	3.4 s	160 s	130 min
Ice	80 ms[a]	4.5 s[a]	190 s[a]	110 min[a]
Na–K (56 % Na)[c]	39 ms[a]	2.2 s[a]	92 s[a]	54 min[a]
Water[c]	160 ms[a]	9.2 s[a]	440 s	290 min

[a]Film resistance controls, Bi < 0.1. All other values in the table fall in the mixed control (particle conduction and film resistance) regime
[b]() Values assuming no film resistance. These values are greatly in error
[c]Assume no circulation in the particle. In the presence of circulation, the relaxation time is much reduced

2. ΔT *among particles.* Because of the large heat capacity of solids (about 1,000 times that of gas, per unit of volume) and because of the rapid circulation of solids in the fluidized bed, we can assume that the particles are uniform in temperature everywhere in the bed.

3. ΔT *between particles and exit gases.* To get a rough order of magnitude estimate of this ΔT, assume plug flow of gas and well-mixed solids in the bed. Referring to Fig. 14.2, a heat balance across the slice of bed of thickness dx then gives

$$-d\dot{q} = \dot{m}_g C_g dT_g = \rho_g u_0 A_t C_g dT_g = h A_{tx}(T_g - T_s)\, dx$$

Replacing values, rearranging, and integrating gives

$$\ln\frac{T_{g,\,\text{in}} - T_s}{T_{g,\,\text{out}} - T_s} = \frac{haL}{\rho_g u_0 C_g} = \frac{\text{Nu}_p}{\text{Pr}\cdot\text{Re}_p}\cdot\frac{6(1-\varepsilon_f)L}{d_p}$$

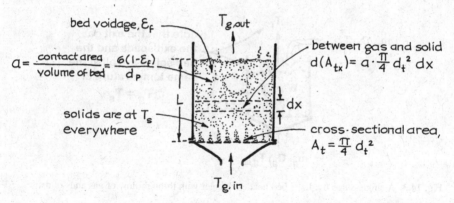

Fig. 14.2 Development of the heat balance between gas and solids in a fluidized bed

Now for a 95 % temperature approach of gas to the temperature of the particles

$$\ln \frac{T_{g,\,in} - T_s}{T_{g,\,out} - T_s} = \ln 20 \geq 3$$

and combining the above two expressions gives

$$\frac{L}{d_p} \geq 0.5 \frac{\mathrm{Pr} \cdot \mathrm{Re}_p}{\mathrm{Nu}_p (1 - \varepsilon_f)}$$

Typically for 0.5-mm particles we find

$$\frac{L}{d} > 25.6 \text{ or } L > 1.28 \text{ cm}$$

So, for beds deeper than 1 cm, we can reasonably assume that $T_{g,out} \cong T_s$, and consequently, for beds deeper than this, we can take $h \cong \infty$ between gas and solid or, more properly, $(h_{g-s}\,a) \cong \infty$.

4. To *summarize* the findings of this preliminary analysis, all particles in a fluidized bed are isothermal and at the same temperature. In addition, gas leaves at the temperature of the bed solids and $(h_{g-s}a) \cong \infty$. These then are the assumptions which we will use to characterize the behavior of gas–solid fluidized bed heat exchangers.

14.2   Mixed Flow *G*/Mixed Flow *S* or Single-Stage Fluidized Bed Exchangers

Consider the fluidized bed exchanger of Fig. 14.3. A heat balance over the whole unit gives

$$\boxed{-\dot{m}_g C_g \left(T_1 - T_{g,\,in} \right) = \dot{m}_s C_s \left(T_1 - T_{s,\,in} \right)} \tag{14.1}$$

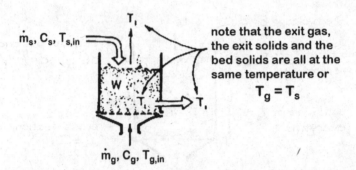

Fig. 14.3 A single-stage fluidized bed heat exchanger with through-flow of gas and solids

Note that no rate equation is used here. This is because we took $(h_{g\text{-}s}a) = \infty$. So on rearranging we find

$$T_1 = \frac{T_{s,\,in} + \phi T_{g,\,in}}{1 + \phi} \qquad (14.2)$$

where ϕ is the heat flow ratio of the two flowing streams, defined as

$$\phi = \frac{\dot{m}_g C_g}{\dot{m}_s C_s} = -\frac{\Delta T_s}{\Delta T_g} = \frac{W/\dot{m}_s}{(1 - \varepsilon_f)(L/u_0)} \cdot \frac{\rho_g C_g}{\rho_s C_s} \qquad (14.3)$$

The efficiencies of heat utilization are then, from equation (14.2),

$$\left. \begin{aligned} \eta_g &= \frac{\Delta T_g}{\Delta T_{max}} = \frac{T_{g,\,in} - T_1}{T_{g,\,in} - T_{s,\,in}} = \frac{1}{1 + \phi} \\ \eta_s &= \frac{\Delta T_s}{\Delta T_{max}} = \frac{T_1 - T_{s,\,in}}{T_{g,\,in} - T_{s,\,in}} = \frac{\phi}{1 + \phi} = 1 - \eta_g \end{aligned} \right\} \qquad (14.4)$$

Efficiencies of heat utilization are always low in single fluidized bed heat exchangers. For example if $\eta = 0.7$ for the gas, it will be 0.3 for the solid (always sum up to 1). To raise the efficiency, we multistage, either crossflow or counterflow. We consider these next.

14.3 Counterflow Stagewise Fluidized Bed Exchangers

Consider N beds of equal size, as shown in Fig. 14.4. A heat balance around each bed gives

$$\left. \begin{aligned} \text{Bed 1}: \quad & T_1 - T_{s,\,in} &=& \quad \phi(T_2 - T_1) \\ \text{Bed 2}: \quad & T_2 - T_1 &=& \quad \phi(T_3 - T_2) \\ & \vdots & \vdots & \qquad \vdots \\ \text{Bed } N: \quad & T_N - T_{N-1} &=& \quad \phi(T_{g,\,in} - T_N) \end{aligned} \right\} \qquad (14.5)$$

Fig. 14.4 A multistage counterflow fluidized bed heat exchanger

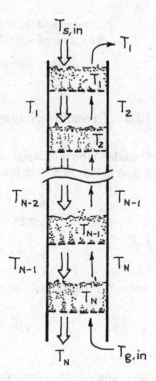

Combining equation (14.5) and eliminating intermediate temperatures give

$$\left.\begin{aligned}\eta_g &= \frac{T_{g,\,\text{in}} - T_1}{T_{g,\,\text{in}} - T_{s,\,\text{in}}} = \frac{\displaystyle\sum_{n=0}^{N-1}\phi^n}{\displaystyle\sum_{n=0}^{N}\phi^n} = \frac{1 + \phi + \phi^2 + \cdots \phi^{N-1}}{1 + \phi + \phi^2 + \cdots \phi^{N-1} + \phi^N}\\[2em]\eta_s &= \frac{T_N - T_{s,\,\text{in}}}{T_{g,\,\text{in}} - T_{s,\,\text{in}}} = \frac{\displaystyle\sum_{n=1}^{N}\phi^n}{\displaystyle\sum_{n=0}^{N}\phi^n} = \frac{\phi + \phi^2 + \cdots \phi^{N-1} + \phi^N}{1 + \phi + \phi^2 + \cdots \phi^{N-1} + \phi^N}\end{aligned}\right\} \quad (14.6)$$

Adjusting flow rates so that $\phi = 1$ (if one stream loses 100°, the other gains 100°) gives

$$\eta_g = \eta_s = \frac{N}{N+1} \quad (14.7)$$

For a large number of stages, $N \to \infty$, countercurrent plug flow is approached and

$$\left.\begin{array}{ll} \text{for equal heat flows, or } \phi \ = 1: & \eta_g = \eta_s = 1 \\ \text{for excess of solids, or } \phi < 1: & \eta_g = 1 \text{ and } \eta_s = \phi \\ \text{for excess of gas, or } \phi > 1: & \eta_g = \dfrac{1}{\phi} \text{ and } \eta_s = 1 \end{array}\right\} \quad (14.8)$$

14.4   Crossflow Stagewise Fluidized Bed Exchangers

Consider N beds of equal size having the same gas flow through each stage, as shown in Fig. 14.5. A heat balance about each bed gives

$$\left.\begin{array}{lll} \text{Bed 1:} & T_1 - T_{s,\text{ in}} = \phi'\left(T_{g,\text{ in}} - T_1\right) \\ \text{Bed 2:} & T_2 - T_1 \ = \phi'\left(T_{g,\text{ in}} - T_2\right) \\ \ \vdots & \quad \vdots & \quad \vdots \\ \text{Bed } N: & T_N - T_{N-1} \ = \phi'\left(T_{g,\text{ in}} - T_N\right) \end{array}\right\} \quad (14.9)$$

where ϕ' is based on the heat flow through each stage or

$$\phi' = \frac{(\dot{m}_g/N) \cdot C_g}{\dot{m}_s C_s} = \frac{\phi_{\text{countercurrent}}}{N} \quad (14.10)$$

Now with equal gas flows through each of the N stages, we have

$$T_{g,\text{ out}} = \frac{T_1 + T_2 + \cdots + T_N}{N} \quad (14.11)$$

Combining equations (14.9) and (14.11) gives the efficiencies of operation as

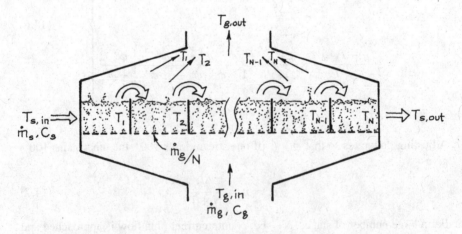

Fig. 14.5 A multistage crossflow fluidized bed heat exchanger

$$\left.\begin{aligned}
\eta_g &= \frac{\Delta T_g}{\Delta T_{\max}} = \frac{T_{g,\,\text{in}} - T_{g,\,\text{out}}}{T_{g,\,\text{in}} - T_{s,\,\text{in}}} = \frac{1}{N\phi'}\left[1 - \frac{1}{(1+\phi')^N}\right] \\
\eta_s &= \frac{\Delta T_s}{\Delta T_{\max}} = \frac{T_N - T_{s,\,\text{in}}}{T_{g,\,\text{in}} - T_{s,\,\text{in}}} = \left[1 - \frac{1}{(1+\phi')^N}\right] = N\phi'\eta_g
\end{aligned}\right\} \qquad (14.12)$$

Comparing crossflow and counterflow operations shows that for any number of stages, N counterflow has the advantage of being more efficient thermally. However, counterflow has the drawbacks of higher pressure drop, more hydraulic problems, especially with downcomers, and more mechanical design complications.

14.5   Countercurrent Plug Flow Exchangers

This contacting pattern, see Fig. 14.6, approximates moving bed and shaft kiln operations, and its Q versus T diagram is shown in Fig. 14.7.

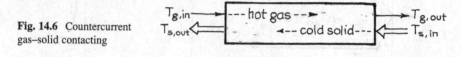

Fig. 14.6 Countercurrent gas–solid contacting

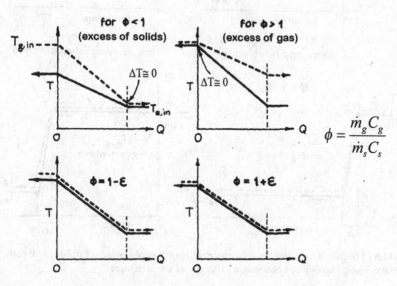

$$\phi = \frac{\dot{m}_g C_g}{\dot{m}_s C_s}$$

Fig. 14.7 Q vs. T diagram for countercurrent plug flow of gas and solid wherein $(h_{g-s}a) = \infty$

With a stream of small particles, the surface area of contact is large, and the particle relaxation time is short compared to the time of passage of particles through the exchanger (see Sect. 14.1 of this chapter). The behavior of such systems is equivalent to having a very large $(h_{g-s}a)$ value, approaching infinity. This means that all the heat exchange takes place in a very narrow zone of the exchanger. Consequently, the temperature versus distance diagram for this operation is as sketched in Fig. 14.8. Note that for close to equal heat flow rates, or $\phi \cong 1$, the location of the temperature front is uncertain. With a slight excess of solids ($\phi = 1 - \varepsilon$), the front slowly migrates to the gas inlet, point A in Fig. 14.8. With a slight excess of gas, it slowly migrates to the solid inlet, point B in Fig. 14.8.

Compare the corresponding sketches of Figs. 14.7 and 14.8 to satisfy yourself that they make sense.

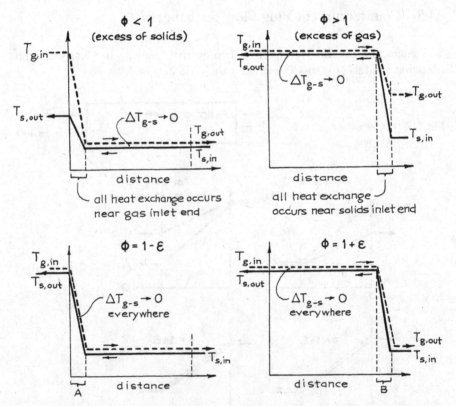

Fig. 14.8 Temperature profiles for countercurrent plug flow of gas and solid. For small particles, all heat exchange takes place near one or other end of the exchanger

A heat balance over the whole exchanger gives

$$\left.\begin{array}{ll} \text{for equal heat flows, or } \phi = 1: & \eta_g = \eta_s = 1 \\ \text{for an excess of solids, or } \phi < 1: & \eta_g = 1 \text{ and } \eta_s = \phi \\ \text{for an excess of gas, or } \phi > 1: & \eta_g = \dfrac{1}{\phi} \text{ and } \eta_s = 1 \end{array}\right\} \quad (14.13)$$

Comments

1. These expressions show that the stream with lower heat flow achieves 100 % heat utilization; the stream in excess does not.
2. For very small particles the heat exchange zone is very narrow, and the temperature of gas and solid is very close to each other nearly everywhere. For larger particles the heat exchange zone broadens, and for very large particles one may have to account for nonuniformity in temperature within particles. This type of problem can be treated by a direct extension of the analysis of Chap. 15.

14.6 Crossflow of Gas and Solids

The way we analyze this contacting pattern depends on the temperature distribution of the flowing solids. There are three extremes that we may want to consider. In all of these cases, we assume a relatively short relaxation time or $(h_{g-s}a) \to \infty$.

14.6.1 Well-Mixed Solids/Plug Flow Gas

The analysis is analogous to that in Sect. 13.V with UA replaced by ha. And for $ha \to \infty$, equations (13.29) and (13.30) reduce to

$$\left.\begin{array}{l} T_{g,\,out} = T_{s,\,out} = \dfrac{\dot{m}_g C_g T_{g,\,in} + \dot{m}_s C_s T_{s,\,in}}{\dot{m}_g C_g + \dot{m}_s C_s} \\[3mm] \eta_g = \dfrac{1}{\phi + 1} \\[3mm] \eta_s = \dfrac{\phi}{\phi + 1} \end{array}\right\} \quad (14.14)$$

Fluidized beds of large particles approximate this extreme.

14.6.2   Solids Mixed Laterally but Unmixed Along Flow Path/Plug Flow Gas

This extreme is representative of a thin stream of solids being contacted crosswise by gas, as shown in Fig. 14.9.

This contacting pattern is identical to the crossflow stagewise fluidized bed exchanger, treated in Sect. 14.4, but with an infinite number of stages. So letting $N \rightarrow \infty$ as $\phi' \rightarrow 0$ in equation (14.12) gives

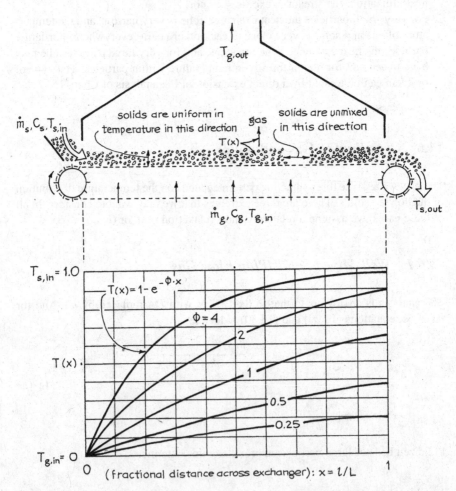

Fig. 14.9 Crossflow exchanger with a thin layer of solids and its temperature distribution at various gas–solid flow ratios, ϕ

$$\left. \begin{array}{l} \eta_g = \dfrac{\Delta T_g}{\Delta T_{g,\,max}} = \dfrac{1}{\phi}[1 - e^{-\phi}] \\[2mm] \eta_s = \dfrac{\Delta T_s}{\Delta T_{s,\,max}} = 1 - e^{-\phi} \end{array} \right\} \qquad (14.15)$$

14.6.3 Solids Unmixed/Plug Flow Gas

This extreme represents a thick stream of solids contacted crosswise by gas, as shown in Fig. 14.10. For plug flow of both hot gas and cold solids, a sharp temperature front exists as shown in Fig. 14.11; thus, the efficiency of heat utilization is

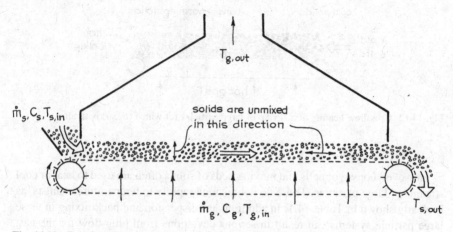

Fig. 14.10 Crossflow exchanger with a thick layer of solids (unmixed vertically)

$$\left. \begin{array}{l} \text{for excess of solids,}\ \phi < 1: \quad \eta_g = 1 \text{ and } \eta_s = \phi \\[2mm] \text{for excess of gas,}\ \phi > 1: \quad \eta_g = \dfrac{1}{\phi} \text{ and } \eta_s = 1 \end{array} \right\} \qquad (14.16)$$

14.7 Comments

The whole treatment of this chapter assumes a short temperature relaxation time for the solids. This assumption is quite reasonable for fluidized beds of fine particles with their very large surface-to-volume ratios.

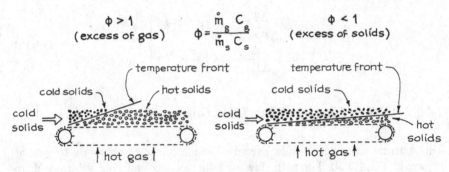

Fig. 14.11 Temperature distribution in a crossflow exchanger with a thick layer of solids and $(h_{g-s}a) = \infty$

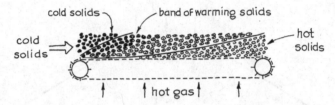

Fig. 14.12 Crossflow heating of a stream of large particles for which $(h_{g-s}a)$ is small

However, conveyor belts and moving beds of solids often are used to heat or cool streams of large particles. These do not respond rapidly to temperature changes, as is clearly shown in Table 14.1. In addition, gas dispersion and backmixing in these large particle systems can result in serious deviations from plug flow for the gas. Both these factors cause a blurring and broadening of the temperature front in these exchangers, as sketched in Fig. 14.12.

The extent of this broadening depends on the temperature relaxation time of the solids and the extent of gas dispersion compared to the residence time of gas and solid in the gas–solid exchanger. This effect is strongly affected by an increase in particle size. The analysis of this situation is not easy but can be developed following the analysis of Chap. 15.

Example 14.1 Counterflow Multistage Fluidized Bed Exchanger
We plan to cool a continuous stream of hot solids from 820 °C to 220 °C with cold gas at 20 °C in staged counterflow fluidized beds.

(a) Determine the number of stages needed; see drawing below.
(b) Find the temperature of the flowing streams in the exchanger.

(continued)

(continued)

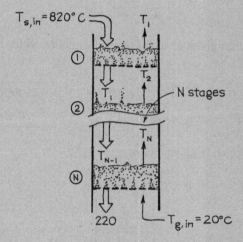

$T_{s,in} = 820°C$

T_1

①

T_2

T_1

N stages

②

T_N

T_{N-1}

Ⓝ

220

$T_{g,in} = 20°C$

Data. Adjust the flow rates of gas and solid so as to obtain the same thermal utilization for the two streams, or $\eta_g = \eta_s$, or $\phi = 1$.

Solution

(a) From the information for the stream of solids:

$$\eta_s = \frac{\Delta T_s}{\Delta T_{s,max}} = \frac{220 - 820}{20 - 820} = 0.75$$

For equal thermal utilization, $\eta_s = \eta_g$, equation (14.7) gives

$$0.75 = \frac{N}{N+1}$$

Thus, the number of stages required is

$$\boxed{N = 3}$$

(b) For the gas stream:

$$\eta_g = 0.75 = \frac{\Delta T_g}{\Delta T_{g,\,max}} = \frac{T_1 - 20}{820 - 20}$$

(continued)

Hence, the exit gas temperature

$$T_1 = 620°C$$

Equation (14.5) then gives the temperature distribution within the exchanger. Thus, with $\phi = 1$,

$$620 - 820 = 1\,(T_2 - 620) \quad \text{or} \quad T_2 = \boxed{420°C}$$
$$420 - 620 = 1\,(T_3 - 420) \quad \text{or} \quad T_3 = \boxed{220°C}$$

Thus, the final sketch is shown below.

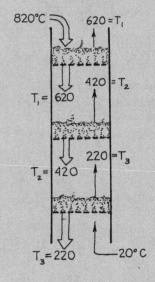

Example 14.2 Crossflow Multistage Fluidized Bed Exchanger

Repeat Example 14.1 with just one change—use a crossflow rather than a counterflow exchanger; see below.

(continued)

(continued)

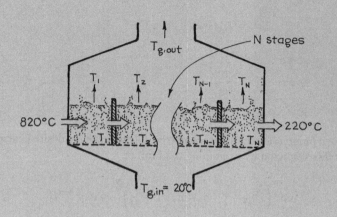

Solution

For equal thermal utilization we have, as with Example 14.1,

$$\eta_s = 0.75, \quad \eta_g = 0.75, \quad -\frac{\Delta T_g}{\Delta T_s} = 1, \quad \frac{\dot{m}_g C_g}{\dot{m}_s C_s} = 1$$

Now since ϕ' refers to the flow through each stage,

$$\phi' = \frac{(\dot{m}_g/N)C_g}{\dot{m}_s C_s} = \frac{1}{N} \cdot \frac{\dot{m}_g C_g}{\dot{m}_s C_s} = \frac{1}{N}$$

Then, equation (14.12) gives

$$\eta_s = 0.75 = 1 - \frac{1}{(1+\phi')^N} = 1 - \frac{1}{[1+(1/N)]^N}$$

or

$$\left(\frac{N+1}{N}\right)^N = 4$$

(continued)

Solve by trial and error:

Guess N	$\left(\frac{N+1}{N}\right)^N$
1	2
3	2.37
9	2.58
99	2.70

The above progression does not seem to be able to reach the desired value of "4." We verify this suspicion by examining the limit or

$$\lim_{N\to\infty} \left(\frac{N+1}{N}\right)^N = \lim_{N\to\infty} \left(1+\frac{1}{N}\right)^N = e = 2.718$$

This shows that one cannot get a solution to equation (ii). This means that

It is impossible to get the required 75 % heat utilization in any crossflow exchanger, even though this can be done in 3-slage counterflow exchangers.

Problems on Direct-Contact Gas–Solid Nonstoring Exchangers

14.1. A stream of fine solids is to be cooled from 820 °C to 100 °C by cold gas heating from 20 °C to 500 °C in a counterflow multistage fluidized bed exchanger. How many ideal stages are needed?

14.2. A crossflow multistage fluidized bed exchanger is to cool a stream of fine solids from 820 °C to 220 °C by contact with gas which heats from 20 °C to 420 °C. How many ideal stages are needed?

14.3. Suppose the equipment for example 14.1 is built and is working smoothly. Then, one day we are told that the temperature of the incoming solids will be raised to 1,020 °C. We still want the solids to leave at 220 °C, and we don't want to modify the equipment, add stages, etc. What do you suggest we do?

14.4. We wish to heat a stream of solids (1 t/h, $C_p = 1{,}000$ J/kg K) from 0 °C to 800 ° C in a 4-stage direct-contact crossflow fluidized bed exchanger using hot air at 1,000·°C. What is the required volumetric flow rate of air?
Data: For incoming air, $\pi = 120$ kPa, $C_p = 32$ J/mol K.

Our factory needed to cool a continuous stream of 880 ° C solids with a stream of 40 ° C air. For this purpose we built a single-stage fluidized bed exchanger as shown in (a), and for the flow rates of our process, we find that the solids leave at 320 °C. This is not bad, but it is not good enough. Let us try to improve the operations.

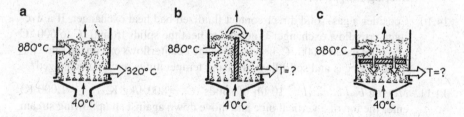

14.5. One thought is to put a vertical baffle right down the middle of the unit as shown in (b). Keeping all flow rates unchanged, find the temperature of solids leaving the exchanger.

14.6. Another thought is to insert a horizontal baffle in the exchanger, as shown in (c). Again, with flow rates unchanged, find the temperature of the leaving solids.

14.7. At present a stream of hot gas (1,000 °C) contacts a stream of cold solids (0 °C) in a single-stage fluidized bed exchanger. The stream of solids leaves at 500 °C. This is not good enough so let us add baffles so as to divide the exchanger into four equal parts, as shown below, keeping all flows unchanged. With this modification find the temperature of the leaving solids and the leaving gas.

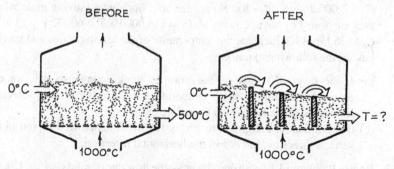

14.8. Hot gas at 1,000 °C enters a three-stage direct-contact crossflow fluidized bed exchanger and contacts a stream of solids which enter at 0 °C. The solids leave at 512 °C. We now add vertical baffles so as to divide each stage into two, thus ending up with six crossflow stages.

(a) By making this change what happens to the temperature of the leaving solids?

(b) How does this change affect the contacting efficiency for the two streams? To answer calculate the contacting efficiency before and after the change.

14.9.   Consider a gas–solid direct-contact fluidized bed heat exchanger. If a four-stage crossflow exchanger can heat solids from 20 °C to 820 °C using hot entering gas at 1,020 °C, how many stage counterflows (with the same entering temperatures for gas and solids) can do the same job?

14.10. Consider a gas–solid direct-contact fluidized bed heat exchanger. If a five-stage crossflow exchanger is needed to heat the solids from 0 °C to 600 °C using hot gas at 1,000 °C, how many stage counterflows can do the same job if the inlet gas and solid flow rates and temperatures remain unchanged?

14.11. *Moving bed contactor.* 10 t/h of solids ($C_p = 800$ J/kg K, $T_{s,in} = 1,000$ K) enter the top of a vertical pipe and move down against an upflowing stream of gas ($C_p = 800$ J/kg K, $T_{g,in} = 500$ K). Plot the temperature distribution of gas and solids in the pipe, and show on the plot the outlet temperature of the two streams.

(a) Gas flow is 10 t/h.
(b) Gas flow is 20 t/h.
(c) Gas flow is 8 t/h.
(d) Find the mean residence time of gas $\bar{t}_g$ and of solids $\bar{t}_s$ in the contactor for the flows of part (a).

Assume perfect countercurrent plug flow contacting, fairly small particles, and a bed voidage $\varepsilon = 0.4$.

14.12. *Moving grate contactor.* Cold crushed solids (small particles, $\dot{m}_s = 10$ kg/s, $C_p = 1,000$ J/kg K, $T = 300$ K) are fed to a horizontal moving grate where they are heated by hot upflowing air ($\pi = 116,000$ Pa, $T = 600$ K, $v = 12\,\mathrm{m}^3$/s, $C_p = 36$ J/mol K). Estimate the temperature of the leaving solids and leaving gas for the following extremes:

(a) If the layer of solids on the grate is thick enough so that we can reasonably assume a sharp temperature front in the vertical direction as well as in the horizontal direction.
(b) If the solids on the grate are gently fluidized, hence well mixed in the vertical direction, but not in the horizontal direction.

14.13. Repeat Problem 14.12 with one change—the flow rate of solids is $\dot{m}_s = 15$ kg/s.

14.14. Repeat Problem 14.12 with one change—the flow rate of solids is $\dot{m}_s = 5$ kg/s.

14.15. To improve the efficiency of a shale processing plant, fresh shale is to be preheated by hot waste combustion gases from the process. To do this, a layer of finely crushed fresh shale rock (0 °C) is transported along a long horizontal porous conveyor belt while the hot gas (640 °C) is forced upward through the bed at a high enough velocity to just fluidize the solids.

At present the shale leaves at 480 °C. What would be the outlet temperature of the solids if the gas flow rate were:

(a) Raised 20 %?
(b) Lowered 20 %, in which case the fluidized bed collapses into a moving fixed bed of solids?

Related Reading

D. Kunii, O. Levenspiel, *Fluidization Engineering*, 2nd edn. (Butterworth, Boston, 1991)

Chapter 15
Heat Regenerators: Direct-Contact Heat Storing Exchangers Using a Batch of Solids

"The theories of the regenerator are among the most difficult and involved that are encountered in engineering."

M. Jacob

Because solids, on a volume basis, have a very large heat capacity compared to gases, they can effectively be used as an intermediary in the transfer of heat from one gas to another. This requires a two-step operation. In the first step hot gas gives up its heat to cold solids. The solids heat up, and then in the second step, the solids release this heat to a second cold gas. For continuous operations regenerators are used in pairs, as shown in Fig. 15.1.

This type of storage exchanger is used primarily when heat has to be exchanged between enormous amounts of gas, as in the steel and other metallurgical industries, or when the gases are dirty and dust laden and liable to plug up a recuperator, as is the case with flue gases in coal-burning electric power stations, or when one of the gases is too hot or reactive for the materials of construction of a recuperator, as is the case with gases from glassmaking furnaces.

Regenerators can also be designed for continuous operations as shown in Fig. 15.2.

In turn, let us take up the two major classes of regenerators: first, the fixed solid devices (the packed bed, the rotating wheel, the monolith unit) and then the well-mixed solid devices (the single-stage and the multistage fluidized bed).

© Springer Science+Business Media New York 2014
O. Levenspiel, *Engineering Flow and Heat Exchange*,
DOI 10.1007/978-1-4899-7454-9_15

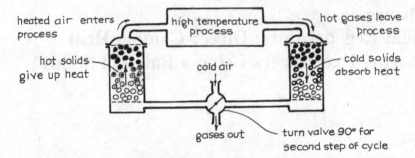

Fig. 15.1 A pair of cyclic regenerators for recovering heat from waste gas

15.1   Packed Bed Regenerators: Preliminary

15.1.1   Spreading of a Temperature Front

These units are usually packed with large solids—bricks, rocks—so that the pressure drop for gas flow does not become excessive and so that fine solids suspended and entrained by the gas do not plug up the unit.

When hot gas enters an initially cold bed of solids, a hot temperature front of gas moves down the bed trailed by a hot front of solids, as shown in Fig. 15.3. Three phenomena lead to the spreading of these hot fronts:

- Deviation from plug flow of gas in the packed bed—some fluid moving faster, some moving slower. This behavior is characterized by the axial dispersion coefficient for the gas D, a sort of diffusion coefficient.
- Film resistance to heat transfer between gas and solid. Since the particles are large, the interfacial area and the heat transfer coefficient ha can be very much lower than for beds of fine solids. The term ha characterizes this resistance.
- Resistance to heat flow into the particles. With large solids such as bricks and rocks, the characteristic time for the heating of the particles can be large. The thermal diffusivity of the solids $k_s/\rho_s C_s$ characterizes this resistance.

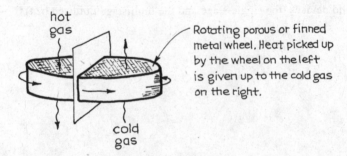

Fig. 15.2   Continuously operating rotating-wheel regenerator

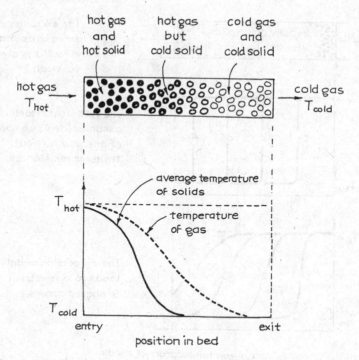

Fig. 15.3 Advancing temperature front in gas and in solid in a packed bed regenerator

15.1.2   Models for the Temperature Spread

We have three levels of analysis for fixed bed regenerators, as shown in Fig. 15.4:

1. The *flat front approximation* of Fig. 15.4a is the simplest model. It assumes ideal plug flow of gas and immediate equalizing of gas and solid temperature. This is a crude approximation, but still useful for baseline estimates of behavior.
2. The *dispersion approach* of Fig. 15.4b describes each of the three spreading factors by a diffusion-like phenomenon. This leads to a symmetrical S-shaped temperature–distance curve for solids characterized by its variance σ^2. Assuming independence of the three spreading phenomena, we can add the variances to give

$$\sigma^2_{\text{overall}} = \sigma^2_{\substack{\text{gas} \\ \text{dispersion}}} + \sigma^2_{\substack{\text{film} \\ \text{resistance}}} + \sigma^2_{\substack{\text{particle} \\ \text{conduction}}} \qquad (15.1)$$

This approach should reasonably approximate the real temperature distribution in a not-too-short regenerator.

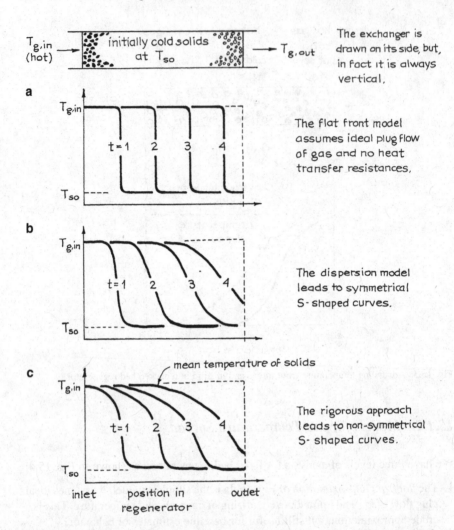

Fig. 15.4 Temperature of solids in a packed bed regenerator according to different models from the simplest to the most complicated

3. The *rigorous analysis* which accounts properly for all three spreading phenomena should yield unsymmetrical S-shaped curves sketched in Fig. 15.4c. This analysis is extremely difficult and has not as yet been done. The many approaches and partial solutions to this problem are presented and discussed by Hausen (1983) and Jakob (1957). Other related references are presented by McAdams (1954) and Kern (1950).

We will take up the flat front and dispersion approaches in analyzing both single-pass and periodic operations of heat regenerators. We do not attempt the rigorous approach.

15.1.3   Measure of Thermal Recovery Efficiency

Suppose hot gas at $T_{h,\text{in}}$ enters a cold regenerator at T_c for a length of time t. We define the *efficiency of heat capture by solids* or the *efficiency of heat removal from the gas* for this period as

$$\eta_h = \left(\frac{\begin{array}{c} \text{heat taken up by cold solids} \\ \text{in time } t \\ \hline \text{maximum possible take up} \\ \text{in time } t \end{array}}{} \right) = \left(\frac{\begin{array}{c} \text{heat lost by hot gas} \\ \text{in time } t \\ \hline \text{maximum possible} \\ \text{heat loss in time } t \end{array}}{} \right) \quad (15.2)$$

$$= \left(\frac{\overline{\Delta T_h}}{\Delta T_{\max}} \right)_{\text{of gas}} = \left(\frac{T_{h,\text{in}} - \overline{T}_{h,\text{out}}}{T_{h,\text{in}} - T_c} \right)_{\text{of gas}}$$

Similarly, for cold gas at $T_{c,\text{in}}$ entering a hot regenerator T_h for a time period t,

$$\eta_h = \left(\frac{\begin{array}{c} \text{heat lost by solids} \\ \text{in time } t \\ \hline \text{maximum possible} \\ \text{heat loss in time } t \end{array}}{} \right) = \left(\frac{\begin{array}{c} \text{heat gained by cold gas} \\ \text{in time } t \\ \hline \text{maximum possible heat} \\ \text{gain in time } t \end{array}}{} \right) \quad (15.3)$$

$$= \left(\frac{\overline{\Delta T_c}}{\Delta T_{\max}} \right)_{\text{gas}} = \left(\frac{\overline{T}_{c,\text{out}} - T_{c,\text{in}}}{T_h - T_{c,\text{in}}} \right)_{\text{gas}}$$

Let us relate these efficiencies. For this consider hot fluid entering an initially cold regenerator. Making a heat balance at time t gives

$$\begin{array}{c} \text{Fraction of} \\ \text{solids heated} \end{array} = \left(\frac{\begin{array}{c} \text{heat introduced} \\ \text{by gas in time } t \\ \hline \text{heat needed to} \\ \text{heat up all the} \\ \text{solids} \end{array}}{} \right) \quad (15.4)$$

$$= \frac{\dot{m}_h C_h (T_{h,\text{in}} - T_c) t}{W_s C_s (T_{h,\text{in}} - T_c)} = \frac{\dot{m}_h C_h t}{W_s C_s}.$$

The characteristic time needed to heat all the solids is then

$$\hat{t}_h = \frac{W_s C_s}{\dot{m}_h C_h} = \frac{\rho_s (1 - \varepsilon) C_s}{G_h C_h} \cdot L \quad (15.5)$$

Similarly, for the cooling of initially hot solids we have

$$\hat{t}_c = \frac{W_s C_s}{\dot{m}_c C_c} = \frac{\rho_s (1-\varepsilon) C_s}{G_c C_c} \cdot L \qquad (15.6)$$

When $\hat{t}_h = \hat{t}_c$ we have what is called *symmetric operations*. For unequal flow rates of hot and cold gases, the characteristic heating and cooling times will differ, or $\hat{t}_h \neq \hat{t}_c$, and we have *unsymmetric operations*.

We will show that symmetric operations are simpler to analyze, always give a higher heat exchange efficiency, and should therefore always be used in practice. We treat this situation.

15.1.4   Periodic Cocurrent and Countercurrent Operations

Periodic operations can be run in two ways. In cocurrent operations the cold fluid and the hot fluid enter one after the other at the same end of the regenerator. In countercurrent operations the hot fluid enters at one end, the cold fluid at the other end of the regenerator. These two modes are shown in Fig. 15.5.

It is not obvious which of these contacting patterns is better. The simple flat front model says that both can be equally good. However, as we will show, the dispersion model predicts that countercurrent operations have a higher efficiency. We will analyze both modes of operations.

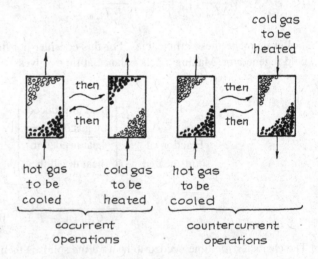

Fig. 15.5 Switching pattern for periodic cocurrent and periodic countercurrent operations of a packed bed regenerator

15.2 Packed Bed Regenerators: Flat Front Model

15.2.1 Cocurrent Operations with $\hat{t}_h = \hat{t}_c = \hat{t}$

Let hot and cold fluid flow through the two side-by-side regenerators at the same rate, and by this we mean that the temperature fronts move through the regenerators at the same speed. In symbols this says that $\dot{m}_h C_h = \dot{m}_c C_c$ or that $\hat{t}_h = \hat{t}_c$. Let us call this time $\hat{t}$.

If the switching time is chosen to be $\hat{t}_h = \hat{t}_c$, then the heat given up by the hot gas is just enough to heat the regenerator. This represents the most efficient heat interchange scheme, or

$$\eta_h = \eta_c = 100\ \%\quad \text{for } t_{sw} = \hat{t} \tag{15.7}$$

If the switching time is shorter than $\hat{t}_h = \hat{t}_c$, then

$$\eta_h = \eta_c = 2 - \frac{\hat{t}}{t_{sw}}\quad \text{for } t_{sw} = \text{between } \tfrac{2}{3}\hat{t} \text{ and } \hat{t} \tag{15.8}$$

If the switching time is longer than $\hat{t}_h = \hat{t}_c$, then

$$\eta_h = \eta_c = \frac{\hat{t}}{t_{sw}}\quad \text{for } t_{sw} > \hat{t} \tag{15.9}$$

Figure 15.6 shows one pair of regenerators for the two undesirable situations above. These results show that one should always use cocurrent switching times equal to $\hat{t}_h = \hat{t}_c$. Unsymmetric operations can never give 100 % exchanger efficiency and should be avoided.

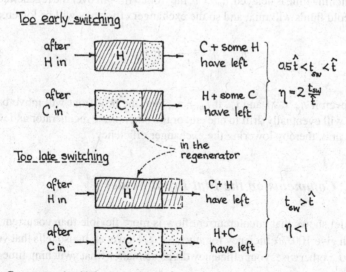

Fig. 15.6 Cocurrent pair of heat regenerators

15.2.2   Countercurrent Operations with $\hat{t}_h=\hat{t}_c=\hat{t}$

By referring to sketches of Fig. 15.7, we see that if $t_{sw} \leq t$, then the temperature front remains in the regenerator, moving from right to left, but never spilling out of either end. For this situation

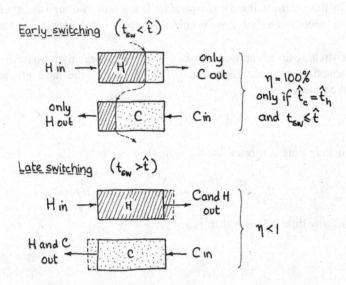

Fig. 15.7 Countercurrent pair of heat regenerators

$$\eta_h = \eta_c = 100 \% \quad \text{for } t_{sw} \leq \hat{t} \tag{15.10}$$

If the switching time is delayed, $t_{sw} > \hat{t}$, the front will spill over the regenerator exits, hot and cold fluids will mix, and so the exchanger efficiency will be lowered to

$$\eta_h = \eta_c = \frac{\hat{t}}{t_{sw}} \quad \text{for } t_{sw} \geq \hat{t} \tag{15.11}$$

Finally, even if $t_{sw} < \hat{t}_h$ and $\hat{t}_c$, if $\hat{t}_h \neq \hat{t}_c$, then the heat front will move back and forth but will eventually drift to one end or the other of the regenerator and will spill from the unit, thereby lowering the exchanger efficiency.

15.2.3   Comments on the Flat Front Model

This model shows that countercurrent flow is more flexible than cocurrent flow in that it can give 100 % efficiency for any $t_{sw} < \hat{t}$. However, it also tells that you must use $\hat{t}_h = \hat{t}_c$; otherwise, your efficiency drops no matter what switching time is used.

Finally, this model is a useful first approximation. In the more realistic dispersion model, which we treat next, we will see that the predicted flat front efficiencies are higher than would be found in practice.

15.3   Packed Bed Regenerators: Dispersion Model

Consider in turn:

- The contribution of the three heat transfer resistances to the spreading of the advancing temperature front
- The thermal efficiency of one-pass operations
- The thermal efficiency of periodic cocurrent operations
- The thermal efficiency of periodic countercurrent operations

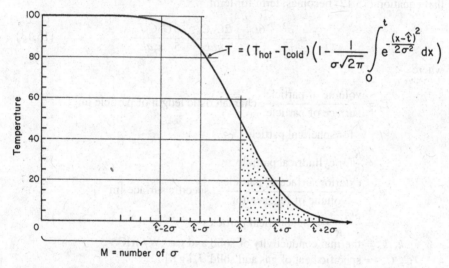

$$T = (T_{hot} - T_{cold}) \left(1 - \frac{1}{\sigma\sqrt{2\pi}} \int_{0}^{t} e^{-\frac{(x-\hat{t})^2}{2\sigma^2}} \, dx \right)$$

Fig. 15.8 Shape of the advancing hot temperature front, according to the dispersion model

15.3.1   Evaluation of σ^2, the Quantity Which Represents the Spreading of the Temperature Front

The dispersion (or diffusion type) model leads to a symmetrical S-shaped advancing temperature front which represents the integral of the Gaussian distribution function. This S-shaped curve is characterized completely by a single quantity, the variance σ^2. Figure 15.8 shows how the temperature front spreads with σ.

From diffusion theory, if the individual contributions to the spreading act independently, then one can add variances for the individual contributions or

$$\left(\begin{array}{c}\text{Spreading of}\\ \text{temperature}\\ \text{front}\end{array}\right)^2 = \left(\begin{array}{c}\text{spreading caused}\\ \text{by deviation from}\\ \text{plug flow}\end{array}\right)^2 + \left(\begin{array}{c}\text{spreading caused}\\ \text{by surface heat}\\ \text{transfer resistance}\end{array}\right)^2$$

$$+ \left(\begin{array}{c}\text{spreading caused}\\ \text{by resistance to}\\ \text{conduction in solid}\end{array}\right)^2$$

or in symbols

$$\sigma^2 = \sigma^2_{\substack{\text{axial gas}\\ \text{dispersion}}} + \sigma^2_{\substack{\text{film}\\ \text{resistance}}} + \sigma^2_{\substack{\text{particle}\\ \text{heating}}} \qquad (15.12)$$

If it is assumed in addition that no heat travels along the solids in the direction of gas flow (reasonable for a packed bed of spherical or randomly packed nonmetallic particles, but possibly not reasonable for a compact exchanger or monolith structure), then Levenspiel (1984), using the results of Sagara et al. (1970), has shown that equation (15.12) becomes, term for term,

$$M^2 = \frac{\sigma^2}{\hat{t}^2} = \frac{6L_p}{L} + \frac{2G_0C_g}{haL} + \frac{6}{5}\frac{G_0C_gL_p}{k_saL} \qquad (15.13)$$

where

$L_p = \dfrac{\text{volume of particle}}{\text{surface of particle}}$, characteristic length of particle [m]

$\quad = \dfrac{R}{3}$ for spherical particle

$\quad = \dfrac{R}{2}$ for cylindrical particle

$a = \dfrac{\text{exterior surface of particles}}{\text{volume of regenerator}}$, specific surface $[\text{m}^{-1}]$

$\quad = \dfrac{3(1-\varepsilon)}{R}$ for spherical particles

$k_s, k_g = $ thermal conductivity of solid and gas [W/m K]

$C_s, C_g = $ specific heat of gas and solid [J/kg K]

$G_0 = u_0\rho$, superficial mass velocity of gas [kg/m² s]

$M = $ measure of the spreading temperature front in the vessel, see equations (15.12) and (15.13)

For spherical particles equation (15.13) becomes

$$M^2 = \frac{\sigma^2}{\hat{t}^2} = \frac{d_p}{L} + \frac{1}{3(1-\varepsilon)}\cdot\frac{G_0C_gd_p}{hL} + \frac{1}{30(1-\varepsilon)}\cdot\frac{G_0C_gd_p^2}{k_sL} \qquad (15.14)$$

This expression shows that as the exchanger is made longer, the relative spread of the temperature front becomes smaller; hence, the efficiency of the unit more closely approaches the flat front ideal.

Let us now see how to evaluate the thermal efficiency for various operating patterns knowing the value of σ^2 from equation (15.13) or (15.14).

15.3.2   One-Pass Operations; Dispersion Model

Consider a step input of hot fluid into a cold regenerator. After time $\hat{t}$ the temperature distribution of solids in the regenerator will be as shown in Fig. 15.9.

The efficiency of heat capture by the solids after time $\hat{t}$ can be found by referring to Fig. 15.9. Thus,

$$\eta_{\text{single pass}} = \left(\begin{array}{c} \text{fractional} \\ \text{heat recovery} \end{array} \right) = 1 - \frac{\text{dotted area in Fig. 15.9}}{\text{area } ABCD}$$

$$= \frac{\text{hatched area in Fig.15.9}}{\text{area } ABCD} \qquad (15.15)$$

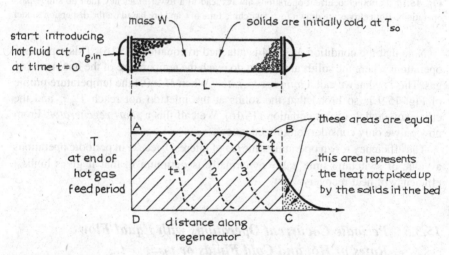

Fig. 15.9 Temperature of solids in a regenerator in a one-pass operation, according to the dispersion model

Noting that the S-shaped curve of Fig. 15.9 represents the Gaussian integral shown in Fig. 15.8 and that the dotted area is calculated to be 0.4σ, the single-pass efficiency is then

$$\eta = 1 - \frac{0.4\sigma M}{\sigma} = 1 - 0.4M \quad \text{for } M \leq 0.4 \qquad (15.16)$$

The upper line of Fig. 15.10 then shows how the single-pass efficiency depends on the value of M.

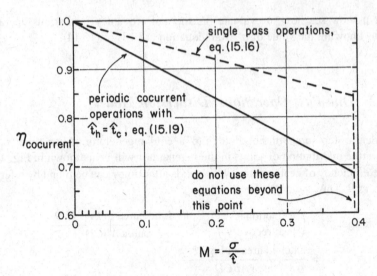

Fig. 15.10 Periodic cocurrent operations always lead to a lower efficiency than do single-pass operations. Both curves are drawn for a switching time of $\hat{t}$ and are based on the dispersion model

Note that the condition $M \leq 0.4$ is attached to equation (15.16). This represents operations where the solids at the inlet do reach the temperature of the hot incoming gas. This is what we call a *long regenerator*. When $M > 0.4$ the temperature profile of Fig. 15.9 is so broad that the solids at the inlet do not reach $T_{g,\,in}$ and the efficiency deviates from equation (15.16). We call this a *short regenerator*. From now on we only consider long regenerators.

The efficiency for an operating time $\hat{t}$ is of interest because in periodic operations a value of switching time $t_{sw} = \hat{t}$ normally is optimum in that it gives highest efficiency of heat recovery.

15.3.3 Periodic Cocurrent Operations with Equal Flow Rates of Hot and Cold Fluids or $\hat{t}_h = \hat{t}_c = t_{sw}$ Dispersion Model

Suppose we choose to switch from hot to cold fluid at time $\hat{t}$ and back again after an additional time $\hat{t}$. The temperature distribution will then be somewhat as shown in Fig. 15.11, and reflection shows that the inefficiency in heat absorption during heating is represented by the shaded areas A and B. Thus, the efficiency is given by

$$\eta_{\text{heating}} = 1 - \left(\frac{\text{area } A + \text{area } B}{\text{area } WXYZ} \right)_{\text{Fig. 15.11}}$$

From equations (15.15) and (15.16) this becomes

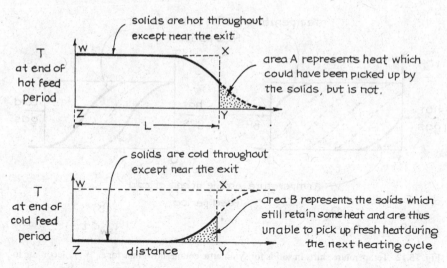

Fig. 15.11 Temperature of solids for symmetric periodic cocurrent flow with $\hat{t}_h = \hat{t}_c = t_{sw}$, according to the dispersion model

$$\eta_{\text{heating}} = 1 - 0.8M \tag{15.17}$$

and for these symmetric operations

$$\eta_{\text{cooling}} = \eta_{\text{heating}} \tag{15.18}$$

An analysis of the situation leads to the following conclusions:

1. The highest efficiency always occurs when the switching time t_{sw} is chosen to be $\hat{t}$.
2. Whenever $M \leq 0.4$ the succeeding temperature fronts of Fig. 15.11 do not affect each other. Thus, the previous front is swept out before the next one comes along. In this situation reflection shows that the efficiency is given by

$$\eta_{\text{periodic cocurrent}} = 2\eta_{\text{single pass}} - 1 \tag{15.19}$$

3. Figure 15.10 shows how the efficiency of single-pass and periodic cocurrent operations depends on the value of M.

15.3.4 Periodic Countercurrent Operations with Equal Flow Rates of Hot and Cold Fluids or $\hat{t}_h = \hat{t}_c$ Dispersion Model

Starting with hot fluid entering a cold regenerator, successive temperature fronts will look somewhat as shown in Fig. 15.12.

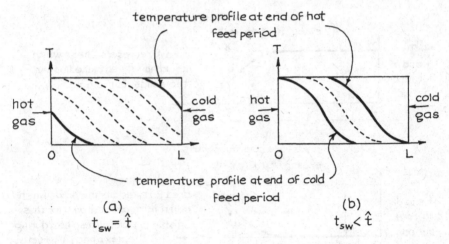

Fig. 15.12 Temperature shifts in solids for symmetric counterflow, thus for $\hat{t}_h = \hat{t}_c$, according to the dispersion model

As the temperature profile shifts back and forth, from right to left, dispersion will make the front spread. However, in partially heating a cold particle, it is the outside shell which heats first, and it is this which first cools with cold gas. This results in a self-healing of the spreading front or a tendency to approach a flat front profile.

The overall effect of these two opposing effects is something which cannot be evaluated by the analyses of today. Probably the best compromise at this time, and until we know better, is to assume that the temperature front remains unchanged from the one pass to the next. We use this assumption throughout for the counter-current operations.

Again, noting that the temperature profiles follow the integral of the Gaussian distribution, the thermal efficiency is evaluated with the aid of Fig. 15.13 and the following calculation procedure:

(a)   Determine $\hat{t}_h = \hat{t}_c = \hat{t}$ from equations (15.5) and (15.6).
(b)   *Set $t_{sw} \leq \hat{t}$.*
(c)   Determine $\sigma_h = \sigma_c = \sigma$ from equation (15.13) or (15.14).
(d)   Calculate $\sigma_{sw} = \sigma(t_{sw}/\hat{t})^{1/2}$.
(e)   *Calculate $P = (\hat{t} - t_{sw})/2\sigma_{sw}$.*
(f)   Calculate $Q = t_{sw}/\sigma_{sw}$.

The efficiency is then given by

$$\eta = 1 - \left(\frac{\text{area } E \text{ and area } F}{\text{area } WXYZ}\right)_{\text{Fig. 15.13}} \tag{15.20}$$

This is a function of P and Q, as displayed in Fig. 15.14.

Note that if one can make t_{sw} smaller than $\hat{t} - 5\sigma_{sw}$, then one can get almost complete heat recovery, as long as the temperature profile stays balanced between

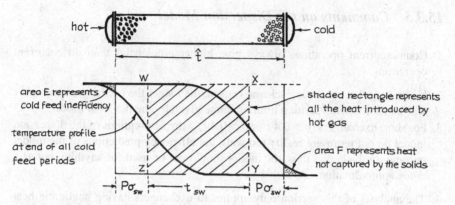

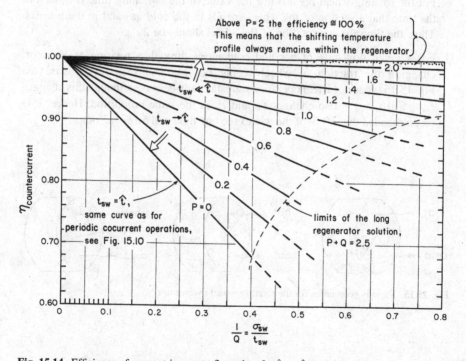

Fig. 15.13 Graph which shows how the efficiency of heat recovery is determined for symmetric counterflow, thus for $\hat{t}_h = \hat{t}_c$

the two ends of the regenerator. However, as t_{sw} is allowed to approach $\hat{t}$, the efficiency drops to that of one-pass operations, and for $t_{sw} > \hat{t}$ the efficiency drops rapidly toward zero.

Comparing Fig. 15.14 with Fig. 15.10 shows that symmetric countercurrent operations can be much more efficient than symmetric cocurrent operations; just use a shorter switching time.

Fig. 15.14 Efficiency of symmetric counterflow, thus for $\hat{t}_h = \hat{t}_c$

15.3.5 Comments on the Dispersion Model

1. Countercurrent operations always give higher efficiencies than do cocurrent operations.

2. No matter what contacting scheme is used, the longer the exchanger, the higher will be its efficiency, with a limiting value given by the flat front model.

3. For short exchangers ($\sigma > 0.4\,\hat{t}$ or $\sigma > 0.4\,t_{sw}$) the assumptions of the dispersion model do not represent reality too well, the efficiency predictions of the model become too high, and thus, the model should not be used for anything but as a crude approximation.

4. The analysis of this section only applies to exchangers having negligible heat conduction through the solids in the direction of gas flow. Thus, it applies reasonably to packed bed regenerators and to monolith regenerators made of ceramic or other poorly conducting solids.

For metallic monoliths where conduction along the metal is significant, the efficiencies calculated in this chapter should be lowered somewhat, to give behavior somewhere between the packed bed and the fluidized bed (see next section).

5. For dusty gases countercurrent operations have an additional advantage in that they help avoid plugging of the regenerator.

6. For the rotating-wheel regenerator the value of the switching time is related to the time that a section of the wheel spends in the cold gas and in the hot gas. Thus, the time for one rotation of the wheel should be $2\,t_{sw}$.

7. For a given flow rate of hot and cold gases through a rotating regenerator, changing the fraction of wheel seeing hot gas and seeing cold gas (see Fig. 15.15) does not affect operations substantially because although this changes $t_{sw,\,h}$ and $t_{sw,\,c}$, it also changes $\hat{t}_h$ and $\hat{t}_c$ in the same proportion. Hence, it is recommended that 50 % of the wheel see one fluid and 50 % the other.

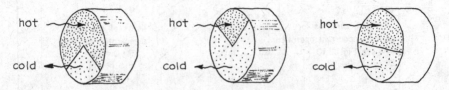

Fig. 15.15 Various geometries for the rotating-wheel regenerator

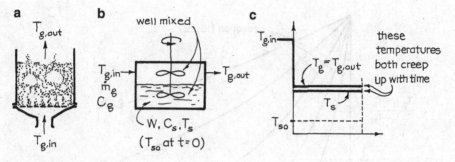

Fig. 15.16 The heating of a batch of solids in a fluidized bed regenerator, (**a**) sketch of unit, (**b**) contacting model, and (**c**) temperatures in the unit

15.4   Fluidized Bed Regenerators

Fluidized beds of fine particles are characterized by good mixing of solids and $Ua \rightarrow \infty$ (see Chap. 14). Thus, we may take the solids to be uniform in temperature at any time and the leaving gas to be at the temperature of the solids, as sketched in Fig. 15.16. A heat balance around the whole vessel becomes

$$\dot{m}_g C_g \left(T_{g,\ in} - T_s \right) = W_s C_s \frac{dT_s}{dt} \tag{15.21}$$

Separating and integrating then give

$$\frac{T_{g,\ in} - T_s}{T_{g,\ in} - t_{s0}} = \frac{\Delta T}{\Delta T_0} = e^{-t/\hat{t}} \quad \text{where } \hat{t} = \frac{W_s C_s}{\dot{m}_g C_g} \tag{15.22}$$

Graphically this temperature progression is shown in Fig. 15.17. Note that equation (15.22) is a special case of the corresponding recuperator expression, equation (13.24), in which $Ua \rightarrow \infty$.

Let us consider the thermal efficiency of fluidized bed recuperators.

15.4.1   Efficiency of One-Pass Operations

Suppose cold gas enters a hot fluidized bed. At the start the gas leaves hot and the efficiency is 100 %. But with time the temperature of solids and exit gas will decrease as shown in Fig. 15.17 and the point efficiency of operations will likewise decrease. Referring to Fig. 15.18 the average efficiency for an operating period t_{sw} is then

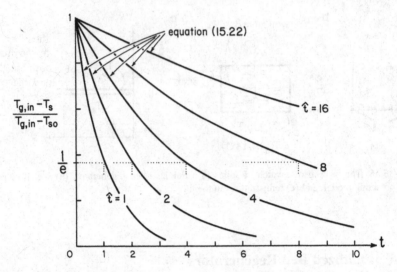

Fig. 15.17 Temperature–time curves when a batch of hot fluidized solids is cooled by a cold gas stream

$$\bar{\eta} = \frac{\text{dotted area in Fig. 15.20}}{\text{area } ABCD} = \frac{\int_0^{t_{sw}} e^{-t/\hat{t}} \, dt}{t_{sw}} \tag{15.23}$$

Performing the integration gives

$$\bar{\eta} = \frac{\hat{t}}{t_{sw}} \left(1 - e^{-t_{sw}/\hat{t}} \right) \tag{15.24}$$

Figure 15.19 shows that the efficiency starts at 100 % at the beginning of the run and then slides to zero; thus, the shorter the operating time, the more efficient is the regenerator for one-pass operations.

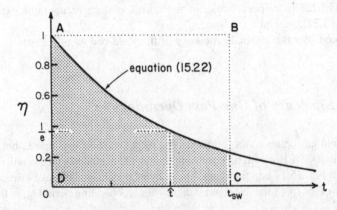

Fig. 15.18 Efficiency of heat recovery at different times during a one-pass operation of a fluidized bed regenerator

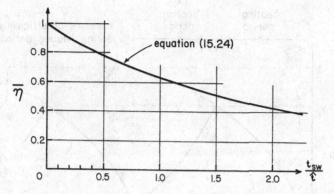

Fig. 15.19 Average heat recovery efficiency in a one-pass operation of a fluidized bed regenerator

15.4.2 *Efficiency of Periodic Operations*

First, consider the situation where $\hat{t}_h = \hat{t}_c$. Here the temperature of the fluidized bed regenerator varies with time as shown in Fig. 15.20, and for optimum operations with given t_{sw}, symmetry suggests that we should keep

$$1 - y_2 = y_1, \quad \text{in Fig. 15.20}$$

The efficiency of operations is then obtained from integration of equation (15.23), using the limits t_1 and t_2. This gives

$$\bar{\eta} = \frac{\hat{t}}{t_{sw}}(y_1 - y_2)$$

where

$$y_1 = e^{-t_1/\hat{t}}, \quad y_2 = e^{-t_2/\hat{t}}, \quad \text{and } t_{sw} = t_2 - t_1$$

Combining the above five expressions, eliminating y_1, y_2, t_1, t_2, and simplifying then give

$$\bar{\eta} = \frac{\hat{t}}{t_{sw}}\left(\frac{e^{t_{sw}/\hat{t}} - 1}{e^{t_{sw}/\hat{t}} + 1}\right) \qquad (15.25)$$

From the geometry of Fig. 15.20, we can see that when the switching time approaches zero, the efficiency reaches a maximum of 50 %, and as the switching time is increased, the efficiency decreases toward zero.

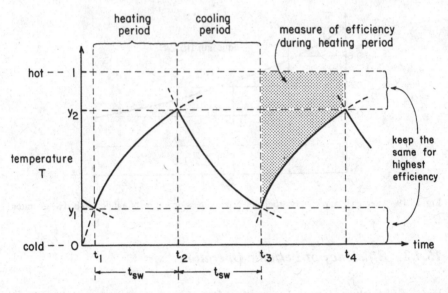

Fig. 15.20 Temperature changes for symmetric period operations $(\hat{t}_h = \hat{t}_c)$ of a fluidized bed regenerator

15.4.3   Comments on Fluidized Bed Regenerators

The analysis of this section leads to the following conclusions:

1. Single-stage periodic operation with $\hat{t}_h = \hat{t}_c$ has efficiencies approaching a maximum of 50 % for very frequent switching times. For longer switching times the efficiency drops toward zero.
2. Efficiencies for single-stage fluidized bed operations are very much lower than for packed bed operations; hence, these units are not used in practice.
3. Multistage fluidized bed operations will give higher efficiencies than single-stage operations. However, the large power requirement needed to fluidize the large mass of solids and the still lower efficiency when compared to a well-designed packed bed regenerator are reasons for not using such units in practice.

Example 15.1 The Great Paperweight Disaster
Idiot!! Can't you read!! The order said, "One vermillion paperweight," not one million paperweights. And here we've already made 999,239 of these beauties. What will we do with the extra 999,238 that we have in our storehouse? You'd better come up with an answer quick or else you lose your job and the West Coast Paperweight Co. goes out of business.

   One possibility is to sell them as packing ($\varepsilon = 0.4$) for a pair of unique, deluxe extra-high-efficiency heat regenerators, each 1 m^2 in cross section.

(continued)

Example 15.1 (continued)

Hot and cold gases would then be passed successively through these units at a superficial gas velocity of 4 m/s measured at 20 °C and 1 atm.

(a) For cocurrent operations find the desired switching time $t_{sw} = \hat{t}$ and the efficiency of operations.
(b) For counterflow operations find the efficiency of operations for a switching time $t_{sw} = \hat{t}$.
(c) For counterflow operations find the efficiency of operations for a switching time $t_{sw} = 0.75\, \hat{t}$.

Data: For the nearly spherical crown glass paperweights,

$$d_p = 0.05 \text{ m}$$
$$\phi = 0.94, \text{ sphericity}$$
$$k_s = 1.066 \text{ W/m K}$$
$$C_s = 714 \text{ J/kg K}$$
$$\rho_s = 2{,}500 \text{ kg/m}^3$$

Take equal quantities of hot and cold gases, and to keep things simple, assume that they have the same properties as air at 20 °C. From the appendix we find these properties to be

$$\mu = 1.8 \times 10^{-5} \text{ kg/m s}$$
$$C_g = 1{,}013 \text{ J/kg K}$$
$$\rho_g = 1.2 \text{ kg/m}^3$$
$$k_g = 0.026 \text{ W/m K}$$
$$(mw) = 0.0289 \text{ kg/mol}$$

Solution

We will solve this problem with the dispersion model, but first some preliminaries.

1. Height of paperweight-filled regenerators:

$$L = \left(\frac{999{,}238\,pw}{2}\right)\left[\frac{\pi}{6}(0.05)^3 \frac{\text{m}^3 \text{ solid}}{pw}\right]\left(\frac{1 \text{ m}^3 \text{ reg}}{0.6 \text{ m}^3 \text{ solid}}\right)\left(\frac{1}{1 \text{ m}^2 \text{ c.s.}}\right) = 54.5 \text{ m each}$$

(continued)

2. Mass flow rate and superficial mass velocity of gas:

$$\dot{m} = \left(4\frac{m}{s}\right)\left(\frac{1\,m^3\,\text{vol}}{1\,m\,\text{height}}\right)\left(\frac{1\,\text{mol}}{0.0224\,m^3}\right)\left(\frac{0.0289\,\text{kg}}{\text{mol}}\right)\left(\frac{273}{293}\right) = 4.8\,\text{kg/s}$$

and

$$G_0 = \frac{\dot{m}}{A} = 4.8\,\text{kg/m}^2\,\text{s}$$

3. Heat transfer coefficient between gas and pw packing is given by equation (9.37). Evaluating the needed dimensionless groups gives

$$\text{Re}_p = \frac{d_p u_0 \rho_g}{\mu} = \frac{(0.05)(4)(1.2)}{1.8 \times 10^{-5}} = 13,333$$

$$\text{Pr} = \frac{C_g \mu}{k_g} = \frac{(1,013)(1.8 \times 10^{-5})}{(0.026)} = 0.7013$$

Thus, equation (9.37) becomes

$$\frac{h(0.05)}{0.026} = 2 + 1.8(13,333)^{1/2}(0.7013)^{1/3}$$

giving
$$h = 97.06 \text{ W/m}^2 \text{ K}$$

4. Calculate $\hat{t}$, the time needed to cool or to heat the solids. From equation (15.5),

$$\hat{t} = \frac{W_s C_s}{\dot{m}_g C_g} = \frac{[(54.5)(2,500)(0.6)](714)}{(4.80)(1,013)} = 12,000 \text{ s}$$
$$= 2 \text{ h } 20 \text{ min}$$

5. Calculate the spreading of the temperature front. From equation (15.14),

$$M^2 = \frac{\sigma^2}{\hat{t}^2} = \frac{d_p}{L} + \frac{1}{3(1-\varepsilon)} \cdot \frac{G_0 C_g d_p}{hL} + \frac{1}{30(1-\varepsilon)} \cdot \frac{G_0 C_g d_p^2}{k_s L}$$

Replacing all values gives

(continued)

$$M^2 = 0.0009 + 0.0255 + 0.0116 = 0.0381$$

or

$$M = \frac{\sigma}{\hat{t}} = 0.1952$$

6. Single-pass efficiency. From equation (15.16) or Fig. 15.10,

$$\eta_{\text{single pass}} = 1 - 0.4(0.1952) = 0.9219, \text{ or } 92\%$$

We are now ready to solve the problem.

(a) Cocurrent flow with $t_{sw} = \hat{t}$. From equation (15.19) or Fig. 15.10,

$$\eta_{\text{cocurrent}} = 2(0.9219) - 1 = 0.8438, \text{ or } 84\%$$

(b) Counterflow with $t_{sw} = \hat{t}$. From the procedure leading to Fig. 15.14,

$$\sigma_h = \sigma_c = \sigma_{sw} = M\hat{t} = 0.1952(12{,}000) = 2{,}342 \text{ s}$$

$$P = \frac{\hat{t} - t_{sw}}{2\sigma_{sw}} = 0$$

$$\frac{1}{Q} = \frac{\sigma_{sw}}{t_{sw}} = \frac{2{,}343}{12{,}000} = 0.1952$$

Then, from Fig. 15.14

$$\eta_{\text{counter}} = 0.8438, \text{ or } 84\% \qquad \text{(same as for single pass)}$$

(c) Counterflow with $t_{sw} = 0.75\hat{t} = 0.75(12{,}000) = 9{,}000$ s. Following the procedure of Section III.D, we have

$$\sigma_h = \sigma_c = (0.1952)(12{,}000) = 2{,}342 \text{ s}$$
$$\sigma_{sw} = 2{,}342 \left(\frac{9{,}000}{12{,}000} \right)^{1/2} = 2{,}028 \text{ s}$$
$$P = \frac{12{,}000 - 9{,}000}{2(2{,}028)} = 0.7396$$
$$\frac{1}{Q} = \frac{2{,}028}{9{,}000} = 0.2253$$

Then, from Fig. 15.13

$$\eta_{\text{counter}} \cong 94\%$$

Note that counterflow with a smaller switching time gives a higher exchange efficiency.

Problems on Regenerators

15.1. The geometry for the regenerator of example 15.1 is not satisfactory because it is too tall and skinny. If we made the regenerator one-quarter the height and double the diameter but maintained the same volumetric flow rate of gases, what would be the cocurrent and counterflow efficiencies of this modified design for the cases of example 15.1?

15.2. We want four times the flow rate of gases of example 15.1. For this we plan to double the diameter and reduce the height of the regenerator of example 15.1, but keep u_0 unchanged at 4 m/s. What would be the cocurrent and counterflow efficiencies of this operation for the cases of example 15.1?

15.3. Two packed bed regenerators are used in periodic swing operations to recover heat from hot exchange gases at 800 °C and to preheat inlet air at 100 °C. Each generator is 30 m long and 4-m diameter. Solid pebbles used are approximately spherical and 6-cm diameter. Find the heat recovery efficiency:

(a) For cocurrent operations at the optimum switching time. Use the dispersion model.
(b) For cocurrent operations at the optimum switching time. Use the flat front model.
(c) For cocurrent operations with a 7.2-h switching time. Use the flat front model.
(d) For cocurrent operations with an 18-h switching time. Use the flat front model.

Data : Solids	Mean gas properties, at 450 °C
$\rho_s = 3{,}900 \text{ kg/m}^3$	$\mu = 3 \times 10^{-5} \text{ kg/m} \cdot \text{s}$
$C_s = 1{,}000 \text{ J/kg K}$	$C_g = 1{,}020 \text{ J/kg K}$
$k_s = 0.5 \text{ W/m} \cdot \text{k}$	$k_g = 0.05 \text{ W/m} \cdot \text{k}$
$\varepsilon = 0.4$	$\rho = 0.5 \text{ kg/m}^3$
$d_p = 6 \text{ cm}$	Use $t_{sw, c} = t_{sw, h}$ throughout

Note. This problem is related to that treated by Dudukovic and Ramachandran, *Chem. Eng.*, pg. 70 (June 10, 1985).

15.4. Repeat the previous problem with the change of using countercurrent flow instead of cocurrent flow. Find the heat recovery efficiency.

(a) For $t_{sw} = \hat{t}_h = \hat{t}_c$, using the dispersion and then the flat front models.

(b) For a switching time $t_{sw} = 7.2$ h, again using the dispersion model and then the flat front model.

15.5. The laboratory's high-temperature wind tunnel experiment requires an air-flow of 30 m/s at 540 °C and 1 atm through a square test section of 0.3 m × 0.3 m. One way to do this is to take in ambient air at 20 °C and heat it continually.

(a) Find how many household electric heaters, rated at 1,500 W, would be needed to provide this hot air continually during the run.
Alternatively, we can leisurely heat up a packed bed of rocks or other solid beforehand as shown below, and at the start of the run, blow through the required ambient air at 20 °C, this representing a one-pass operation.

(b) What length of packed bed and how many tons of steel balls would be required to store the heat needed for a 15-min wind tunnel experiment? Assume the flat front approximation for this calculation.

(c) How long can we run a hot-air wind tunnel experiment with the stored heat from an 8-m-high bed of steel balls? Note that at 2.5 σ away from the mean, the S-shaped dispersion curve is within 1 % of its asymptote.

Data: Take 300 °C to be the average conditions of air in the packed bed; see diagram for additional data.

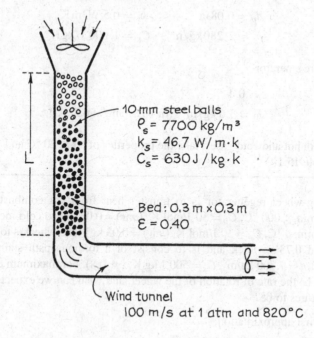

10 mm steel balls
$\rho_s = 7700$ kg/m^3
$k_s = 46.7$ W/m·k
$C_s = 630$ J/kg·k

Bed: 0.3 m × 0.3 m
$\varepsilon = 0.40$

Wind tunnel
100 m/s at 1 atm and 820°C

15.6. Repeat Problem 15.5, but instead of using steel balls in the regenerator, let us use uniformly sized crushed rock from the local rock quarry.
Data: For rock,

$$d_p = 0.05 \text{ m} \qquad k_s = 0.80 \text{ W/kg K}$$
$$\rho_s = 2,190 \text{ kg/m}^3 \quad \varepsilon = 0.48$$
$$C_s = 800 \text{ J/kg K}$$

15.7. A pair of regenerators 32 m high and 3 m in diameter filled with uniformly sized and close to spherical basaltic beach stones is to be used to transfer heat from hot waste gases leaving a process to cold incoming air. For equal flows of hot and cold gases and for the following operating conditions and properties of materials, find:

(a) The relative contributions of the three resistances to heat transfer
(b) The switching time to use and the efficiency of heat recovery for the best cocurrent operations
(c) The efficiency of heat recovery for countercurrent operations with a switching time $t_{sw} = 2$ h

Data: For the solid,

$$d_p = 0.08 \text{ m} \qquad k_s = 0.5 \text{ W/m K}$$
$$\rho_s = 2,280 \text{ kg/m}^3 \quad C_s = 1,000 \text{ J/kg K}$$

In the regenerator

$$\varepsilon = 0.4$$
$$G_0 = 3.6 \text{ kg/m}^2 \text{ s (or } u_0 \cong 3 \text{m/s at } 20\,°\text{C).}$$

For both hot and cold gases, take the properties of air at 20 °C and 1 atm; see example 15.1.

A rotating-wheel regenerator is to transfer heat from hot combustion gases [6,000 mol/min, 1,000 °C, $C_p = 30$ J/mol K, $(mw) = 0.03$ kg] to cold incoming air [6,000 mol/min, 0 °C, $C_p = 30$ J/mol K, $(mw) = 0.03$ kg]. The wheel is to be 1 m in diameter and 0.73 m thick and is to consist of a tortuous path stainless steel honeycomb ($\rho_s = 7,700$ kg/m^3, $C_s = 500$ J/kg K, $\varepsilon = 0.8$). For maximum efficiency, what should be the rate of rotation of the wheel, and what can we expect the outlet gas temperatures to be?

15.8. As a first approximation:

(1) Assume no heat flow resistance from gas to metal (or $h \rightarrow \infty$) and into the metal sheet.
(2) Assume plug flow of gas through the regenerator.

(3) Assume that the regenerator is so built that no heat flows from one location in the metal to another.

15.9. Replace assumption (3) with the following:

(3') Assume that there is no resistance to heat flow in the metal along a flow channel; thus, all the metal at a given angle θ in the drawing below is at the same temperature.

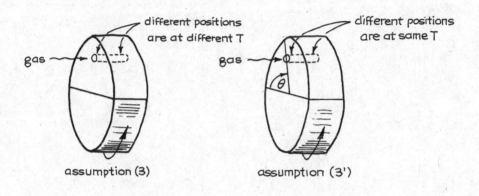

assumption (3)   assumption (3')

15.10. A pair of identical fluidized beds is to be used to recover heat from hot waste gas leaving a process and transfer it to fresh incoming gas. What thermal recovery efficiency may be expected from this pair of regenerators if the switching time is set at:

(a) 15 min?
(b) 30 min?

Data: Weight of sand in each regenerator $= 570$ kg. For both hot and cold gases, take the properties of air at 20 °C. Flow rate of hot and cold gases $= 0.5$ kg/s.

References

H. Hausen, *Heat Transfer in Counterflow, Parallel-Flow, and Cross-Flow* (trans. from German by M.S. Sayer), (McGraw-Hill, New York, 1983)

M. Jakob, *Heat Transfer*, vol. 2, Chapter 35, (Wiley, New York, 1957)

D.Q. Kern, *Process Heat Transfer* (McGraw-Hill, New York, 1950)

O. Levenspiel, Design of long heat regenerators by use of the dispersion model, Chem. Eng. Sci. **38**[12], 2035–2045 (1983)

W.H. McAdams, *Heat Transmission*, 3rd edn. (McGraw-Hill, New York, 1954)

M. Sagara, P. Schneider, J.M. Smith, The determination of heat-transfer parameters for flow in packed beds using pulse testing and chromatographic theory. Chem. Eng. J. **1**, 47 (1970)

Chapter 16
Potpourri of Problems

Here are some problems which use ideas from more than one chapter:

16.1 *Quenching high-temperature reactions.* Let us use the fluidized bed to freeze
   a fast multistep chemical reaction and thereby obtain valuable reaction
   intermediates. For this introduce the hot exit gases from a jet reactor with
   all its reaction intermediates immediately into the bottom of a cold vigorously
   fluidized bed, thereby using this gas to fluidize the batch of solids, as shown in
   the drawing on the next page. The bed is cooled by heat transfer to the walls of
   a cooling section. The rest of the bed is well insulated.
   Determine the temperature of the solids in the bed.
   Data: For solids,

$$\rho_s = 1{,}000 \text{ kg/m}^3$$
$$d_p = 100\mu m$$
$$C_s = 800 \text{ J/kg K}$$

© Springer Science+Business Media New York 2014
O. Levenspiel, *Engineering Flow and Heat Exchange*,
DOI 10.1007/978-1-4899-7454-9_16

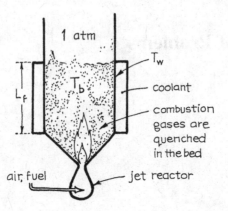

For gas,

$$C_g = 1,000 \text{ J/kg K, average value}$$
$$\rho_g = 1.273 \text{ kg/m}^3, \text{ at 273 K and 1 atm}$$
$$u_0 = 0.1 \text{ m/s, measured at 273 K and 1 atm}$$
$$T_g = 1,000 \text{ °C, entering gas}$$

For the bed,

$$d_b = 0.1 \text{ m, } \varepsilon_f = 0.49$$
Height of bed and of cooling section : $L_f = 0.4 \text{ m}$
Temperature of wall of cooling section : $T_w = 0 \text{ °C}$
For heat loss at wall : $h_w = 191 \text{ W/m}^2\text{K}$

16.2 In the jet reactor of Problem 16.1, heat is removed by transfer to the walls.
Consider a different arrangement wherein heat is removed by passing
(or flowing) solids at a steady rate through the reactor. If solids enter at
20 °C and at a rate such that a vessel volume of solids is replaced every
160 s, find the temperature of gas and solids leaving the reactor

(a) If the reactor walls are perfectly insulated
(b) If the reactor walls are cooled as in Problem 16.1

16.3 *Solar hot-water heaters*. Each of our horizontal rooftop solar collectors is a
shallow black-painted water-filled pan about 1 m on a side, well insulated on
the sides and bottom by the equivalent of 3 mm of Styrofoam, and covered
with a thin plastic film to stop evaporation. The heated water in these units is
circulated to a large heat reservoir in the basement of our home, and all
works well.
At night, however, the water, in the collectors, gets cold and if allowed to
circulate into the basement reservoir would cool it somewhat. Hence, the flow
is shut off at night and the water stays stagnant in the pan all night. Determine
the night temperature below which this water is likely to freeze.

Data: The emissivity of the collector is 0.95. The temperature of the night sky in our climate is $-73\,°C$. For Styrofoam $k = 0.035$ W/m K. Assume the worst possible conditions—still air.

16.4 *Dry scrubber for dusty gases.* Combustion Power Company uses a moving bed of granular solids as the filter medium to remove fine solids from hot gas without cooling the gas appreciably. The unit operates by having the gas move radially inward across an annulus of downflowing solids.

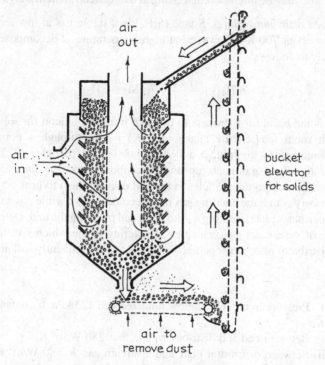

The solids trap the dust which is then removed on a moving grate. The clean solids are then recirculated to repeat the filtering action. If dusty gas ($C_g = 50$ J/mol K) enters at 200 mol/s, 800 K, and 120 kPa, if solids ($C_s = 1,000$ J/kg K) circulate at 1 kg/s, and if the solids cool 100 °C in the conveyor and bucket elevator, determine the temperature of the leaving gas.

16.5. *Geothermal water for heating cities.* Prodded by geologist John Hook of Salem, Oregon, Northwest Natural Gas Company is considering drilling wells on the slopes of Mt. Hood and piping this hot water (74 °C) at 1.6 m³/s in a 1.1-m-i.d. pipe insulated with 15 mm of Styrofoam to Portland, 70 km away and 760 m lower in elevation.

If too much heat is lost en route and if the water arrives colder than 65 °C, then the whole project might as well be scrapped. Calculate the temperature of the water when it reaches Portland.

(a) Account for heat loss to the surroundings only.
(b) Also include in your calculations the heat generated by friction in the pipes.

For winter conditions take the surroundings to be at 0 °C.
NOTE: YOU may be interested in looking at the related Problems, 2.13 and 2.14.

16.6 *Plugged distributor plates.* Silane (SiH_4) is a stable gas at low temperature (below about 700 K). However, at high temperature, it decomposes spontaneously as follows:

$$SiH_4(g) \xrightarrow{\ T> 700\ K\ } Si_\downarrow + 2H_2$$

This is the basis for a process to produce ultrapure silicon for solar cells in which room temperature silane (300 K) passes through a porous metal distributor plate to fluidize a bed of hot silicon particles (973 K). The incoming silane gas heats up and decomposes, and the fine silicon dust formed fuses onto the particles in the bed causing them to grow.
One worry is that the entering gas may become too hot while passing through the distributor plate, forming a solid there and plugging the unit. Determine the range of superficial gas velocities of entering silane which will insure that the distributor plate never gets above 650 K and consequently will not plug up.

Data:

(i) Pressure in chamber below fluidized bed 123 kPa. C_g (silane) = 60 J/mol K.
(ii) Between bed and distributor plate: $h = 280$ W/m^2 K.
(iii) Between distributor plate and upstream gas: $h = 40$ W/m^2 K.
(iv) It is reasonable to assume that the metal distributor plate is isothermal.

16.7 *The cooling of spent shale.* The US Bureau of Mines in Albany, Oregon, is looking into a process for recovering aluminum salts from spent shale rock by water leaching. If this hot solid is dispersed directly into water, terribly noxious gases evolve and the aluminum salts are transformed into insoluble compounds. However, if the solids are first cooled before dispersing in water, then no gases form and the aluminum salts remain in their soluble form and thus may possibly be recovered economically.
Consider the cooling of 30,000 t/day of hot spent shale from 600 °C to 120 °C by having them settle through a square tower penetrated by a whole host of

50-mm-o.d. horizontal tubes on 0.1-m equilateral triangle centers. These tubes are fed hot water at 100 °C and produce steam at 100 °C. Find:

(a) The cross section of tower to be used

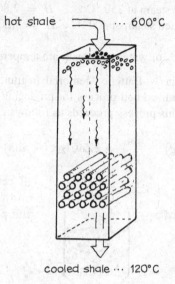

hot shale ··· 600°C

cooled shale ··· 120°C

(b) The number of heat transfer tubes and the height of heat transfer section needed for this cooling operation
(c) The steam production rate

Data: For the stream of flowing solids,

$$\overline{C}_p = 950\,\mathrm{J/Kg\,K} \qquad u_0 = 0.1\,\mathrm{m/s}$$
$$\rho_{\mathrm{bulk}} = 870\,\mathrm{kg/m^3} \quad U_{to\,\mathrm{tubes}} = 150\,\mathrm{W/m^2\,K}$$
$$\varepsilon_{\mathrm{bulk}} = 0.46$$

Also assume that the particles are small enough so that their temperature relaxation time is short enough for them to be isothermal and at the same temperature at any level in the exchanger.

16.8 Repeat the previous problem with the following changes:

1. The bare heat transfer tubes are replaced by finned tubes, which triples the surface area of the tubes, but lowers the overall U to 120 W/m² K, based on the outside surface area of tubes.
2. Water at 24 °C and 40 MPa enters at the bottom row of the tubes and superheated steam at 400 °C leaves at the top of the exchanger.

Data:

Cold water at 24 °C :	$H = 100 \, \text{kJ/kg}$
Boiling water at 250 °C :	$H = 1,100 \, \text{kJ/kg}$
Saturated steam at 250 °C :	$H = 2,800 \, \text{kJ/kg}$
Superheated steam at 400 °C :	$H = 3,200 \, \text{kJ/kg}$

Assume that the C_p of water and steam are temperature independent.

16.9   *Silicon for solar cells.* Battelle Memorial Institute of Columbus, Ohio, is looking into a fluidized bed process for producing solar grade silicon. The reaction step for this process proceeds as follows:

$$2\,\text{Zn(g)} + \text{SiCl}_4(\text{g}) \xrightarrow[\text{1181 K}]{\text{above}} 2\,\text{ZnCl}_2(\text{g}) + \text{Si(s)} \quad \Delta H_r = -113 \, \text{kJ}$$

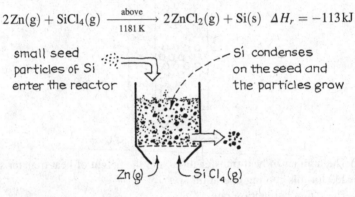

and the reactor is to look somewhat as sketched. Now, lower temperatures give higher equilibrium conversions and are favored, but zinc condenses at 1,181 K, and if this happens a fog forms which ruins operations. Thus, they would like to keep the temperature above 1,181 K, but as close to it as they dare.

With this in mind, Battelle plans to keep the fluidized solids at 1,200 K and the reactor walls at 1,185 K (for cooling purposes). A stoichiometric gas feed is to enter the 1-m high reactor at 0.15 m/s, 1,200 K, and 1.1 atm. At this temperature equilibrium conversion is 72 %; however, the expected conversion is 60 %. If $h = 160 \, \text{W/m}^2 \, \text{K}$ between bed and wall, calculate the largest diameter which can be used for the reactor, and find the silicon production rate (kg/h) that can be obtained from it.

Data: For silicon, $(mw) = 0.028 \, \text{kg/gm atom}$.

16.10   *Oil from shale.* The United States has vast deposits of shale, a solid which contains up to about 10 % organic material. There is more oil in these shale deposits than in all the petroleum reserves in the world. Crush and heat this rock and a part of this organic material comes off as volatile hydrocarbons.

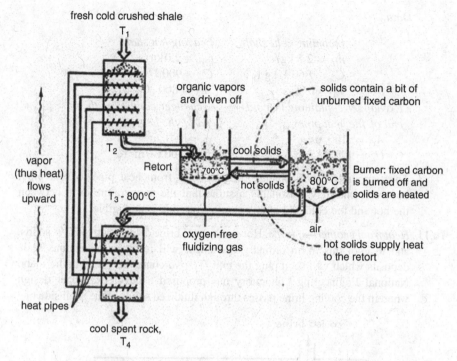

The rest remains with the rock as fixed carbon which can then be burned to give off heat.

Processes designed to recover these useful volatiles face one big problem. Because so much rock accompanies each kilogram of volatiles, a significant fraction of the recoverable energy is used just to heat the rock and is lost when the hot waste rock is discharged.

Consider the process shown in the diagram:

1. It achieves countercurrent heat exchange between the streams of hot waste solids and cold fresh solids by use of heat pipe exchangers.
2. It is able to use the not very useful fixed carbon in the shale to provide heat for the process.
3. It keeps nitrogen away from the volatiles, thus not diluting the desired product gas.
4. It uses gravity flow for all the solids.
5. It also uses gravity flow of condensate in all the heat pipes, leading to higher heat fluxes.

These are all attractive features.

For this process calculate the temperatures T_2 and T_4, and determine the fraction of what would have been wasted sensible heat which is recovered by the exchangers.

Data:

<table>
<tr><td>Incoming cold shale</td><td>Leaving hot shale</td></tr>
<tr><td>$\dot{m}_c = 2.5\ \text{kg/s}$</td><td>$\dot{m}_h = 2.0\ \text{kg/s}$</td></tr>
<tr><td>$C_c = 960\ \text{J/kg K}$</td><td>$C_h = 900\ \text{J/kg K}$</td></tr>
<tr><td>$T_1 = 0\ ^\circ\text{C}$</td><td>$T_3 = 800\ ^\circ\text{C}$</td></tr>
</table>

Exchanger containing the cold	*Exchanger containing the hot*
end of the heat pipes	*end of the heat pipes*
$A_c = 100\ \text{m}^2$	$A_h = 80\ \text{m}^2$
$U_c = 120\ \text{W/m}^2\text{K}$	$U_h = 100\ \text{W/m}^2\text{K}$

The temperature will change progressively from heat pipe to heat pipe; however, one may reasonably assume that the temperature drop between the hot and the cold end of any single heat pipe is negligible.

16.11 *Heat from geothermal brine.* Hot geothermal brine contains up to 30 % solids, and to cool this in an ordinary exchanger will result in horrendous scale deposits which will soon plug the unit. To overcome this problem the Idaho National Engineering Laboratory has proposed a novel exchanger design wherein the cooling brine passes through fluidized solids in the shell side of

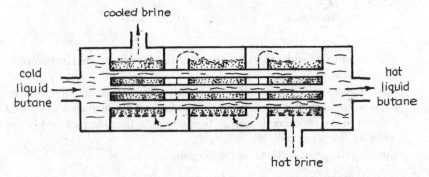

the recuperator, as sketched above. Since the total surface area of solids is so much greater than that of the tubes, the salts should deposit preferentially on the particles while the tubes are kept burnished by the buffing action of the particles. In the commercial unit, brine will flow in the shell side and liquid butane in the tube side; however, in our test runs with such a unit, we use water for both streams. From the data presented below, determine the outside-of-tube heat transfer coefficient h_0 for the following experimental conditions:

Data:

$$\text{Inside of tubes} \begin{cases} \dot{m}' = 1\,\text{l/s} \\ h_{i,\text{estimated}} = 3{,}600\,\text{W/m}^2\text{K} \end{cases} \qquad \begin{cases} T'_{\text{in}} = 5\ ^\circ\text{C} \\ T'_{\text{out}} = 50\ ^\circ\text{C} \end{cases}$$

$$\text{In the shell} \begin{cases} \dot{m}'' = 1\,\text{l/s} \\ T''_{\text{in}} = 70\ ^\circ\text{C} \end{cases}$$

Total area of tubes passing through the three fluidized beds: $A = 12\ \text{m}^2$

[From C. A. Allen and E. S. Grimmett, INEL report UC-66d, April 1978]

16.12 *Freezing of people*. In certain experimental emergency procedures, the anesthetized patient is plunged into ice water and his whole body is cooled to 30 °C. If the heat transfer coefficient between patient and water is 32 W/m^2 K and his skin area is 2.3 m^2, estimate how long it should take to cool an 80-kg patient of medium build down to the desired state. Of course the patient is alive throughout in a sort of hibernating state, breathing slowly, heart beating, and blood flowing.

(a) Assume no heat generation in the body due to metabolic action.
(b) Assume heat generation of 3,200 kJ/day or 37 W.

Clearly state any other assumptions that you need to make.

16.13 *Hot-water heaters*. G. F. Montgomery of the National Bureau of Standards wrote an article entitled "Product Technology and the Consumer," in the December 1977 issue of *Scientific American*, in which he compared a conventional hot-water heater with an improved energy saving model.

(a) For the same amount of insulation around the tanks, what can you say about the difference in efficiency of the pairs of alternatives sketched in the diagram?

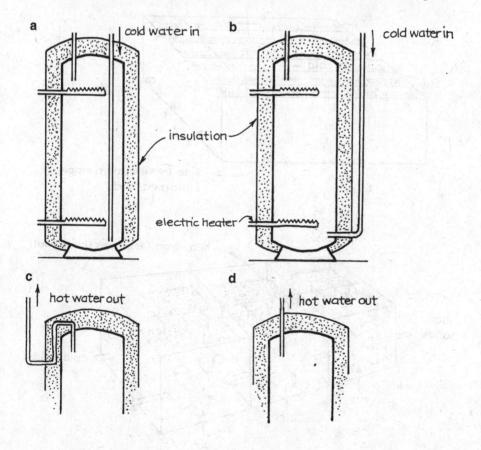

(b) Which of the cold water inlet (a) or (b) and which of the hot water outlet (c) or (d) represent the improved energy saving design?

16.14. *Solid to solid counterflow heat exchanger*. In a faraway gathering (IMM Conference, Sydney, Australia, 1981) O. E. Potter introduced a novel fluidized solid heat exchanger consisting of alternate counterflowing channels of hot and cold solids, somewhat as sketched below. Consider a small two-channel test unit as shown in the drawing on the next page, each channel consisting of a three-stage fluidized bed, with a common finned wall to facilitate heat transfer between hot and cold streams. If the hot solids enter at 600 °C and cold solids at 60 °C, find the temperature and the heat recovery efficiency of the leaving streams.

Data:

$$\text{Hot solids} \begin{cases} \dot{m}_h = 3 \text{ t/h} \\ C_h = 720 \text{ J/kg K} \end{cases} \quad \text{Cold solids} \begin{cases} \dot{m}_c = 2.7 \text{ t/h} \\ C_c = 800 \text{ J/kg K} \end{cases}$$

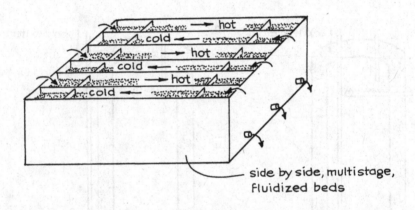

side by side, multistage, fluidized beds

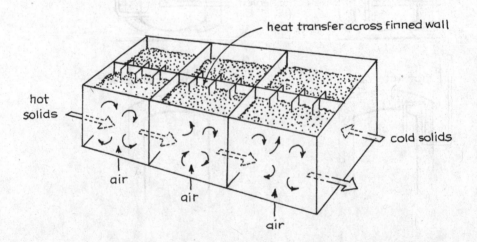

heat transfer across finned wall

hot solids

cold solids

air   air   air

Area on hot side (includes all three stages and all finned surfaces) $= 18$ m^2. Same area on cold side. Mean heat transfer coefficient of all surfaces: $h_h = h_c = 400$ W/m^2 K. Ignore the resistance to heat transfer of the metal dividing wall and the fins.

16.15. *Heating solids with a gas.* At present a stream of solids (0 °C) is heated by hot combustion gases (900 °C) in two equal-sized rotating-shaft exchangers in series, in which gas and solids approximate countercurrent plug flow. Unfortunately, contacting is poor in these exchangers because gas flows over the top of most of the solids, and as a result the outlet temperatures of gas and solid are both 450 °C. This is not good enough and so it has been suggested that a fluidized bed be placed between these rotating shaft exchangers, as sketched below. What size of fluidized bed would be most effective, and with this modification what would be the outlet temperature of the solids? Ignore all heat losses to the surroundings.

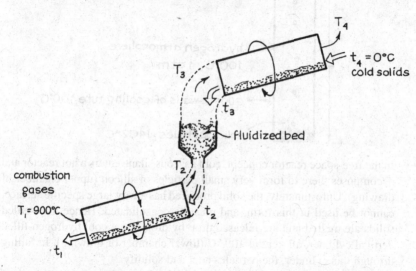

16.16. *Producing very pure silicon.* A number of processes have been proposed for producing ultrapure silicon from the thermal decomposition of silane:

$$SiH_4(g) \xrightarrow{\text{heat}} Si_{\downarrow} + 2H_2$$

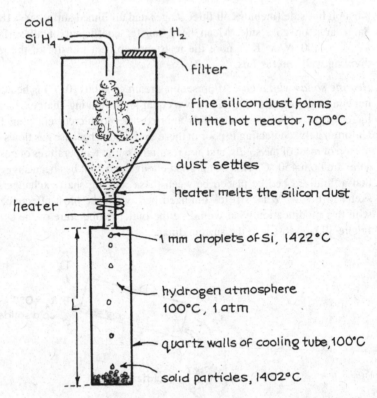

cold
Si H₄ →

→ H₂

filter

fine silicon dust forms
in the hot reactor, 700°C

dust settles

heater melts the silicon dust

heater

1 mm droplets of Si, 1422°C

hydrogen atmosphere
100°C, 1 atm

L

quartz walls of cooling tube, 100°C

solid particles, 1402°C

In the free-space reactor concept, cold gaseous silane enters a hot reactor and decomposes there to form very small particles of silicon (upper chamber of drawing). Unfortunately, the solid produced has a very large specific surface, cannot be used in this form, and must be consolidated. Hence, the settled solids are melted and are released drop by drop into a cool hydrogen-filled vertical cylinder, all kept at 100 °C (lower chamber of drawing). In falling through this cylinder, the particles cool and solidify.

Please come up with a rough but conservative first estimate on the length of cylinder to be used if the 1-mm silicon droplets leaving the nozzle are 10 °C above their melting point and if the particles are to be 10 °C below their melting point when they reach the bottom of the cylinder.

Data: For silicon,

ρ (both solid and liquid) $\cong 2,300$ kg/m³ λ (melting) $= 1.10 \times 10^6$ J/kg

$\quad c_p$(liquid) $= 1,010$ J/kg K T (melting) $= 1,412$ °C

$\quad c_p$ (solid) $= 713$ J /kg K k_l and $k_s \cong 1$ W/m K

For the quartz tube walls, $T = 100$ °C and emissivity $= 1$. For a conservative estimate assume that the particles very quickly reach their terminal velocity.

Solid–solid heat exchange using a third solid. Figure 12.7 shows Shell Company's SPHER process designed to transfer heat from a stream of finely divided hot solids to a stream of finely divided cold solids. This is to be done in two fluidized beds using a stream of circulating metallic balls to soak up the heat in the lower unit and then release it in the upper unit.

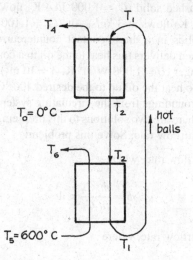

(a) Make a Q vs. T diagram to represent this system.

(b) Determine the efficiency of heat removal from the hot solids, η_h, and the efficiency of heat uptake by cold solids, η_c.

16.17 Assume plug flow of all streams.

16.18 Assume mixed flow of all streams.

16.19 Assume mixed flow of fine solids and plug flow of the large steel balls.

Data: The flow rates and properties of the flowing streams are as follows:

Hot solids	Cold solids
$T_5 = 600\ °C$	$T_3 = 0\ °C$
$\dot{m}_h = 25\ kg/s$	$\dot{m}_c = 25\ kg/s$
$C_h = 1{,}000\ J/kg\ K$	$C_c = 1{,}000\ J/kg\ K$
Circulating	
metal balls	In the beds
$\dot{m}_b = 50\ kg/s$	$U = 200\ W/m^2 of\ balls\ K$
$C_b = 500\ J/kg\ K$	$\bar{t}_{fines} = 80\ s$ in each bed
$\rho_b = 5{,}000\ kg/m^3$	$\bar{t}_{balls} = 25\ s$ in each bed
$d_b = 6\ mm$	

Ignore the heat contribution of the fluidizing gases in these two exchangers, and label the temperature of the various streams as shown.

Indirect heat exchange between liquid and solid. A stream of oil is to be heated from 0 °C to 400 °C using the heat from a stream of hot 1,000 °C solids. Direct contact exchange is ruled out because the solid would contaminate the oil, so let us consider using air as an intermediary for the transfer of heat from solid to oil.

Rather finely crushed solid ($C_s = 1,000$ J/kg K) flows at 5 kg/s and the oil ($C_o = 3,000$ J/kg K) flows at 3 kg/s. Air ($C_a = 1,100$ J/kg K) picks up heat from the hot solids in a direct-contact countercurrent flow raining–solid exchanger and then delivers this heat to the oil in a countercurrent concentric pipe heat exchanger ($U = 1,000$ W/m^2 K, $A = 10$ m^2). What circulation rate of air is needed to heat the oil up to the desired 400 °C? Assume that no heat is lost to the surroundings from the circulating system.

It turns out that there are two solutions to this problem, one at low airflow rate and one at high airflow rate. Solve this problem:

16.20 For the lower airflow rate, where

$$\phi = \frac{\dot{m}_a C_a}{\dot{m}_s C_s} < 1$$

16.21 For the higher airflow rate, where

$$\phi = \frac{\dot{m}_a C_a}{\dot{m}_s C_s} > 1$$

Label temperatures according to the flow diagram below.

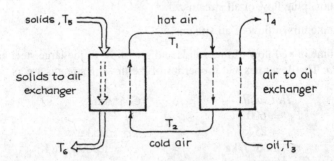

16.22 In the above two problems, instead of recirculating the air why not have fresh incoming air ($T_0 = 0$ °C) contact the solids, then the oil, and then be discarded. With this modification, all else remaining unchanged, find the minimum flow rate of air needed to heat the oil to 400 °C.

16.23 *Design of an atmospheric fluidized bed combustor.* We wish to make a preliminary estimate of the dimensions of an atmospheric fluidized bed combustor for a utility scale power plant (1,000 $MW_{electrical}$). The bed will consist of a fluidized mixture of coal, ash, and limestone with immersed heat transfer tubes (5 cm o.d. on 15-cm centers) for steam generation. Boiling water enters the tubes at 204 °C and steam at 538 °C leaves the tubes.

(a) Find the number of 4 m × 25-m beds needed for such a plant.
(b) Find the number of rows of heat transfer tubes needed to heat the steam.
(c) Find the amount of coal needed/hour to fuel the plant.

Data:

Bed conditions : $T = 843$ °C; air enters at 100 °C, coal at 20 °C.
c_p of all gases $= 33.4$ J/mol K
Boiling water at 204 °C : $\quad H = 870,000$ J /kg
Steam at 204 °C : $\quad\quad H = 2,795,000$ J/kg
Steam at 538 °C : $\quad\quad H = 3,555,000$ J/kg

Assume that coal is pure carbon, that stoichiometric air is used, and that the following reaction goes to completion:

$$C + O_2 \longrightarrow CO_2 \quad \Delta H_r = -393,396 \text{ J}$$

Between the tubes and the bed, $U = 250$ W/m² K. At the bed temperature the gas velocity through the bed, $u_0 = 2.5$ m/s. Efficiency of conversion of heat to steam and to electricity is 40 %.

16.24 *Production of zinc strips.* Ridmoss Corp. pulls a continuous strip of zinc from a melt and then cools, cuts, and coils these strips for sale as moss killers for roofs of houses. The rate-limiting step of this production operation is the cooling of the strips.
What production rate can we expect to obtain (m/s) if the continuous strip (5 cm wide, 400 μm thick) passes vertically upward through the 2-m-high cooler where cold 0 °C air is blown past the strip? The strip enters the cooler at 400 °C and leaves at a mean temperature of 100 °C.
Data: The heat transfer coefficient from air to the zinc strip is estimated to be $h = 50$ W/m² · K.

$$\text{and for zinc}: \quad k = 103.8 \text{ W/m} \cdot \text{K}$$
$$\rho_s = 7,140 \text{ kg/m}^3$$
$$C_s = 426 \text{ J/kg} \cdot \text{K}$$

16.25 In the previous problem the production rate of zinc rolls is to be speeded, so we are considering replacing the air cooler with a fluidized bed cooler in

which the strip passes through and across a 1-m-wide fluidized bed of very
small copper spheres ($d_p = 100\,\mu m$) kept at 0 °C.
What production rate of these zinc strips (m/s) can we expect with this
cooler?

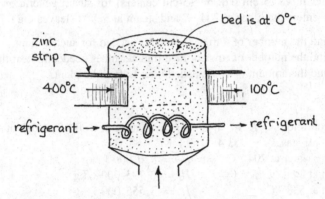

16.26 Consider one modification to the above problem. Here we do not know the
temperature of the bed; we just know that the refrigerant in its copper tube
enters the fluidized bed at $-20\,°C$, passes through its coiled tube, and leaves
at 0 °C.
With this change find the production rate (m/s) of the zinc strip.
Data: The exterior surface of the refrigerant coil is 0.124 m² and its h in the
fluidized bed is the same as for the zinc strip.

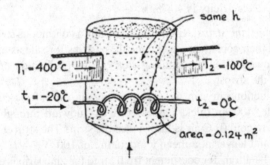

16.27 *What did Watts invent?* Spherical lead shot was in great demand by the
military and by hunters; however, no one was able to make it. Egg shaped,
yes, with seams, yes, but really spherical, no. The one great breakthrough
came in 1782 when an English plumber, William Watts, received a patent for
a method to produce shot that was "solid throughout, perfectly globular in
form, and without the dimples, scratches and imperfections which other shot,
heretofore manufactured, usually have on their surface..." His invention is
quite simple: melt lead in an iron pot, add a pinch of arsenic as lubricant,
pour through a sieve, and let the resulting droplets fall far enough to solidify

completely, eventually plunging into a tank of water which serves only to cushion the fall.

The essentials of this method remain unchanged to this day, and about 30 drop towers exist worldwide, including five in the United States. But I wonder how tall these towers need to be. Please calculate the height of a shot tower needed to produce 4.8-mm shot, the largest size produced today.

Data: Lead ($mw = 0.207$ kg/mol) melts at 327 °C with a latent heat of 24,740 J/kg.

Liquid lead just above its freezing point falls through the screen.

Air in the tower is at 20 °C.

For further readings on shot towers, see *American Heritage of Invention and Technology*, 52, Spring/Summer, 1990.

16.28 The showers at the local gym put out discrete 2-mm droplets of water, and I have the distinct impression that the water hitting my knees is much cooler than the water hitting my shoulders, 1 m higher up. I wonder whether this really is so or whether it is an illusion. Would you please estimate this temperature difference if the shower room is at 25 °C and if the droplets hitting my shoulder are at 50 °C. Ignore the contribution of evaporation to cooling for this problem.

16.29 Heat is to be transferred between two equal-sized batches of water, one at 100 °C and other at 0 °C. What is the very coolest to which you can cool the hot water if $h_1 = h_2 = U = \infty$? With a sketch show how to do this.

16.30 *Cost of pumping Vita-Cola.* Illinois Cola plan to pump their specially formulated and deliciously tasty mixture of fruit juices from Honduras to Chicago called Vita-Cola, bottle it there, and sell to the waiting public.

They plan to pump Vita-Cola in a clean, smooth 5,000-km-long 0.1-m-i.d. pipeline at 0.5 m/s, which they are now building, and you should also know that Vita-Cola is a power law fluid with a shear stress of 1/3

$$\tau = 6.3 \ (du/dy)$$

and density close to that of water.

If the pumping system is 50 % efficient, what would be the power needed to pump this mixture, $W = J/s$?

Appendix
Dimensions, Units, Conversions, Physical Data, and Other Useful Information

A.1 SI Prefixes

Factor	Prefix name	Symbol
10^{12}	tera	T
10^{9}	giga	G
10^{6}	mega	M
10^{3}	kilo	k
10^{-3}	milli	m
10^{-6}	micro	μ
10^{-9}	nano	n
10^{-12}	pico	p

A.2 Length

The standard unit of length is the meter.

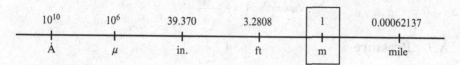

2.54 cm = 1 in., 30.48 cm = 1 ft

© Springer Science+Business Media New York 2014
O. Levenspiel, *Engineering Flow and Heat Exchange*,
DOI 10.1007/978-1-4899-7454-9

A.3   Volume

The standard unit of volume is the cubic meter

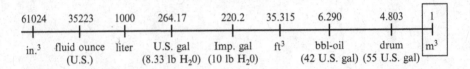

61024	35223	1000	264.17	220.2	35.315	6.290	4.803	1
in.^{3}	fluid ounce (U.S.)	liter	U.S. gal (8.33 lb H$_2$0)	Imp. gal (10 lb H$_2$0)	ft^3	bbl-oil (42 U.S. gal)	drum (55 U.S. gal)	m^3

A.4   Mass

The standard unit of mass is the kilogram

35.274	2.2046	1	0.0011023	0.001	0.0009842
ounce, avoirdupois	lb	kg	short ton (2000 lb)	metric ton	long ton (2240 lb)

A.5   Newton's Law

$$F = ma$$
$$a = g = 9.80665 \ \text{m/s}^2 \quad \text{at "standard" sea leavel}$$

(In fact, g varies from 9.77 to 9.82 at different locations on the earth's surface.)

A.6   Force

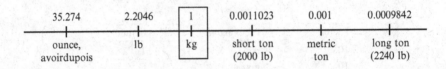

$$(\text{Force}) = (\text{mass})(\text{acceleration})$$

the newton: $1 \ \text{N} = 1 \ \dfrac{\text{kgm}}{\text{s}^2}$

A.7   Pressure

$$(\text{Pressure}) = (\text{force}) / (\text{area})$$

the pascal: $1 \ \text{Pa} = 1 \ \dfrac{\text{N}}{\text{m}^2} = 1 \dfrac{\text{kg}}{\text{m s}^2}$

$1 \text{ atm} = 760 \text{ mm Hg} = 14.7 \frac{\text{lb}_f}{\text{in.}^2} = 29.92 \text{ in. Hg} = 101{,}325 \text{ Pa} \cong 34 \text{ ft H}_2\text{O}$

$1 \text{ Torr} = 1 \text{ mm Hg} = 133.3 \text{ Pa} = \frac{1}{760} \text{ atm}$

$1 \text{ bar} = 10^5 \text{ Pa (close to 1 atm)}$

$1 \text{ in. H}_2\text{O} = 248.86 \text{ Pa} \cong 250 \text{ Pa}$

A.8 Work, Heat, and Energy

$$(\text{Work}) = (\text{force})(\text{distance})$$

$$\text{the joule: } 1 \text{ J} = 1 \text{ Nm} = 1 \, \frac{\text{kg m}^2}{\text{s}^2}$$

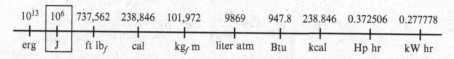

$$10^{13} \quad 10^6 \quad 737{,}562 \quad 238{,}846 \quad 101{,}972 \quad 9869 \quad 947.8 \quad 238.846 \quad 0.372506 \quad 0.277778$$

erg | J | ft lb$_f$ | cal | kg$_f$ m | liter atm | Btu | kcal | Hp hr | kW hr

$1 \text{ cal} = 4.184 \text{ J}$

$1 \text{ Btu} = 778 \text{ ft lb}_f = 252 \text{ cal} = 1{,}055 \text{ J}$

$1 \text{ lit atm} = 101.325 \text{ J}$

A.9 Power

$$(\text{Power}) = (\text{work or energy})/(\text{time})$$

$$\text{the watt: } 1 \text{ W} = 1\frac{\text{J}}{\text{s}} = 1\frac{\text{Nm}}{\text{s}} = 1\frac{\text{kg m}^2}{\text{s}^3}$$

$$1 \text{ kW} = 239 \frac{\text{cal}}{\text{s}} = 1.341 \text{ Hp}$$

$$1 \text{ Hp} = 550 \frac{\text{ft lb}_f}{\text{s}} = 33{,}000 \frac{\text{ft lb}_f}{\text{min}}$$

A.10 Molecular Weight

In SI units, which are used throughout,

$$(mw) = \left(\frac{kg}{mol}\right)$$

$$(mw)_{O_2} = 0.032\,\frac{kg}{mol}$$

$$(mw)_{air} = 0.0289\,\frac{kg}{mol}\,,\text{etc.}$$

$$\text{no. of mols}\quad kg$$
$$n = \frac{m}{(mw)}$$
$$kg\,/\,mol$$

A.11   Ideal Gas Law

$$pV = nRT \quad \text{or} \quad \frac{p}{\rho} = \frac{RT}{(mw)} \quad \text{or} \quad pV = \frac{mRT}{(mw)}$$

Gas constant

$$R = 8.314\,\frac{J}{mol\,K} = 8.314\,\frac{m^3 Pa}{mol\,K}$$

Ideal Gas Law Constant

In flowing systems:

$$G = u\rho = \frac{\dot{n}(mw)}{A} = \frac{p\dot{v}(mw)}{ART} = \frac{pu(mw)}{RT} \quad \left[\frac{kg}{m^2 \cdot s}\right]$$

A.12   Density

$$(\text{Density}) = (\text{mass})/(\text{volume}), \quad \text{or } \rho = \left[\frac{kg}{m^3}\right]$$

For an ideal gas

$$\rho = \frac{p(mw)}{RT}\,\frac{air}{20°C,}\,\frac{(101,325)(0.0289)}{(8.314)(293)} = 1.20\,\frac{kg}{m^3} \qquad 1\ \text{atm}$$

For water

Temperature, °C	ρ, kg/m^3
0–12	1,000
13–18	999
19–23	998
24–27	997
28–30	996
40	992
50	988
60	983
70	978
80	972
90	965
100	958

See Appendix A.21 for more ρ values.

A.13 Viscosity

$$\text{Dynamic or absolute viscosity} \quad \mu = \left[\frac{kg}{m\,s}\right] = [Pa \cdot s]$$

For a Newtonian

$$(\text{Shear stress}) \propto \left(\frac{\text{velocity}}{\text{gradient}}\right) \quad \text{or} \quad \tau = \mu \left(\frac{du}{dy}\right) \quad \left[\frac{kg}{m \cdot s^2} = Pa\right]$$

0.672	1	10	1,000	2,420
$\dfrac{1b_m}{ft\,s}$	$\dfrac{kg}{m\,s}$	$\dfrac{gm}{cm\,s}$	cp	$\dfrac{1b_m}{ft\,hr}$
	(Poiseville, Pl)	(poise p)	(centipoise)	

$$= Pa \cdot s$$

$$= \left(\frac{kg}{m \cdot s^2}\right)s$$

$$= \left(\frac{kg}{m \cdot s}\right)$$

$$\boxed{1Pl = 10p = 1000cp}$$

Poiseville   10Pa   poise

For liquid water (20 °C): $\mu = 10^{-3}$ kg/ms
For gases (20 °C): $\mu \cong 10^{-5}$ kg/ms
For air (20 °C): $\mu \cong 1.8 \times 10^{-5}$ kg/ms

For liquid water

Temperature, °C	μ, kg/ms
0	1.79×10^{-3}
5	1.52×10^{-3}
10	1.31×10^{-3}
15	1.14×10^{-3}
20	1.00×10^{-3}
25	0.894×10^{-3}
30	0.801×10^{-3}
40	0.656×10^{-3}
50	0.549×10^{-3}
60	0.469×10^{-3}
70	0.406×10^{-3}
80	0.357×10^{-3}
90	0.317×10^{-3}
100	0.284×10^{-3}

The following listing shows the wide range of viscosities of familiar Newtonian fluid

Fluid	μ, kg/ms
Gases	
H2 (20 °C)	0.876×10^{-5}
Steam (100 °C)	1.25×10^{-5}
CO2 (20 °C)	1.48×10^{-5}
Air (0 °C)	1.71×10^{-5}
(20 °C)	1.83×10^{-5}
(100 °C)	2.17×10^{-5}
Liquids	
Gasoline (20 °C)	0.6×10^{-3}
H2O (20 °C)	1.0×10^{-3}
C2H5OH (20 °C)	1.2×10^{-3}
Kerosene (20 °C)	2.0×10^{-3}
Whole milk (0 °C)	4.3×10^{-3}
Sucrose solutions (20 °C)	
20 wt%	2.0×10^{-3}
40 wt%	6.2×10^{-3}
60 wt%	58×10^{-3}
70 wt%	486×10^{-3}
SAE 10W-30 motor oil	
(−18 °C)	1.2–2.4
(99 °C)	9×10^{-3}–12×10^{-3}
Olive oil (20 °C)	84×10^{-3}
SAE 30W motor oil (20 °C)	100×10^{-3}

(continued)

(continued)

Fluid	μ, kg/ms
Heavy machine oil (20 °C)	660×10^{-3}
Glycerin (20 °C)	860×10^{-3}
Molasses, very heavy (20 °C)	6.6
Clover honey (20 °C)	10–50
Pitch (0 °C)	5.1–10^{10}

The viscosity of liquids is practically independent of pressure; that of gases increases slowly with a rise in pressure, not even doubling at the critical pressure. With an increase in temperature liquids become *less* viscous, gases become *more* viscous. For the viscous characteristics of non-Newtonians see Chap. 5.

A.14 Kinematic Viscosity

$$\nu = \frac{\mu}{\rho} = \left[\frac{m^2}{s}\right]$$

$$1\frac{m^2}{s} = 10^4 \frac{cm^2}{s} = 10^4 \text{ stoke} = 10^6 \text{centistoke}$$

A.15 Thermal Conductivity

$$k = \left[\frac{W}{(m^2 c.s.)(K/m \text{ length})}\right] = \left[\frac{W}{m\,K}\right]$$

$$1\frac{W}{mK} = 0.00239 \frac{cal}{s\ cm°C} = 0.578 \frac{Btu}{h\ ft\ °F}$$

$k \cong$ independent of pressure

For water (20 °C): $k = 0.597$ W/m K
For air (20 °C): $k = 0.0257$ W/m K
For steam (100 °C): $k = 0.0251$ W/m K

See Appendix A.21 for more k values

A.16 Specific Heat

$$C_p = \left[\frac{J}{kg\ K} \right]$$

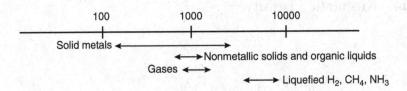

$$1\frac{J}{kg\ K} = 239 \times 10^{-6}\frac{cal}{g°C} = 239 \times 10^{-6}\frac{Btu}{lb°F}$$

For water (20 °C): $C_p = 4{,}184$ J/kg K $= 1$ cal/g °C $= 1$ Btu/lb °F
For air (20 °C): $C_p = 1{,}013$ J/kg K $= 29.29$ J/mol K
                               $= 0.24$ cal/g °C $= 7$ cal/mol C
For steam (100 °C): $C_p = 2{,}063$ J/kg K $= 37.13$ J/mol K

See Appendix A.21 for more C_p values.

A.17 Thermal Diffusivity

$$\alpha = \frac{k}{\rho C_p} = \left[\frac{m^2}{s} \right]$$

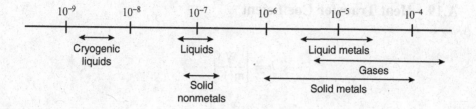

$$1\frac{m^2}{s} = 10.76\,\frac{ft^2}{s} = 38,750\,\frac{ft^2}{h}$$

For air (20 °C): $\alpha = 2.12 \times 10^{-5}\ m^2/s$
For steam (100 °C): $\alpha = 2.05 \times 10^{-5}\ m^2/s$
For water (20 °C): $\alpha = 1.43 \times 10^{-7}\ m^2/s$

α values for other substances can be found from $k/\rho C_p$ values in Appendix A.21.

A.18   Thermal Radiative Properties

$$\varepsilon,\ \alpha = [\text{dimensionless}]$$

ε and α for room temperature radiation

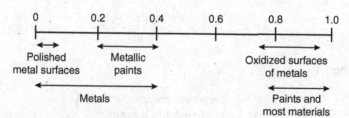

$\sigma = 5.67 \times 10^{-8}\ W/m^2\ K^4$
  $= 0.1713 \times 10^{-8}\ Btu/h\ ft^2\ °F^4$, the radiation constant

See Chap. 12 for a short table of emissivities and absorptivities.

A.19   Heat Transfer Coefficient

$$h = \left[\frac{W}{m^2\,K}\right]$$

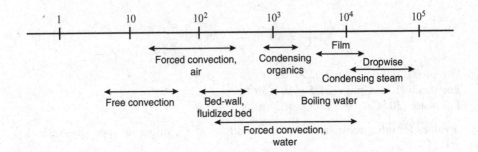

$$1\frac{W}{m^2\,K} = 2.39 \times 10^{-5}\frac{cal}{s\,cm^2\,{}^\circ C} = 0.1761\frac{Btu}{h\,ft^2\,{}^\circ F}$$

A.20   Dimensionless Groups

Archimedes number:

$$Ar = \left(d_p^*\right)^3 = \frac{d_p^3 \rho_g \left(|\rho_s - \rho_g|\right) g}{\mu^2}$$

Biot number:

$$Bi = \frac{hL}{k} = \frac{\left(\begin{array}{c}\text{interior resistance to heat}\\ \text{transfer in a particle}\end{array}\right)}{\left(\begin{array}{c}\text{resistance to heat transfer}\\ \text{at surface of a particle}\end{array}\right)}$$

Drag coefficient for falling particles:

$$C_D = \frac{\text{drag force}, F_d}{\left(\rho_g u^2/2\right) \cdot \left(\pi d_p^2/4\right)}$$

Darcy friction factor for flow in pipes:

$$f_D = 4f_f$$

Friction factor for packed beds:

$$f_f = \frac{\varepsilon^3}{1-\varepsilon} \cdot \frac{\sum F d_p}{u_0^2 L} = \frac{\left(\begin{array}{c} \text{frictional} \\ \text{energy loss} \end{array} \middle/ \begin{array}{c} \text{kg of} \\ \text{fluid} \end{array}\right)}{\left(\begin{array}{c} \text{kinetic} \\ \text{energy loss} \end{array} \middle/ \begin{array}{c} \text{kg of} \\ \text{fluid} \end{array}\right)}$$

Fanning friction factor for flow in pipes:

$$f_F = \frac{\tau_w}{\rho u_0^2/2} = \frac{\left(\begin{array}{c} \text{frictional} \\ \text{drag} \end{array} \middle/ \begin{array}{c} \text{area of} \\ \text{pipe wall} \end{array}\right)}{\left(\begin{array}{c} \text{kinetic} \\ \text{energy} \end{array} \middle/ \text{m}^3 \text{of fluid}\right)} = \frac{\left(\begin{array}{c} \text{frictional loss} \\ \text{during flow} \end{array}\right)}{\left(\begin{array}{c} \text{kinetic energy} \\ \text{of fluid} \end{array}\right)}$$

Fourier number:

$$\text{Fo} = \frac{\alpha t}{L^2} = \frac{kt}{\rho C_p L^2}$$

Graetz number:

$$\text{Gz} = \text{Re} \cdot \text{Pr} \cdot \left(\frac{d}{L}\right) \dots \text{used in forced convection}$$

Grashof number:

$$\overbrace{-\frac{1}{V}\left(\frac{\partial V}{\partial T}\right)_p \overset{\text{ideal}}{\underset{\text{gas}}{=}} \frac{1}{T}}^{\text{Coefficient of volumetric expansion}}$$

$$\text{Gr} = \frac{L^3 \rho^2 g \beta \Delta T}{\mu^2} = \frac{\text{buoyancy force}}{\text{viscous force}}$$

Hedstrom number:

$$\text{He} = \frac{\tau_0 d^2 \rho}{\eta^2}$$

Knudsen number:

$$\mathrm{Kn} = \left(\frac{\pi C_p}{2C_v}\right)^{1/2} \cdot \frac{\mathrm{Ma}}{\mathrm{Re}} = \frac{\left(\begin{array}{c}\text{mean free path}\\ \text{of molecules}\end{array}\right)}{\left(\begin{array}{c}\text{diameter of}\\ \text{flow channel}\end{array}\right)}$$

Mach number:

$$\mathrm{Ma} = \frac{u}{c} = \frac{\text{velocity of gas}}{\text{speed of sound}}$$

Nusselt number:

$$\mathrm{Nu} = \frac{hd}{k} = \frac{\left(\begin{array}{c}\text{total heat}\\ \text{transfer}\end{array}\right)}{\left(\begin{array}{c}\text{molecular}\\ \text{heat transfer}\end{array}\right)} = \frac{\left(\begin{array}{c}\text{conduction and}\\ \text{convection}\end{array}\right)}{\left(\begin{array}{c}\text{conduction}\\ \text{alone}\end{array}\right)}$$

Prandtl number:

$$\mathrm{Pr} = \frac{C_p \mu}{k} = \frac{\left(\begin{array}{c}\text{molecular}\\ \text{momentum transfer}\end{array}\right)}{\left(\begin{array}{c}\text{molecular}\\ \text{heat tranfer}\end{array}\right)} = \frac{\left(\begin{array}{c}\text{viscous dissipation}\\ \text{of energy}\end{array}\right)}{(\text{heat conduction})}$$

$= 0.66 - 0.75$ for air, A, CO_2, CH_4, CO, H_2, He, N_2 and other common gases
$\cong 1.06$ for steam
$= 10 - 1{,}000$ for most liquids
$= 0.006 - 0.03$ for most liquid metals

Reynolds number:

$$\mathrm{Re} = \frac{du\rho}{\mu} = \frac{\text{inertial force}}{\text{viscous force}}$$

A.21 Tables of Physical Properties of Materials

Solids: Metals and Alloys

	T	k	ρ	C_p	$\alpha \times 10^6$
	°C	W/m K	kg/m³	J/kg K	m²/s
Aluminum	20	204.2	2,707	896	84.2
Copper	20	384	8,954	385	112

(continued)

(continued)

	T	k	ρ	C_p	$\alpha \times 10^6$
	°C	W/m K	kg/m^3	J/kg K	m^2/s
Iron	20	72.7	7,897	452	20.4
Lead	20	34.7	11,393	130	23.4
Silver	20	406.8	10,524	235	164.5
Stainless steel	20	16.3	7,820	460	4.53

Solids: Nonmetals

	T	k	ρ	C_p	$\alpha \times 10^6$
	°C	W/m K	kg/m^3	J/kg K	m^2/s
Brick (building)	20	0.233–0.291	800–1,500	—	—
Cardboard, corrugated	20	0.064	—	—	—
Chalk	50	0.9304	2,000	897	0.529
Coal	20	0.1861	1,400	1,306	0.102
Concrete	20	1.279	2,300	1,130	0.492
Corkboard	20	0.0443	160	—	—
Glass	20	0.7443	2,500	670	0.444
Ice	0	2.25	920	2,261	1.08
Leather	30	0.1593	1,000	—	—
Rubber	0	0.1628	1,200	1,392	0.0975
Sand (bulk properties, $\varepsilon = 0.42$)	20	0.33	1,500	800	0.275
Snow, dry packed	< 0	0.4652	560	2,093	0.397
Wood:					
Oak across grain	20	0.207	800	1,759	0.147
Oak with grain	20	0.3629	800	—	—
Pine across grain	20	0.107	448	—	—
Pine with grain	20	0.2559	448	—	—

Plastics

	k	ρ	C_p
	W/m K	kg/m^3	J/kg K
ABS	0.16–0.27	1,020–1,200	1,510–1,550
Nylon	0.17–0.34	1,030–1,140	1,380–1,670
Polycarbonate	0.19–0.22	910–1,250	1,170–1,260
Polyethylene	0.42–0.49	913–968	2,090–2,300
Polyester	0.33–0.91	1,100–2,010	1,340–2,090
PVC	0.14–0.19	1,240–1,550	1,050
Polystyrene foam	0.03–0.04	16–32	—
Polyurethane foam	0.04	122	—

Gases at 1 atm

| | T | ρ | C_p | $\mu \times 10^6$ | $k \times 10^3$ |
	°C	kg/m³	J/kg K	kg/m s	W/m K
Air	0	1.293	1,005	17.2	24.4
	20	1.205	1,005	18.1	25.9
	100	0.946	1,009	21.9	32.1
	200	0.745	1,019	25.8	39.2
	300	0.615	1,046	29.5	46.0
	400	0.526	1,069	33.7	52.0
	500	0.456	1,093	36.2	57.4
	1,000	0.277	1,185	49.0	80.7
N_2	0	1.250	1,030	16.7	24.3
	100	0.916	1,034	20.7	31.5
	500	0.442	1,105	33.9	55.8
	1,000	0.268	1,202	47.4	72.3
O_2	0	1.429	913	19.4	24.7
	100	1.050	934	24.1	32.9
	500	0.504	1,047	40.0	61.5
	1,000	0.306	1,122	56.5	85.8
CO	0	1.250	1,038	16.6	23.2
	100	0.916	1,043	20.7	30.1
	500	0.442	1,130	34.4	54.1
	1,000	0.268	1,231	48.7	80.6
CO_2	0	1.977	816	14.0	14.6
	100	1.447	913	18.2	22.8
	500	0.698	1,156	33.9	54.9
	1,000	0.423	1,290	51.5	86.3
SO_2	0	2.926	607	12.1	8.4
	100	2.140	662	16.1	12.3
	500	1.033	808	31.3	30.7
	1,000	0.626	867	49.2	57.6
Flue gases	0	1.295	1,043	15.8	22.8
	100	0.950	1,068	20.4	31.3
	500	0.457	1,185	34.8	65.6
	1,000	0.275	1,306	48.3	109.0
H_2	0	0.0899	14,070	8.4	172.1
	100	0.0657	14,480	10.3	219.8
	500	0.0317	14,660	16.8	387.3
	1,000	0.0192	15,520	23.7	571.0
NH_3	0	0.771	2,043	9.4	21.0
	100	0.564	2,219	13.0	34.0
	500	0.272	2,918	28.1	103.6
	1,000	0.165	3,710	47.9	222.1
Steam	100	0.598	2,135	12.0	23.7
	500	0.284	2,135	28.6	68.4
	1,000	0.172	2,483	52.4	140.7

k ratios . . . Gases	$k = Cp/Cv = [-]$
Monatomic (He, Ne)	$k = 5/3 = 1.67$
Diatomic (O_2, N_2, H_2)	$k = 7/5 = 1.4$
Triatomic (CO_2, SO_2)	$k \cong 1.3$

Liquids

	T	ρ	Cp	$\mu \times 10^6$	k
	°C	kg/m³	J/kg K	kg/m s	W/m K
Water	0	1,002	4,216	1,792	0.5524
	20	1,000	4,178	1,006	0.5978
	40	995	4,178	654	0.6280
	60	985	4,183	471	0.6513
	80	974	4,195	355	0.6687
	100	961	4,216	282	0.6804
	200	867	4,505	139	0.6652
	300	714	5,728	96	0.5396
NH_3	20	612	4,798	219	0.521
CO_2	20	772	5,024	70.3	0.0872
Glycerine, $C_3H_5(OH)_3$	20	1,264	2,387	1.49×10^6	0.2861
SO_2	20	1,386	1,365	290	0.1989
Freon-22, CCl_2F_2	20	1,330	963	263	0.072
Ethylene glycol, $C_2H_4(OH)_2$	20	1,116	2,382	21,398	0.2489
Mercury	20	13,550	139	1,545	7.91

Sources

C.P. Kothandaraman, S. Subramanyan, *Heat and Mass Transfer Desk Book* (Wiley, New York, 1977)
W.H. McAdams, *Heat Transmission*, 3rd edn. (McGraw-Hill, New York, 1954)
J.H. Perry, *Chemical Engineers' Handbook*, 3rd edn. (McGraw-Hill, New York, 1950)
Plastic Desk-Top Data Bank, Cordura, 1980
Y.S. Touloukian and co-workers, *Thermophysical Properties of Matter,* in 13 volumes, (Plenum, New York. 1970–1977)

© Springer Science+Business Media New York 2014
O. Levenspiel, *Engineering Flow and Heat Exchange*,
DOI 10.1007/978-1-4899-7454-9

Author Index

© Springer Science+Business Media New York 2014
O. Levenspiel, *Engineering Flow and Heat Exchange*,
DOI 10.1007/978-1-4899-7454-9

Subject Index

© Springer Science+Business Media New York 2014
O. Levenspiel, *Engineering Flow and Heat Exchange*,
DOI 10.1007/978-1-4899-7454-9

Printed in the United States
By Bookmasters